Advances in Fuzzy Systems — Applications and Theory – Vol. 24

Nonlinear Integrals and Their Applications in Data Mining

ADVANCES IN FUZZY SYSTEMS — APPLICATIONS AND THEORY

Published

Vol. 9: Fuzzy Topology
(*Y. M. Liu and M. K. Luo*)

Vol. 10: Fuzzy Algorithms: With Applications to Image Processing and Pattern Recognition
(*Z. Chi, H. Yan and T. D. Pham*)

Vol. 11: Hybrid Intelligent Engineering Systems
(*Eds. L. C. Jain and R. K. Jain*)

Vol. 12: Fuzzy Logic for Business, Finance, and Management
(*G. Bojadziev and M. Bojadziev*)

Vol. 13: Fuzzy and Uncertain Object-Oriented Databases: Concepts and Models
(*Ed. R. de Caluwe*)

Vol. 14: Automatic Generation of Neural Network Architecture Using Evolutionary Computing
(*Eds. E. Vonk, L. C. Jain and R. P. Johnson*)

Vol. 15: Fuzzy-Logic-Based Programming
(*Chin-Liang Chang*)

Vol. 16: Computational Intelligence in Software Engineering
(*W. Pedrycz and J. F. Peters*)

Vol. 17: Nonlinear Integrals and Their Applications in Data Mining
(Z. Y. Wang, R. Yang and K.-S. Leung)

Vol. 18: Factor Space, Fuzzy Statistics, and Uncertainty Inference *(Forthcoming)*
(*P. Z. Wang and X. H. Zhang*)

Vol. 19: Genetic Fuzzy Systems, Evolutionary Tuning and Learning of Fuzzy Knowledge Bases
(O. Cordón, F. Herrera, F. Hoffmann and L. Magdalena)

Vol. 20: Uncertainty in Intelligent and Information Systems
(Eds. B. Bouchon-Meunier, R. R. Yager and L. A. Zadeh)

Vol. 21: Machine Intelligence: Quo Vadis?
(Eds. P. Sincák, J. Vascák and K. Hirota)

Vol. 22: Fuzzy Relational Calculus: Theory, Applications and Software (With CD-ROM)
(K. Peeva and Y. Kyosěv)

Vol. 23: Fuzzy Logic for Business, Finance and Management (2nd Edition)
(*G. Bojadziev and M. Bojadziev*)

Advances in Fuzzy Systems — Applications and Theory – Vol. 24

Nonlinear Integrals and Their Applications in Data Mining

Zhenyuan Wang

University of Nebraska at Omaha, USA

Rong Yang

Shen Zhen University, China

Kwong-Sak Leung

Chinese University of Hong Kong, China

World Scientific

NEW JERSEY • LONDON • SINGAPORE • BEIJING • SHANGHAI • HONG KONG • TAIPEI • CHENNAI

Published by

World Scientific Publishing Co. Pte. Ltd.
5 Toh Tuck Link, Singapore 596224
USA office: 27 Warren Street, Suite 401-402, Hackensack, NJ 07601
UK office: 57 Shelton Street, Covent Garden, London WC2H 9HE

British Library Cataloguing-in-Publication Data
A catalogue record for this book is available from the British Library.

NONLINEAR INTEGRALS AND THEIR APPLICATIONS IN DATA MINING
Advances in Fuzzy Systems – Applications and Theory — Vol. 17

ISBN-13 978-981-281-467-8
ISBN-10 981-281-467-1

Printed in Singapore by Mainland Press Pte Ltd.

To our families

Preface

The theory of nonadditive set functions and relevant nonlinear integrals, as a new mathematics branch, has been developed for more than thirty years. Starting from the beginning of the nineties of the last century, several monographs were published. The first author of this monograph and Professor George J. Klir (The State University of New York at Binghamton) have published two books, *Fuzzy Measure Theory* (Plenum Press, New York, 1992) and *Generalized Measure Theory* (Springer-verlag, New York, 2008) on this topic. These two books cover most of their theoretical research results with colleagues at the Chinese University of Hong Kong in the area of nonadditive set functions and relevant nonlinear integrals. Since the 1980s, nonadditive set functions and nonlinear integrals have been successfully applied in information fusion and data mining. However, only a few applications are involved in the above-mentioned books. As a supplement and in-depth material, the current monograph, *Nonlinear Integrals and Their Applications in Data Mining*, concentrates on the applications in data analysis. Since the number of attributes in any database is always finite, we focus on our fundamentally theoretical discussion of nonadditive set function and nonlinear integrals, which are presented in the first several chapters, on the finite universal set, and abandon all convergence and limit theorems.

As for the terminology adopted in the current monograph, words like *monotone measure* is used for a set function that is nonnegative, monotonic, and vanishing at the empty set. It has no fuzziness in the meaning of Zadeh's fuzzy sets. Unfortunately, its original name is *fuzzy measure* in literature. Word "fuzzy" here is not proper. For example, words "fuzzy-valued fuzzy measure defined on fuzzy sets" causes confusion to some people. Such a revision is the same as made in book *Generalized Measure Theory*. However, in this monograph, we prefer to use *efficiency measure* to name a set function that is nonnegative and vanishing at the empty set, rather than using *general measure*. This is more convenient and intuitive, and leaves more space for further generalizing the domain or the range of the set functions. Hence, similar to the classical case in measure theory [Halmos 1950], the set functions that vanish at the empty set and may assume both nonnegative and negative real values are naturally named as *signed efficiency measures*. The signed efficiency measures were also called *non-monotonic fuzzy measures* by some scholars. Since, in general, the efficiency measures are non-monotonic too, to distinguish the set functions satisfying only the condition of vanishing at the empty set from the efficiency measures and to emphasize that they can assume both positive and negative values as well as zero, we prefer to use the current name, signed efficiency measures, for this type of set functions with the weakest restriction. Thus, in this monograph, we discuss and apply three layers of set functions named monotone measures, efficiency measures, and signed efficiency measures respectively.

The contents of this monograph have been used as the teaching materials of two graduate level courses at the University of Nebraska at Omaha since 2004. Also, some parts of this monograph have been provided to a number of master degree and Ph.D. degree graduate students in the University of Nebraska at Omaha, the University of Nebraska at Lincoln, the Chinese University of Hong Kong, and the Chinese Academy Sciences, for preparing their dissertations.

This monograph may benefit the relevant research workers. It is also possible to be used as a textbook of some graduate level courses for both mathematics and engineering major students. A number of exercises on the basic theory of nonadditive set functions and relevant nonlinear integrals are available in Chapters 2–5 of the monograph.

Several former graduate students of the first author provided some algorithms, examples, and figures. We appreciate their valuable contributions to this monograph. We also thank the Department of Computer Science and Engineering of the Chinese University of Hong

Kong, the Department of System Science and Industrial Engineering of the State University of New York at Binghamton and, especially, the Department of Mathematics, as well as the Art and Science College of the University of Nebraska at Omaha for their support and help.

Zhenyuan Wang
Rong Yang
Kwong-Sak Leung

Contents

List of Tables

List of Figures

Chapter 1

Introduction

The traditional aggregation tool in information treatment is the weighted average, or more general, the weighted sum. That is, if the numerical information received from diverse information sources $x_1, x_2, \cdots, x_n$ are $f(x_1), f(x_2), \cdots, f(x_n)$ respectively, then the synthetic amount, weighted sum y, of the information is calculated by

$$y = w_1 f(x_1) + w_2 f(x_2) + \cdots + w_n f(x_n), \tag{1.1}$$

where $w_1, w_2, \cdots, w_n$ are the weights of $x_1, x_2, \cdots, x_n$, respectively. When $0 \le w_i \le 1$ for $i = 1, 2, \cdots, n$ and $\sum_{i=1}^{n} w_i = 1$, the weighted sum shown in (1.1) is called the weighted average. In databases, these information sources $x_1, x_2, \cdots, x_n$ are regarded as attributes and $f(x_1), f(x_2), \cdots, f(x_n)$ are their observations (or say, their records), respectively. An observation can be considered as a function defined on the finite set consisting of these involved information sources. Thus, the weighted sum, essentially, is the Lebesgue integral defined on the set of information sources and is a linear aggregation model. The linear models have been widely applied in information fusion and data mining, such as in multiregression, multi-objective decision making, classification, clustering, Principal Components Analysis (PCA), and so on. However, using linear methods need a basic assumption that there is no interaction among the contributions from individual attributes towards a certain target, such as the objective attribute in regression problems or the classifying attribute in classification problems. This interaction is totally different from the correlationship in statistics. The latter is used to describe the relation between the appearing values of two considered attributes and is not related to any target attribute.

To describe the interaction among contributions from attributes towards a certain target, the concept of nonadditive set functions, such as λ-measures (called λ-fuzzy measure during the seventies and eighties of the last century), belief measures, possibility measures, monotone measures, and efficiency measures have been introduced. The systematic investigation on nonadditive set functions started thirty five years ago. At that time, they were called *fuzzy measures*. Noticeably, the traditional aggregation tool, the weighted sum, fails when the above-mentioned interaction cannot be ignored and some new types of integrals, such as the Choquet integral, the upper integral and the lower integral, should be adopted. In general, these integrals are nonlinear and are generalizations of the classical Lebesgue integral in the sense that they coincide with the Lebesgue integral when the involved nonadditive measure is simply additive. The fuzzy integral, which was introduced in 1974, is also a special type of nonlinear integrals with respect to so-called fuzzy measures. Since the fuzzy integral adopts the maximum and minimum operators, but not the common addition and the common multiplication, most people do not prefer to use the fuzzy integral in real problems. Currently, the most common nonlinear integral in use is the Choquet integral. It has been widely applied in information fusion and data mining, such as the nonlinear multiregressions and the nonlinear classifications, successfully. However, the corresponding algorithms are relatively complex. Only the traditional algebraic methods are not sufficient to solve most data mining problems based on nonlinear integrals. Some newly introduced soft computing techniques, such as the genetic algorithm and the pseudo gradient search, which are presented in Chapter 7 of this monograph, must be adopted.

In most real problems, there are only finitely many variables. For example, in any real database, there are only finitely many attributes. So, the part of fundamental theory in this monograph is focused on the discussion of the nonadditive set functions and the relevant nonlinear integrals defined on a finite universal set. The readers who are interested in the convergence theorems of the function sequences and integral sequences with respect to nonadditive set functions may refer to monographs *Fuzzy Measure Theory* (Plenum press, New York, 1992) and *Generalized Measure Theory* (Springer-verlag, New York, 2008).

The current monograph consists of eleven chapters, After the Introduction, Chapters 2 to 5 devote to the fundamental theory on sets, fuzzy sets, set functions, and integrals. Chapters 6 to 11 discuss the applications of the nonlinear integrals in information fusion and data mining, as well as the relevant soft computing techniques. The relation among these chapters is illustrated in Figure 1.1.

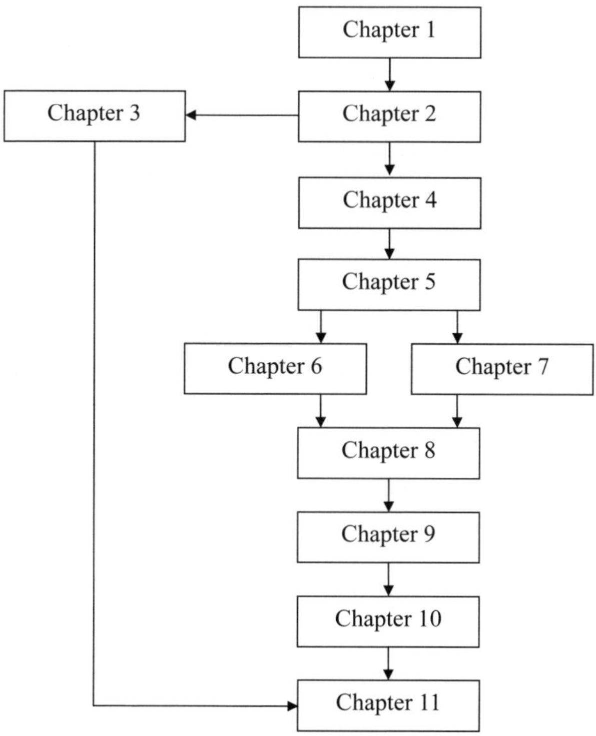

Fig. 1.1 The relation among chapters.

Chapter 2

Basic Knowledge on Classical Sets

2.1 Classical Sets and Set Inclusion

A *set* is a collection of objects that are considered in a particular circumstance. Each object in the set is called a *point* (or an *element*) of the set. Usually, sets are denoted by capital English letters such as *A*, *B*, *E*, *F*, *U*, *X*; while points are denoted by lower case English letters such as *a*, *b*, *x*, *y*. As some special sets, the set of all real numbers is denoted by *R*, and the set of all nonnegative integers is denoted by *N*. For any given set and any given point, the point either belongs to the set or does not belong to the set. "Point *x belongs* to set *A*" is denoted as $x \in A$. In this case, we also say "*A contains x*" or "*x* is in *A*". "Point *x does not belong* to set *A*" is denoted as $x \notin A$. For this, we may also say "*A does not contain x*" or "*x* is *not in A*".

The set consisting of all points considered in a given problem is called the *universal set* (or the *universe of discourse*) and is denoted by *X* usually. The set consisting of no point is called the *empty set* and denoted by ∅. Any set is called a *nonempty set* if it is not empty, i.e., it contains at least one point. A set consisting of exactly one point is called a *singleton*. Any set of sets is called a *class*. The class consisting of no set is the empty class. It is, in fact, the same as the empty set.

A set can be presented by listing all points (without any duplicates) belonging to this set or by indicating the condition satisfied exactly by the points in this set. For example, the set consisting of all nonnegative integers not larger than 5 can be expressed as $\{0, 1, 2, 3, 4\}$ or $\{x \mid 0 \le x < 5, x \in N\}$.

It should be emphasized that any set should not contain some duplication of a point. For instance, $\{2, 1, 2, 3\}$ is not a proper notation of a set since integer 2 appears in the pair of braces twice. After deleting the duplication (but keeping only one of them), $\{2, 1, 3\}$ is a legal notation of the set consisting of integers 1, 2, and 3. The appearing order of points in the notation of sets is not important. For instance, $\{2, 1, 3\}$ and $\{1, 2, 3\}$ denote the same set that consists of integers 1, 2, and 3.

Sets can be used to describe crisp concepts. Also, they represent events in probability theory.

Definition 2.1 Set A is *included* by set B, denoted by $A \subseteq B$ or $B \supseteq A$ iff $x \in A$ implies $x \in B$. In this case, we also say "*B includes A*" or "*A* is a *subset* of *B*".

Example 2.1 In an experiment of randomly selecting a card from a complete deck consisting of 52 cards, there are 52 outcomes. Let the universal set X be the set of these 52 outcomes. Equivalently, X can be regarded as the set of these 52 cards directly. Event "the selected card is a heart", denoted by H, is a subset of X. We can write $H = \{\text{hearts}\} \subseteq X$ simply if there is no confusion. Here, set H describes crisp concept of suit "heart".

Obviously, in a given problem, any set A is included by X, i.e., $A \subseteq X$, while the empty set is included by any set A, i.e., $\varnothing \subseteq A$.

Definition 2.2 Set A is *equal* to set B, denoted by $A = B$, iff $A \subseteq B$ and $B \subseteq A$. If A is not equal to B, we write $A \neq B$.

Definition 2.3 If set A is a subset of set B and $A \neq B$ (i.e., $\exists x \in B$ such that $x \notin A$), then A is called a *proper subset* of B and we write $A \subset B$.

Definition 2.4 Given set A, function $\chi_A : X \to \{0, 1\}$ defined by

$$\chi_A(x) = \begin{cases} 1, & \text{if } x \in A \\ 0, & \text{if } x \notin A \end{cases} \qquad \forall x \in X$$

is called the *characteristic function* of A.

It is easy to know that $A = B$ iff $\chi_A = \chi_B$ (i.e., $\chi_A(x) = \chi_B(x), \forall x \in X$) and $A \subseteq B$ iff $\chi_A \le \chi_B$ (i.e., $\chi_A(x) \le \chi_B(x)$ or $\chi_A(x) = 1 \Rightarrow \chi_B(x) = 1, \forall x \in X$). Similarly, $A \subset B$ iff $\chi_A \le \chi_B$ and there exists at least one point x in X such that $x \in B$ but $x \notin A$ (i.e., $\chi_A(x) \le \chi_B(x)$ and $\exists x \in X$ such that $\chi_B(x) = 1, \chi_A(x) = 0$).

Example 2.2 Let X be the set of all real numbers, i.e., $X = R$. Interval $[1, 2]$ is a subset of interval $[1, 5)$. We have

$$\chi_{[1,\,2]}(x) = \begin{cases} 1, & \text{if } 1 \le x \le 2 \\ 0, & \text{otherwise} \end{cases},$$

$$\chi_{[1,\,5)}(x) = \begin{cases} 1, & \text{if } 1 \le x < 5 \\ 0, & \text{otherwise} \end{cases},$$

and $\chi_{[1,2]} \le \chi_{[1,5]}$.

Example 2.3 Let $X = \{a, b, c\}$, $A = \{a\}$, and $B = \{b\}$. Then, neither $A \subseteq B$ nor $B \subseteq A$. In fact, we have

$$\chi_A(x) = \begin{cases} 1, & \text{if } x = a \\ 0, & \text{if } x \ne a \end{cases},$$

$$\chi_B(x) = \begin{cases} 1, & \text{if } x = b \\ 0, & \text{if } x \ne b \end{cases},$$

and neither $\chi_A \leq \chi_B$ nor $\chi_B \leq \chi_A$.

2.2 Set Operations

Let *X* be the universal set, and let *A* and *B* be subsets of *X*.

Definition 2.5 The *union* of *A* and *B*, denoted by $A \cup B$, is the set consisting of all points that belong to either *A* or *B* (may be both). That is, $A \cup B = \{x \mid x \in A \text{ or } x \in B\}$.

Definition 2.6 The *intersection* of *A* and *B*, denoted by $A \cap B$, is the set consisting of all points that belong to both *A* and *B*. That is, $A \cap B = \{x \mid x \in A \text{ and } x \in B\}$.

Definition 2.7 The *complement* of *A*, denoted by $\overline{A}$, is the set consisting of all points that do not belong to *A*. That is, $\overline{A} = \{x \mid x \notin A\}$.

Corresponding to the characteristic functions, we have

$$\chi_{A \cup B} = \chi_A \vee \chi_B,$$

$$\chi_{A \cap B} = \chi_A \wedge \chi_B,$$

and

$$\chi_{\overline{A}} = 1 - \chi_A,$$

where symbols "$\vee$" and "$\wedge$" are used to denote the maximum and the minimum operators for real numbers respectively, that is, $a \vee b = \max(a, b)$ and $a \wedge b = \min(a, b)$ for any real numbers *a* and *b*.

Definition 2.8 Two sets *A* and *B* are *disjoint* iff $A \cap B = \varnothing$.

Example 2.4 Rolling a regular die once, the outcome may be any one among 1, 2, 3, 4, 5, and 6. Let $X = \{1, 2, 3, 4, 5, 6\}$. Event "obtaining an even number", denoted by A, is a subset of X, i.e., $A = \{2, 4, 6\} \subset \{1, 2, 3, 4, 5, 6\}$. Event "obtaining a number less than 4", denoted by B, is also a subset of X, i.e., $B = \{1, 2, 3\} \subset \{1, 2, 3, 4, 5, 6\}$. Then, we have $A \cup B = \{1, 2, 3, 4, 6\}$, $A \cap B = \{2\}$, and $\overline{A} = \{1, 3, 5\}$.

The subsets of X with set operations union, intersection, and complement have the properties listed in the following Theorem. The proof of the theorem is directly from the definitions 2.5-2.7 and is omitted.

Theorem 2.1 The operations of union, intersection, and complement of sets satisfy the following laws.

Involution law:	$\overline{\overline{A}} = A$
Commutative laws:	$A \cup B = B \cup A$
Associative laws:	$A \cup (B \cup C) = (A \cup B) \cup C$
	$A \cap (B \cap C) = (A \cap B) \cap C$
Distributive laws:	$A \cap (B \cup C) = (A \cap B) \cup (A \cap C)$
	$A \cup (B \cap C) = (A \cup B) \cap (A \cup C)$
Idempotent laws:	$A \cup A = A$
	$A \cap A = A$
Absorption laws:	$A \cup (A \cap B) = A$
	$A \cap (A \cup B) = A$
Domination laws:	$A \cup X = X$
	$A \cap \varnothing = \varnothing$
Identity laws:	$A \cup \varnothing = A$
	$A \cap X = A$
De Morgan's laws:	$\overline{A \cup B} = \overline{A} \cap \overline{B}$
	$\overline{A \cap B} = \overline{A} \cup \overline{B}$
Law of excluded middle:	$A \cup \overline{A} = X$
Law of contradiction:	$A \cap \overline{A} = \varnothing$

The subsets of X with operators union, intersection, and complement form a class so-called *Boolean algebra*.

Beyond the union, the intersection, and the complement, there are more set operations that can be defined. Among them, one useful set operation is the difference defined as follows.

Definition 2.9 The *difference* of A and B, denoted by $A-B$, is the set consisting of all points that belong to A but not to B. That is, $A-B=\{x \mid x \in A \text{ and } x \notin B\}$.

The difference is not symmetric with respect to sets A and B generally, that is, $A-B \neq B-A$, except $A=B$. Thus, we may define another kind of difference for two given sets as follows.

Definition 2.10 The *symmetric difference* of A and B, denoted by $A \Delta B$, is the set consisting of all points that belong to exactly one of A and B. That is, $A \Delta B=\{x \mid x \in A-B \text{ or } x \in B-A\}$.

For the symmetric difference, we have $A \Delta B=B \Delta A$ for any sets A and B.

Example 2.5 Using X, A, and B in Example 2.4, we have $A-B=\{4,6\}$, $A \Delta B=\{1,3,4,6\}$,

By using De Morgan's law $\overline{A \cup B}=\overline{A} \cap \overline{B}$, we can express the union in terms of the intersection and the complement as follows:

$$A \cup B=\overline{\overline{A} \cap \overline{B}}.$$

Similarly, by using De Morgan's law $\overline{A \cap B}=\overline{A} \cup \overline{B}$, we can express the intersection in terms of the union and the complement as well:

$$A \cap B=\overline{\overline{A} \cup \overline{B}}.$$

The difference can be expressed in terms of the intersection and the complement, that is, $A-B=A \cap \overline{B}$. The symmetric difference of A and B can be expressed by the other operations:

$$A\Delta B = (A-B)\cup(B-A) = (A\cap\overline{B})\cup(B\cap\overline{A})$$
$$= (A\cup B)-(A\cap B) = (A\cup B)\cap(\overline{A}\cup\overline{B})$$

So, we have only two basic set operators: either the intersection and the complement, or the union and the complement.

2.3 Set Sequences and Set Classes

A mapping from the set of positive integers (or the set of first n positive integers) to the power set of the universal set X is called a *set sequence* (or a *finite set sequence*, respectively) and denoted by $\{A_i\}$ simply, where A_i is the i-th term of the set sequence. It should be emphasized that $\{A_i\}$ cannot be regarded as a set of sets since A_i 's are allowed to be repeated but a set is not allowed to have any duplicate of elements.

The union and the intersection can be extended for more than two sets. The union of sets $A_1, A_2, \cdots,$ and A_n, is denoted by $A_1\cup A_2\cup\cdots\cup A_n$, simply, $\bigcup_{i=1}^n A_i$. Similarly, their intersection is denoted by $A_1\cap A_2\cap\cdots\cap A_n$ or $\bigcap_{i=1}^n A_i$. Furthermore, considering infinitely many subsets of X: A_1, A_2, $A_3, \cdots$, denoted by $\{A_i\}$, their union and intersection are defined as follows.

Definition 2.11 The *union* of $\{A_i\}$, denoted by $\bigcup_{i=1}^\infty A_i$ (or $\bigcup_i A_i$ simply if there is no confusion), is the set consisting of all points that belong to A_i for at least one $i = 1, 2, \cdots$. That is,

$$\bigcup_{i=1}^{\infty} A_i = \{x \mid x\in A_i \text{ for at least one } i = 1, 2, \cdots\}.$$

Definition 2.12 The *intersection* of $\{A_i\}$, denoted by $\bigcap_{i=1}^\infty A_i$ (or $\bigcap_i A_i$ simply if there is no confusion), is the set consisting of all points that belong to all A_i for $i = 1, 2, \cdots$. That is,

$$\bigcap_{i=1}^{\infty} A_i = \{x \mid x \in A_i \text{ for all } i = 1, 2, \cdots\}.$$

As for their corresponding characteristic functions, we have

$$\chi_{\bigcup_i A_i} = \sup_i \chi_{A_i}$$

and

$$\chi_{\bigcap_i A_i} = \inf_i \chi_{A_i}$$

where sup and inf represent the supremum and infimum respectively (see Section 2.6).

Definition 2.13 Set sequence $\{A_i\}$ is *disjoint* iff A_i and A_j are disjoint for any $i \neq j$, $i, j = 1, 2, \cdots$.

When $\{A_i\}$ is disjoint,

$$\chi_{\bigcup_i A_i} = \sum_i \chi_{A_i}.$$

If we only consider finitely many (but more than one) sets A_1, A_2, $\cdots$, A_n, the above discussion on the characteristic functions is still valid. We just need to let $A_{n+1} = A_{n+2} = \cdots = \varnothing$ in set sequence $\{A_i\}$. Of course, the above "sup" and "inf" become "max" and "min" respectively.

Definition 2.14 Set sequence $\{A_i\}$ is *nondecreasing* iff $A_1 \subseteq A_2 \subseteq \cdots$; it is *nonincreasing* iff $A_1 \supseteq A_2 \supseteq \cdots$. Both of them are said to be *monotonic*.

If set sequence $\{A_i\}$ is monotonic, the above-mentioned "sup" and "inf" for the characteristic functions become "lim".

Example 2.6 Let X be the set of all real numbers, i.e., $X = R = (-\infty, \infty)$. Taking $A_i = [i, \infty)$, $i = 1, 2, \cdots$, we know that $\{A_i\}$ is a nonincreasing set sequence. Furthermore, $\bigcup_{i=1}^{\infty} A_i = [1, \infty) = A_1$ and $\bigcap_{i=1}^{\infty} A_i = \varnothing$. We also have $\lim_{i\to\infty} \chi_{A_i}(x) = 0$ for every real number x.

Furthermore, these discussions can be generalized again. Let $\{A_t\} = \{A_t \mid t \in T\}$ be *a family of sets* where T is a nonempty index set. We may define the union and the intersection of $\{A_t\}$ as well.

Definition 2.15 The *union* of $\{A_t \mid t \in T\}$, denoted by $\bigcup_{t\in T} A_t$, is the set consisting of all points that belong to A_t for at least one $t \in T$. That is, $\bigcup_{t\in T} A_t = \{x \mid x \in A_t \text{ for at least one } t \in T\}$.

Definition 2.16 The *intersection* of $\{A_t \mid t \in T\}$, denoted by $\bigcap_{t\in T} A_t$, is the set consisting of all points that belong to all A_t for $t \in T$. That is, $\bigcap_{t\in T} A_t = \{x \mid x \in A_t \text{ for all } t \in T\}$.

Generally, given class $\mathscr{C}$, we use $\cup\mathscr{C}$ and $\cap\mathscr{C}$ to denote sets $\{x \mid x \in A \text{ for some } A \in \mathscr{C}\}$ and $\{x \mid x \in A \text{ for every } A \in \mathscr{C}\}$, respectively.

When index set T is well ordered, such as $T = [0, 1]$, we can also use the concepts of monotonicity.

Similar to the set sequences, for the corresponding characteristic functions, we have

$$\chi_{\bigcup_{t\in T} A_t} = \sup_{t\in T} \chi_{A_t}$$

and

$$\chi_{\bigcap_{t\in T} A_t} = \inf_{t\in T} \chi_{A_t} .$$

Thus, some laws discussed in Section 2.2 (Theorem 2.1) can be generalized as follows.

Associative laws: $$\bigcup_{t\in T}\Big(\bigcup_{s\in S_t} A_s\Big) = \bigcup_{s\in \bigcup_{t\in T} S_t} A_s$$

$$\bigcap_{t\in T}(\bigcap_{s\in S_t} A_s) = \bigcap_{s\in\bigcup_{t\in T} S_t} A_s$$

Distributive laws: $$B \cap (\bigcup_{t\in T} A_t) = \bigcup_{t\in T}(B \cap A_t)$$

$$B \cup (\bigcap_{t\in T} A_t) = \bigcap_{t\in T}(B \cup A_t)$$

De Morgan's laws: $$\overline{\bigcup_{t\in T} A_t} = \bigcap_{t\in T} \overline{A_t}$$

$$\overline{\bigcap_{t\in T} A_t} = \bigcup_{t\in T} \overline{A_t}$$

where S_t and T are index sets and we take the convention that $\bigcup_{\varnothing} \cdot = \varnothing$ and $\bigcap_{\varnothing} \cdot = X$.

For given nonempty class $\mathscr{C}$, we say that $\mathscr{C}$ is disjoint if A and B are disjoint whenever $A, B \in \mathscr{C}$ and $A \neq B$.

Similar to the set sequence, it is convenient to allow duplicate sets in a class of sets sometimes.

2.4 Set Classes Closed Under Set Operations

Let X be the universal set. The class of all subsets of X, denoted by $\mathscr{P}(X)$, is called the *power set* of X.

Definition 2.17 A nonempty class is called a *ring*, denoted by $\mathscr{R}$, iff $E \cup F \in \mathscr{R}$ and $E - F \in \mathscr{R}$ $\forall E, F \in \mathscr{R}$.

In other words, a ring is a nonempty class closed under the formation of unions and differences. Any ring is also closed under the formation of intersection, i.e., $E \cap F \in \mathscr{R}$ $\forall E, F \in \mathscr{R}$. In fact, the intersection can be expressed in terms of difference: $E \cap F = E - (E - F)$.

Example 2.7 The class of all finite subsets of X is a ring.

Example 2.8 The class of all finite unions of bounded left closed right open intervals is a ring.

Definition 2.18 A nonempty class is called a *semiring*, denoted by $\mathscr{S}$, iff

(1) $\forall E, F \in \mathscr{S},\ E \cap F \in \mathscr{S}$;
(2) $\forall E, F \in \mathscr{S}$ satisfying $E \subseteq F$, there exists a finite class $\{C_0, C_1, \cdots, C_n\}$ of sets in $\mathscr{S}$, such that $E = C_0 \subseteq C_1 \subseteq \cdots \subseteq C_n = F$ and $D_i = C_i - C_{i-1} \in \mathscr{S},\ \forall i = 1, 2, \cdots, n$.

Example 2.9 The class consisting of all singletons and the empty set is a semiring.

Example 2.10 The class of all bounded left closed right open intervals is a semiring. Similarly, the class of all bounded left open right closed intervals is also a semiring.

Definition 2.19 An *algebra*, denoted by $\mathscr{A}$, is a ring containing X.

Any algebra is closed under the formation of complements since the complement of a set can be expressed by its difference from X.

Example 2.11 The class consists of all sets in a ring and their complements is an algebra. Therefore, by Example 2.7, the class of all finite subsets of X and their complements is an algebra.

Definition 2.20 A nonempty class is called a *σ-ring*, denoted by $\mathscr{R}_\sigma$, iff

(1) $\forall E, F \in \mathscr{R}_\sigma,\ E - F \in \mathscr{R}_\sigma$;
(2) $\bigcup_{i=1}^{\infty} E_i \in \mathscr{R}_\sigma$, when $E_i \in \mathscr{R}_\sigma$ for $i = 1, 2, \cdots$.

In other words, a σ-ring is a nonempty class closed under the formation of countable unions and differences. Also, we can say that a σ-ring is a ring closed under the formation of countable unions. Any

σ-ring is also closed under the formation of countable intersections. In fact, any countable intersection can be expressed in terms of countable unions and differences as follows:

$$\bigcap_{i=1}^{\infty} A_i = \bigcup_{i=1}^{\infty} A_i - \bigcup_{i=1}^{\infty}(\bigcup_{j=1}^{\infty} A_j - A_i).$$

Example 2.12 The class of all countable subsets of X is a σ-ring.

Definition 2.21 A *σ-algebra* (*σ-field*), denoted by $\mathscr{F}$, is a σ-ring containing X.

Any σ-algebra is closed under the formation of any countable (including finite) set operations that we have defined.

Example 2.13 The class of all countable subsets of X and their complements is a σ-algebra.

The power set of X is a σ-algebra; any σ-algebra is a σ-ring as well as an algebra; any σ-ring or algebra is a ring; and any ring is a semi-ring. These relations are illustrated in Figure 2.1.

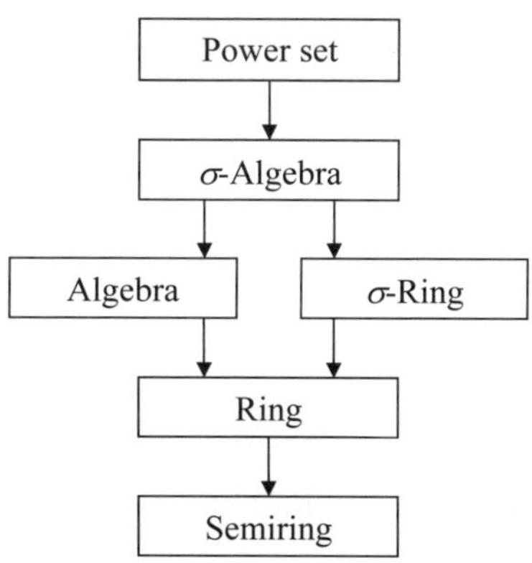

Fig. 2.1 Relations among classes of sets.

For any given semiring $\mathscr{S}$, there exists at least one set E such that $E \in \mathscr{S}$. Since $E \subseteq E$, for any finite class $\{C_0, C_1, \cdots, C_n\}$ of sets in $\mathscr{S}$ satisfying $E = C_0 \subseteq C_1 \subseteq \cdots \subseteq C_n = E$, we must have $C_0 = C_1 = \cdots = C_n = E$, such that $D_i = C_i - C_{i-1} = \emptyset$. This means that the empty set belongs to any semiring. Hence, any ring, any algebra, any σ-ring, and any σ-algebra must contain the empty set.

Theorem 2.2 Let $\mathscr{C}$ be a nonempty class. There exists a unique ring, denoted by $\mathscr{R}(\mathscr{C})$, such that $\mathscr{C} \subseteq \mathscr{R}(\mathscr{C})$ and $\mathscr{C} \subseteq \mathscr{R} \Rightarrow \mathscr{R}(\mathscr{C}) \subseteq \mathscr{R}$ for any ring $\mathscr{R}$. That is, $\mathscr{R}(\mathscr{C})$ is the smallest ring including $\mathscr{C}$.

Proof. Power set $\mathscr{P}(X)$ is a ring including $\mathscr{C}$. Let $\mathbf{C}$ be the set of all rings that include $\mathscr{C}$ and let $\mathscr{R}(\mathscr{C}) = \cap\mathbf{C}$. It is not difficult to verify that $\mathscr{R}(\mathscr{C})$ is still closed under the formations of unions and differences, that is, $\mathscr{R}(\mathscr{C})$ is a ring. Since every ring in $\mathbf{C}$ includes $\mathscr{C}$, so does their intersection $\mathscr{R}(\mathscr{C})$. The uniqueness and being the smallest are guaranteed by the intersection in its definition. □

$\mathscr{R}(\mathscr{C})$ in Theorem 2.2 is called the *ring generated by* $\mathscr{C}$.

Similar conclusions for semiring, algebra, σ-ring, and σ-algebra can also be obtained. We will use symbols $\mathscr{S}(\mathscr{C})$, $\mathscr{A}(\mathscr{C})$, $\mathscr{R}_\sigma(\mathscr{C})$, and $\mathscr{F}(\mathscr{C})$ to denote the semiring, the algebra, the σ-ring and the σ-algebra generated by $\mathscr{C}$, respectively. For any given class $\mathscr{C}$, we have $\mathscr{C} \subseteq \mathscr{S}(\mathscr{C}) \subseteq \mathscr{R}(\mathscr{C}) \subseteq \mathscr{R}_\sigma(\mathscr{C}) \subseteq \mathscr{F}(\mathscr{C})$ and $\mathscr{R}(\mathscr{C}) \subseteq \mathscr{A}(\mathscr{C}) \subseteq \mathscr{F}(\mathscr{C})$.

The ring generated by a semiring can be obtained by collecting all finite disjoint unions of sets in the semiring. The algebra generated by a ring can be obtained by adding the complements of sets in the ring. These can be verified by Examples 2.7-2.11. However, to obtain the σ-ring generated by a ring, the procedure sometimes is very complex.

Example 2.14 The σ-ring generated by either of the semirings shown in Example 2.10 is called the *Borel field* and denoted by $\mathscr{B}$. In fact, it is a σ-algebra. It cannot be obtained by simply taking all countable unions of sets in the semiring and their complements.

2.5 Relations, Posets, and Lattices

Let E and F be nonempty sets.

Definition 2.22 Set $\{(x_1, x_2) | x_1 \in E, x_2 \in F\}$ is called the *product set* of E and F, denoted by $E \times F$.

Example 2.15 Let $X_1 = X_2 = R = (-\infty, \infty)$, the one dimensional Euclidean space (the set of all real numbers, i.e., the real line). Then $X \times Y$ is the two dimensional Euclidean space (the real plane), denoted by R^2, in which each point is an ordered pair of real numbers, (x_1, x_2).

Definition 2.23 A *relation* $\boldsymbol{R}$ *from E to F* is a subset of the product set of E and F, i.e., $\boldsymbol{R} \subseteq E \times F$. According to $\boldsymbol{R}$, if point a in E is related to point b in F, then we write $(a, b) \in \boldsymbol{R}$ or $a\boldsymbol{R}b$. A relation from E to E is simply called a *relation on E*.

Example 2.16 Let Z be the set of all integers. We may define a relation $\boldsymbol{R}_3$ on Z as follows: $a\boldsymbol{R}_3b$ iff $a = b$ (mod 3), i.e., a and b have the same remainder when they are divided by 3.

Example 2.17 Consider Z given in Example 2.16. Symbol $\le$ with the common meaning "less than or equal to" is a relation on Z, denoted by $\boldsymbol{R}_{\le}$. For instance, $(1, 2) \in \boldsymbol{R}_{\le}$, but $(2, 1) \notin \boldsymbol{R}_{\le}$.

Example 2.18 Let X be a nonempty set. The inclusion of sets, $\subseteq$, is a relation on $\mathscr{P}(X)$, i.e., $\{(E, F) | E \subseteq F\}$ is a subset of $\mathscr{P}(X) \times \mathscr{P}(X)$.

Definition 2.24 A relation $\boldsymbol{R}$ on E is:

(1) *reflexive* iff $a\boldsymbol{R}a$ for any $a \in E$;
(2) *symmetric* iff $a\boldsymbol{R}b$ implies $b\boldsymbol{R}a$ for any $a, b \in E$;
(3) *transitive* iff $a\boldsymbol{R}b$ and $b\boldsymbol{R}c$ imply $a\boldsymbol{R}c$ for any $a, b, c \in E$;
(4) *antisymmetric* iff $a\boldsymbol{R}b$ and $b\boldsymbol{R}a$ imply $a = b$ for any $a, b \in E$.

Relation $\boldsymbol{R}_3$ in Example 2.16 is reflexive, symmetric, and transitive. Relations $\boldsymbol{R}_\leq$ and $\subseteq$ in Examples 2.17 and 2.18 are reflexive, transitive, and antisymmetric.

Definition 2.25 A relation $\boldsymbol{R}$ on E is called an *equivalence* relation iff $\boldsymbol{R}$ is reflexive, symmetric, and transitive.

Relation $\boldsymbol{R}_3$ in Example 2.16 is an equivalence relation.

Example 2.19 On $R^2 = (-\infty, \infty) \times (-\infty, \infty)$, for any two points $x = (x_1, x_2)$ and $y = (y_1, y_2)$, define $x \approx y$ iff $x_1^2 + x_2^2 = y_1^2 + y_2^2$. Then relation $\approx$ is an equivalent relation on R^2.

Definition 2.26 Given an equivalence relation $\boldsymbol{R}$ on E and any point $a \in E$, set $\{x|\ x\boldsymbol{R}a\}$ is called the *equivalence class* (with respect to $\boldsymbol{R}$) of a and denoted by $[a]$.

Theorem 2.3 Let $\boldsymbol{R}$ be an equivalence relation on E and $a, b \in E$. Then, $[a] = [b]$ if and only if $a\boldsymbol{R}b$.

Proof. *Necessity*: Since $\boldsymbol{R}$ is an equivalence relation on E, it is reflexive. So, $a\boldsymbol{R}a$ and, therefore, $a \in [a]$. Thus, $[a] = [b]$ means $a \in [b]$.
Sufficiency: Suppose that $a\boldsymbol{R}b$. For any $x \in [a]$, from $x\boldsymbol{R}a$ and the transitivity of $\boldsymbol{R}$, we have $x\boldsymbol{R}b$. This means $x \in [b]$. So, $[a] \subseteq [b]$. By the symmetry of $\boldsymbol{R}$, the reason for $[b] \subseteq [a]$ is totally similar. Thus, $[a] = [b]$. □

Definition 2.27 Let E be a nonempty set. A class of sets $\{E_t \mid t \in T\}$ is called a *partition* of E iff

(1) $E_t \neq \varnothing$ for every $t \in T$;
(2) class $\{E_t \mid t \in T\}$ is disjoint, i.e., $E_t \cap E_s = \varnothing$ for any $t, s \in T$ with $t \neq s$;
(3) $\bigcup_{t \in T} E_t = E$.

Example 2.20 Interval class {[0, 1), [1, 2), [2, 3), [3, 4), [4, 5]} is a partition of interval [0, 5].

Theorem 2.4 Let $\boldsymbol{R}$ be an equivalence relation on E. Then, after deleting the duplicates, class $\{[a] \mid a \in E\}$ is a partition of E.

Proof. (1) For every $a \in E$, since $\boldsymbol{R}$ is reflexive, we have $a \in [a]$, i.e., $[a] \neq \varnothing$.

(2) If there exists a point $x \in [a] \cap [b]$ for two different equivalence class $[a]$ and $[b]$, then from $x\boldsymbol{R}a$, $x\boldsymbol{R}b$, the symmetry and the transitivity of $\boldsymbol{R}$, we have $a\boldsymbol{R}b$. By Theorem 2.3, $[a] = [b]$. Hence, after deleting the duplicates, class $\{[a] \mid a \in E\}$ is disjoint.

(3) For any $a \in E$, there exists $[a] \in \{[a] \mid a \in E\}$ such that $a \in [a]$. So $\bigcup \{[a] \mid a \in E\} = E$. □

Definition 2.28 Let $\boldsymbol{R}$ be an equivalence relation on E. Class $\{[a] \mid a \in E\}$ is called the *quotient set* (or, *quotient space*) of E with respect to $\boldsymbol{R}$.

Example 2.21 In Example 2.16, relation $\boldsymbol{R}_3$ is an equivalence relation on Z, the set of all integers. Equivalence classes $[i] = [i+3]$ for any integer i. Thus, class {[0], [1], [2]} forms a partition of Z, where $[0] = \{\cdots, -6, -3, 0, 3, 6, \cdots\}$, $[1] = \{\cdots, -5, -2, 1, 4, 7, \cdots\}$, and $[2] = \{\cdots, -4, -1, 2, 5, 8, \cdots\}$. Class {[0], [1], [2]} is the quotient set of Z with respect to $\boldsymbol{R}_3$.

Definition 2.29 Relation $\boldsymbol{R}$ on E is called a *partial ordering* if it is reflexive, antisymmetric, and transitive. In this case, $(E, \boldsymbol{R})$ is called a *partial ordered set* (or, *poset*).

In Examples 2.17 and 2.18, $(Z, \leq)$ and $(\mathscr{P}(X), \subseteq)$ are posets.

Example 2.22 On R^n, for any two points $x = (x_1, x_2, \cdots, x_n)$ and $y = (y_1, y_2, \cdots, y_n)$, define $x \leq y$ iff $x_i \leq y_i$ for all $i = 1, 2, \cdots, n$. Then $(R^n, \leq)$ is a poset.

Definition 2.30 A poset (E, $\boldsymbol{R}$) is called a *well* (or, *totally*) *ordered set* or a *chain* iff either $x \le y$ or $y \le x$ for any $x, y \in E$.

In Examples 2.17, (Z, $\le$) is a well ordered set.

In case there is no confusion, we use (P, $\le$) to denote a poset.

Definition 2.31 Let (P, $\le$) be a poset and $E \subseteq P$. A point a in P is called an *upper bound* of E iff $x \le a$ for all $x \in E$. An upper bound a of E is called the *least upper bound* of E (or, *supremum* of E), denoted by sup E or $\vee E$, iff $a \le b$ for any upper bound b of E. A point a in P is called a *lower bound* of E iff $a \le x$ for all $x \in E$. A lower bound a of E is called the *greatest lower bound* of E (or, *infimum* of E), denoted by inf E or $\wedge E$, iff $b \le a$ for any lower bound b of E.

When E consists of only two points, say x and y, we may write $x \vee y$ instead of $\vee \{x, y\}$ and $x \wedge y$ instead of $\wedge \{x, y\}$.

If the least upper bound or the greatest lower bound of a set $E \subseteq P$ exists, then it is unique.

Definition 2.32 A poset (P, $\le$) is called an *upper semilattice* iff $x \vee y$ exists for any $x, y \in P$; A poset (P, $\le$) is called a *lower semilattice* iff $x \wedge y$ exists for any $x, y \in P$; A poset (P, $\le$) is called a *lattice* iff it is both an upper semilattice and a lower semilattice.

Example 2.23 Let X be a nonempty set. Poset ($\mathscr{P}(X)$, $\subseteq$) is a lattice. For any sets $E, F \subseteq X$, $\sup\{E, F\} = E \cup F$ and $\inf\{E, F\} = E \cap F$. However, it is not a well ordered set unless X is a singleton.

2.6 The Supremum and Infimum of Real Number Sets

In this section, we consider the set of all real numbers, called real line sometimes and denoted as R or $(-\infty, \infty)$ directly. Relation $\le$ on R is a full ordering such that $(R, \le)$ is a lattice and, therefore, concepts upper bound, lower bound, supremum, and infimum are also available for any nonempty sets of real numbers.

Example 2.24 Let set E be open interval (a, b). We have $\sup E = b$ and $\inf E = a$.

Example 2.25 Let set E be the set consisting of all real numbers in the sequence $\{a_i\}$, where $a_i = 1 - 2^{-i}$ for $i = 1, 2, \cdots$. Then $\sup E = 1$ and $\inf E = 1/2$.

As a basic property of real numbers sets, the following Proposition is important.

Proposition 2.1 For any nonempty set of real numbers, if it is upper bounded, then its supremum exists; if it is lower bounded, then its infimum exists.

This proposition can be regarded as an axiom and should be always accepted.

Theorem 2.5 Let E be a nonempty set of real numbers. Then for any given $\varepsilon > 0$, there exists $x \in E$ such that $x \geq \sup E - \varepsilon$. Similarly, for any given $\varepsilon > 0$, there exists $x \in E$ such that $x \leq \inf E + \varepsilon$.

Proof. Use a proof by contradiction. Assume that there is no $x \in E$ such that $x \geq \sup E - \varepsilon$. Then $\sup E - \varepsilon$ is an upper bound of *E*. However, $\sup E - \varepsilon < \sup E$. This contradicts the fact that $\sup E$ is the smallest upper bound of *E*. For the infimum, the proof is similar and is omitted. □

From the above theorem directly and the concept of limit, we have the following corollary.

Corollary 2.1 Let E be a nonempty set of real numbers. There exists a sequence $\{a_i\}$ with $a_i \in E$ for $i = 1, 2, \cdots$, such that $\lim_{i \to \infty} a_i = \sup E$. Similarly, there exists a sequence $\{b_i\}$ with $b_i \in E$ for $i = 1, 2, \cdots$, such that $\lim_{i \to \infty} b_i = \inf E$.

Example 2.26 Set E is given in Example 2.25. Sequence $\{a_i\}$ itself has a limit, i.e., $\lim_{i\to\infty} a_i = \sup E = 1$. Taking $b_i = 1/2$ for all $i = 1, 2, \cdots$, we have $\lim_{i\to\infty} b_i = \inf E = 1/2$.

Exercises

Exercise 2.1 Let $X = (-\infty, \infty)$. Explain the following sets and classes in natural language:
(1) $\{x | 0 < x \le 1\}$;
(2) $\{x | x > 0\}$;
(3) $\{0\}$;
(4) $\{\varnothing\}$;
(5) $\{\{x\} | x \in X\}$;
(6) $\{E | E \subset X\}$.

Exercise 2.2 Let $E = [0, 4]$ and $F = [1, 2]$. Find $\chi_{E\cap F}$, $\chi_{E\cup F}$, and χ_{E-F}.

Exercise 2.3 Let $A_i = [1/(i+1),\ 1/i)$, $i = 1, 2, \cdots$. Find $\bigcup_{i=1}^{\infty} A_i$ and $\bigcap_{i=1}^{\infty} A_i$.

Exercise 2.4 Let $A_i = [i, \infty)$, $i = 1, 2, \cdots$. Find $\bigcup_{i=1}^{\infty} A_i$ and $\bigcap_{i=1}^{\infty} A_i$.

Exercise 2.5 Let $A_i = (0, 1/i]$, $i = 1, 2, \cdots$. Find $\bigcup_{i=1}^{\infty} A_i$ and $\bigcap_{i=1}^{\infty} A_i$.

Exercise 2.6 Let the universal set $X = [0, 1]$. Find $\bigcup_{x\in X} \{x\}$ and $\bigcap_{x\in X} \overline{\{x\}}$.

Exercise 2.7 Categorize class $\mathscr{C}$ given in the following descriptions as of a semiring, a ring, an algebra, a σ-ring, a σ-algebra, or none of them:
(1) $X = (-\infty, \infty)$, $\mathscr{C}$ is the class of all bounded, left open, and right closed intervals;
(2) $X = \{1, 2, \cdots\}$, $\mathscr{C} = \{A \mid A \subseteq X, |A| \le 3\}$ where $|A|$ denotes the number of points in set A and called the *cardinality* of A;
(3) X is a nonempty set with $|X| \ge 2$, E is a nonempty proper subset of X, $\mathscr{C} = \{F | E \subset F \subset X\}$;
(4) X is a nonempty set, E is a proper subset of X, $\mathscr{C} = \{F | F \subseteq E\}$;
(5) X is a nonempty set, E is a nonempty subset of X, $\mathscr{C} = \{E\}$.
(6) X is product set $(-\infty, \infty)\times(-\infty, \infty)$, $\mathscr{C} = \{[a, b)\times[c, d) \mid a \le b, c \le d\}$.

Exercise 2.8 Let $X = \{1, 2, \cdots\}$ and $\mathscr{C} = \{\{x\} \mid x \in X\}$. Find $\mathscr{S}(\mathscr{C})$, $\mathscr{R}(\mathscr{C})$, $\mathscr{A}(\mathscr{C})$, $\mathscr{R}_\sigma(\mathscr{C})$, and $\mathscr{F}(\mathscr{C})$.

Exercise 2.9 Let the universal set be the set of all real numbers, i.e., $X = R$, and $\mathscr{S}$ be the set consists of all singletons and the empty set in $\mathscr{P}(X)$. Find $\mathscr{R}(\mathscr{S})$, $\mathscr{R}_\sigma(\mathscr{S})$, $\mathscr{A}(\mathscr{S})$, and $\mathscr{F}(\mathscr{S})$.

Exercise 2.10 Let $\mathscr{C}_1$ and $\mathscr{C}_2$ be nonempty classes satisfying $\mathscr{C}_1 \subseteq \mathscr{C}_2$. Prove that $\mathscr{F}(\mathscr{C}_1) \subseteq \mathscr{F}(\mathscr{C}_2)$. A similar result holds when σ-algebra $\mathscr{F}$ is replaced by semiring $\mathscr{S}$, ring $\mathscr{R}$, algebra $\mathscr{A}$, as well as σ-ring $\mathscr{R}_\sigma$ respectively.

Exercise 2.11 Show that the σ-ring generated by the second semiring shown in Example 2.10 is also the Borel field $\mathscr{B}$.

Exercise 2.12 Let $\mathscr{C}$ be a nonempty class. Prove that $\mathscr{F}(\mathscr{R}(\mathscr{C})) = \mathscr{F}(\mathscr{C})$.

Exercise 2.13 Determine whether (yes or no) each of the following relations on set $A = \{1, 2, 3, 4\}$ is reflexive, symmetric, antisymmetric, and/or transitive.

(1) $\{(1,1), (1,2), (2,1), (2,4), (3,3), (4,4)\}$.

(2) $\{(1,1), (1,3), (1,4), (2,2), (3,1), (3,3), (4,1), (4,4)\}$.

(3) $\{(1,2), (3,4)\}$.

(4) $\{(4,4)\}$.

Exercise 2.14 Let $X = (-\infty, \infty)$ and $\cong$ be the relation on $X \times X$ defined by

$$(x_1, y_1) \cong (x_2, y_2) \quad \text{iff} \quad x_1 + y_1 = x_2 + y_2 .$$

Prove that $\cong$ is an equivalence relation.

Exercise 2.15 Let $X = \{a, b, c, d\}$ and $\mathscr{C} = \{\varnothing, A, \{a, c\}, B, X\}$. Find all possible set pairs A and B such that $\mathscr{C}$ is a chain with respect to set inclusion $\subseteq$.

Exercise 2.16 Let E be the set of all irrational numbers in [0, 1].

(1) Find $\sup E$ and $\inf E$.

(2) Find sequences $\{a_i \mid a_i \in E, i = 1, 2, \cdots\}$ and $\{b_i \mid b_i \in E, i = 1, 2, \cdots\}$ such that $\lim_{i\to\infty} a_i = \sup E$ and $\lim_{i\to\infty} b_i = \inf E$.

Exercise 2.17 Let the universal set be the set of all real numbers, Find the sup and the inf of the following sets.

(1) $A = \{\text{all rational number in } (0, 1)\}$.

(2) $B = \{1, 2, 3\}$.

(3) $C = \{(-1)^i (1 - 1/(i+1)) \mid i = 1, 2, \cdots\}$.

Chapter 3

Fuzzy Sets

3.1 The Membership Functions of Fuzzy Sets

Let X be the universal set. From Chapter 2, we know that any crisp subset (or say, a classical subset) of X, simply called a set if there is no confusion, may be used to describe a crisp concept and is identified by its characteristic function. For a given set, any specified point is either in this set or not in this set, impossible to be both. However, in many real problems, some concepts, called fuzzy concepts, are not so clear. Hence, it is necessary to introduce the concept of *fuzzy subsets* of X (or, simply, *fuzzy sets* if there is no confusion) for describing fuzzy concepts. Similar to the fact that a crisp set is identified by its characteristic function, a fuzzy set is identified by its *membership function*, denoted by $m: X \to [0,1]$. Value $m(x)$ is called the *membership degree* of the fuzzy set at x, where $x \in X$. The characteristic function of sets discussed in Chapter 2 can be regarded as a special case of the membership function of fuzzy sets. So, the concept of fuzzy sets is a generalization of the concept of classical crisp sets.

To simplify the notation, we still used capital letter, such as A, $B, \cdots$, to denote fuzzy sets if there is no confusion. When more than one fuzzy sets are in discussion, we use subscripts to indicate the respective membership function, such as m_A, $m_B, \cdots$ denoting the membership function of fuzzy sets A, $B, \cdots$ respectively. Sometimes, to emphasize that the fuzzy sets are discussed, we use a wave at the top of the symbols, such as $\widetilde{A}$, $\widetilde{B}, \cdots$. When X is the set of real numbers, its fuzzy subsets sometimes may also be denoted by lower case letters with a wave at the top, such as $\tilde{a}$, $\tilde{b}, \cdots$. The set of all fuzzy subsets of X is denoted by $\mathscr{F}(X)$ and called the *fuzzy power set* of X.

Example 3.1 On the age axis, we take $X = [0, 120]$ as the universal set. Concepts "young" and "old" are fuzzy. We may use the following membership functions m_Y and m_O to indicate them respectively.

$$m_Y(x) = \begin{cases} 1 & \text{if } x \leq 25 \\ (40 - x)/15 & \text{if } 25 < x < 40 \\ 0 & \text{if } x \geq 40 \end{cases} \qquad \forall x \in X ;$$

$$m_O(x) = \begin{cases} 0 & \text{if } x \leq 50 \\ (x - 50)/15 & \text{if } 50 < x < 65 \\ 1 & \text{if } x \geq 65 \end{cases} \qquad \forall x \in X \cdot$$

Then, the membership degree of Y at 28 years age is 0.8 while that of O at 45 years age is 0. The graph of membership functions m_Y and m_O are shown in Figures 3.1 and 3.2 respectively.

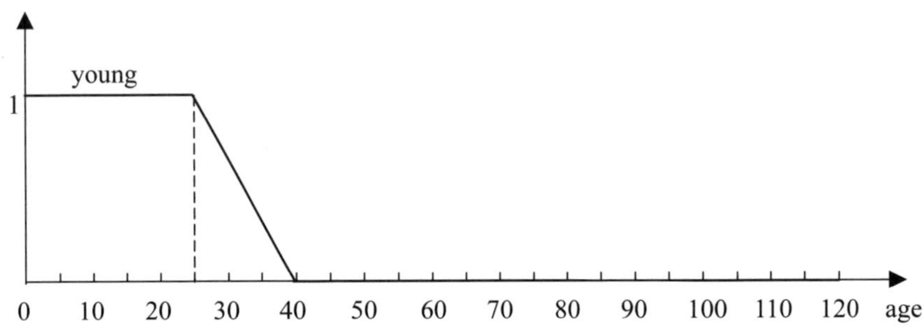

Fig. 3.1 The membership function of Y.

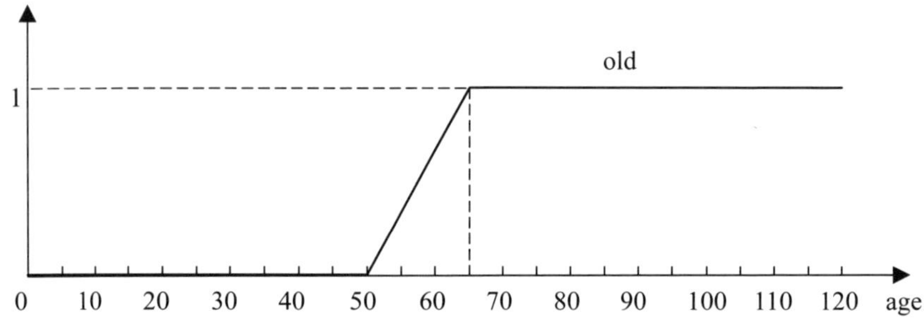

Fig. 3.2 The membership function of O.

Example 3.2 Let $X = \{0, 1, 2, \cdots\}$. Fuzzy concept "around 10" can be expressed as a fuzzy subset of *X*, denoted by *D*, with membership function

$$m_D(x) = \begin{cases} 0.3 & \text{if } x = 10 \pm 2 \\ 0.8 & \text{if } x = 10 \pm 1 \\ 1 & \text{if } x = 10 \\ 0 & \text{otherwise} \end{cases} .$$

Its graph is shown in Figure 3.8(a) on page 43. By using Zadeh's notation, it can also be denoted as

$$m_D = 0.3/8 + 0.8/9 + 1/10 + 0.8/11 + 0.3/12 ,$$

where symbol "/" does not mean "divided by" but means "at". Alternatively, we may intuitively write

$$m_D = \left\{ \frac{0.3}{8}, \frac{0.8}{9}, \frac{1}{10}, \frac{0.8}{11}, \frac{0.3}{12} \right\},$$

where symbol "—" means "at".

Definition 3.1 Let A be a fuzzy subset of X with membership function $m_A(x)$. The *support set* of A, denoted by supp A, is the crisp set described as follows:

$$\text{supp } A = \{x \mid m_A(x) > 0, x \in X\}.$$

The support sets of fuzzy sets Y, O, and D in Examples 3.1 and 3.2 are [0, 40), (50, 120], and {8, 9, 10, 11, 12}, respectively.

3.2 Inclusion and Operations of Fuzzy Sets

Let A and B be fuzzy subsets of universal set X.

Definition 3.2 Fuzzy set A is *included* by fuzzy set B, denoted by $A \subseteq B$, iff $m_A(x) \leq m_B(x) \quad \forall x \in X$. Fuzzy sets A and B are *equal*, denoted by $A = B$, iff $A \subseteq B$ and $B \subseteq A$.

The concepts of inclusion and equality for fuzzy sets are generalizations of the concepts of inclusion and equality for crisp sets given in Definitions 2.1 and 2.2 respectively.

Definition 3.3 The *union* of A and B, denoted by $A \cup B$, is the fuzzy set possessing membership function

$$m_{A \cup B}(x) = m_A(x) \vee m_B(x) = \max[m_A(x), m_B(x)] \quad \forall x \in X.$$

Moreover, if $\{A_t \mid t \in T\}$ be a class of fuzzy sets, where T is an index set, then the union of $\{A_t \mid t \in T\}$, denoted by $\bigcup_{t \in T} A_t$, has a membership function

$$m_{\bigcup_{t \in T} A_t}(x) = \sup_{t \in T} m_{A_t}(x) \quad \forall x \in X.$$

Definition 3.4 The *intersection* of A and B, denoted by $A \cap B$, is the fuzzy set possessing membership function

$$m_{A\cap B}(x) = m_A(x) \wedge m_B(x) = \min[m_A(x), m_B(x)] \quad \forall x \in X .$$

Similarly, if $\{A_t \mid t \in T\}$ be a class of fuzzy sets, where T is an index set, then the intersection of $\{A_t \mid t \in T\}$, denoted by $\bigcap_{t\in T} A_t$, has a membership function

$$m_{\bigcap_{t\in T} A_t}(x) = \inf_{t\in T} m_{A_t}(x) \quad \forall x \in X .$$

Definition 3.5 The *complement* of fuzzy set A, denoted by $\overline{A}$, is the fuzzy set possessing membership function

$$m_{\overline{A}}(x) = 1 - m_A(x) \quad \forall x \in X .$$

The concepts of union, intersection, and complement for fuzzy sets are also generalizations of the corresponding concepts for crisp sets given in Definitions 2.5, 2.6, and 2.7 respectively. Similar to the laws for operations for crisp sets shown in Section 2.2, the following theorem gives the laws of operations for fuzzy sets. Its proof is omitted as well.

Theorem 3.1 The operations of union, intersection, and complement of fuzzy sets satisfy the following laws.

Involution law: $\overline{\overline{A}} = A$

Commutative laws: $A \cup B = B \cup A$, $A \cap B = B \cap A$

Associative laws: $\bigcup_{t\in T}(\bigcup_{s\in S_t} A_s) = \bigcup_{s\in \bigcup_{t\in T} S_t} A_s$

$\bigcap_{t\in T}(\bigcap_{s\in S_t} A_s) = \bigcap_{s\in \bigcup_{t\in T} S_t} A_s$

Distributive laws: $B \cap (\bigcup_{t \in T} A_t) = \bigcup_{t \in T} (B \cap A_t)$

$B \cup (\bigcap_{t \in T} A_t) = \bigcap_{t \in T} (B \cup A_t)$

Idempotent laws: $A \cup A = A$

$A \cap A = A$

Absorption laws: $A \cup (A \cap B) = A$

$A \cap (A \cup B) = A$

Domination laws: $A \cup X = X$

$A \cap \varnothing = \varnothing$

Identity laws: $A \cup \varnothing = A$

$A \cap X = A$

De Morgan's laws: $\overline{\bigcup_{t \in T} A_t} = \bigcap_{t \in T} \overline{A_t}$

$\overline{\bigcap_{t \in T} A_t} = \bigcup_{t \in T} \overline{A_t}$

where S_t and T are index sets.

Note that the following laws hold for crisp sets, but they are not in the above list.

Law of excluded middle: $A \cup \overline{A} = X$
Law of contradiction: $A \cap \overline{A} = \varnothing$.

In fact, these two laws are not true for fuzzy sets generally. So, the fuzzy sets in $\mathscr{F}(X)$ with operators union, intersection, and complement form a *De Morgan algebra* (or say, *soft algebra*), but not a *Boolean algebra* mentioned in Section 2.2.

Example 3.3 In Example 3.1, the complement of "young", read as "not young" and denoted by $\overline{Y}$, has membership function

$$m_{\overline{Y}}(x)=\begin{cases}0 & \text{if } x\le 25\\ (x-25)/15 & \text{if } 25<x<40\\ 1 & \text{if } x\ge 40\end{cases}.$$

The concept "not young and not old", called "middle age" and denoted by *M*, is a fuzzy set possessing membership function m_M that has the form

$$m_M(x)=m_{\overline{Y}\cap\overline{O}}(x)=m_{\overline{Y\cup O}}(x)=1-m_{Y\cup O}(x)=\begin{cases}0 & \text{if } x\le 25 \text{ or } x\ge 65\\ (x-25)/15 & \text{if } 25<x<40\\ 1 & \text{if } 40\le x\le 50\\ (65-x)/15 & \text{if } 50<x<65\end{cases}.$$

The graphs of the membership functions $m_{\overline{Y}}$ and m_M are shown in Figures. 3.3 and 3.4 respectively.

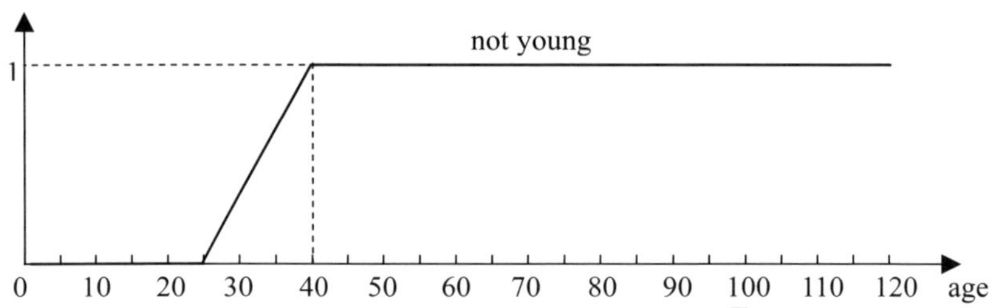

Fig. 3.3 The membership function of $\overline{Y}$.

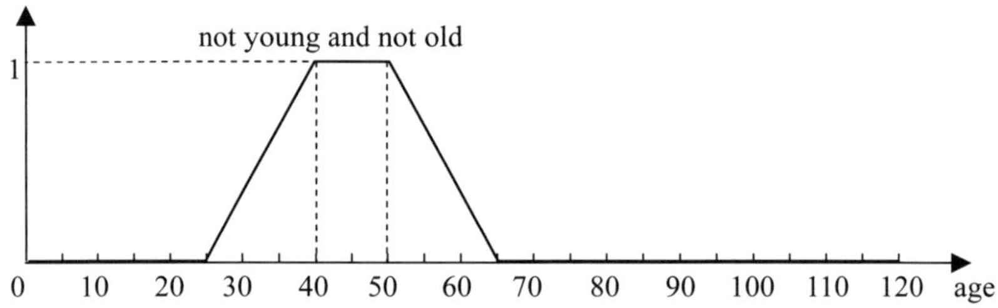

Fig. 3.4 The membership function of *M*.

Definition 3.6 A nonempty class of fuzzy sets $\{A_t \mid t \in T\}$ is called a *fuzzy partition* of fuzzy set A iff

(FP1) $\text{supp}\, A_t \neq \varnothing \quad \forall t \in T$;

(FP2) $\sum_{t \in T} m_{A_t}(x) = m_A(x) \quad \forall x \in X$,

where T is a nonempty index set.

In Definition 3.6, index set T may be infinite, where we can conclude that, for each $x \in X$, there are at most countably many t in T such that $m_{A_t}(x) > 0$. The concept of fuzzy partition is a generalization of the concept of partition for crisp sets shown in Section 2.5. However, conditions (2) and (3) in Definition 2.27 may be violated by a fuzzy partition.

Example 3.4 In Example 3.3, class $\{Y, M, O\}$ is a fuzzy partition of X. We may see that $Y \cap M \neq \varnothing$, $M \cap O \neq \varnothing$, and $Y \cup M \cup O \neq X$.

Example 3.5 The range of the evaluation to each criterion for submitted research papers by an academic journal editor is the interval I = [0, 5]. However, the reviewers, usually, are only required to rate the criteria by the following words: “bad”, “weak”, “fair”, “good”, and “excellent”. These are fuzzy concepts and can be described by fuzzy subsets of I:

$$\widetilde{a}_b,\ \widetilde{a}_w,\ \widetilde{a}_f,\ \widetilde{a}_g,\ \text{and}\ \widetilde{a}_e,$$

with membership functions

$$m_b(t) = \begin{cases} 1 & \text{if } t \in [0, 1] \\ 3 - 2t & \text{if } t \in (1, 1.5] \\ 0 & \text{otherwise} \end{cases},$$

$$m_w(t)=\begin{cases}2t-2 & \text{if } t\in[1,1.5)\\ 1 & \text{if } t\in[1.5,2]\\ 5-2t & \text{if } t\in(2,2.5]\\ 0 & \text{otherwise}\end{cases},$$

$$m_f(t)=\begin{cases}2t-4 & \text{if } t\in[2,2.5)\\ 1 & \text{if } t\in[2.5,3]\\ 7-2t & \text{if } t\in(3,3.5]\\ 0 & \text{otherwise}\end{cases},$$

$$m_g(t)=\begin{cases}2t-6 & \text{if } t\in[3,3.5)\\ 1 & \text{if } t\in[3.5,4]\\ 9-2t & \text{if } t\in(4,4.5]\\ 0 & \text{otherwise}\end{cases},$$

and

$$m_e(t)=\begin{cases}2t-8 & \text{if } t\in[4,4.5)\\ 1 & \text{if } t\in[4.5,5]\\ 0 & \text{otherwise}\end{cases},$$

respectively. Then, $\{\tilde{a}_b, \tilde{a}_w, \tilde{a}_f, \tilde{a}_g, \tilde{a}_e\}$ is a fuzzy partition of *I*. Figure 3.5 shows the membership functions of these five fuzzy sets.

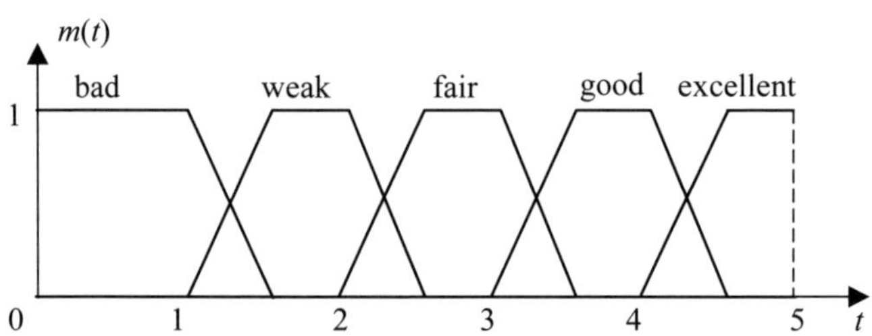

Fig. 3.5 Membership functions of $\tilde{a}_b, \tilde{a}_w, \tilde{a}_f, \tilde{a}_g, \tilde{a}_e$.

Generally, any fuzzy set and its complement form a fuzzy partition of X if none of them is empty.

3.3 α-Cuts

In Section 3.5, we can see that any fuzzy set may be expressed by a class of crisp sets, which are called α-cuts defined as follows.

Definition 3.7 Let A be a fuzzy subset of X. For any $\alpha \in [0, 1]$, crisp set $\{x \mid m_A(x) \geq \alpha, x \in X\}$ is called the *α-cut set* (or, simply, *α-cut*) of A, denoted by A_α; while crisp set $\{x \mid m_A(x) > \alpha, x \in X\}$ is called the *strong α-cut set* (or, simply, *strong α-cut*) of A, denoted by $A_{\alpha+}$.

It is clear that $A_0 = X$, $A_{0+} = \text{supp } A$, and $A_{1+} = \varnothing$.

Example 3.6 In Example 3.1, $Y_{0.5} = [0, 32.5]$ and $Y_{0.5+} = [0, 32.5)$. They can be seen from Figure 3.6.

Example 3.7 In Example 3.2, $D_{0.5} = D_{0.5+} = \{9, 10, 11\}$, but $D_{0.8} = \{9, 10, 11\}$ and $D_{0.8+} = \{10\}$ are different.

Following Theorems 3.2, 3.3, 3.4, and 3.5 show some properties of α-cuts. The first two are direct results from the definitions and, therefore, their proofs are omitted.

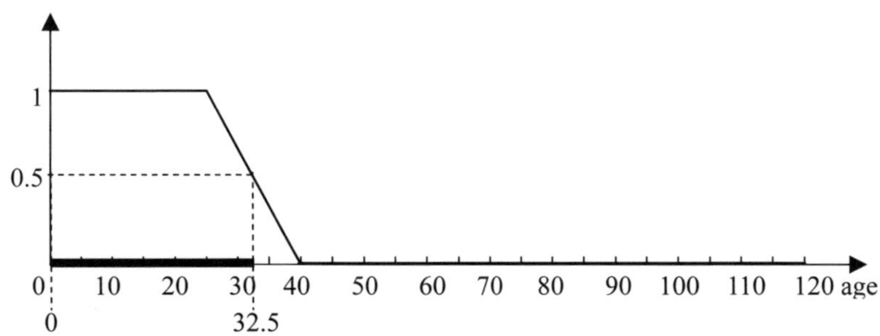

Fig. 3.6 The α-cut and strong α-cut of fuzzy set Y when $\alpha = 0.5$.

Theorem 3.2 For any fuzzy set A,

$$\text{If } \alpha_1 \le \alpha_2 \text{, then } A_{\alpha_2} \subseteq A_{\alpha_1} \text{ and } A_{\alpha_2 +} \subseteq A_{\alpha_1 +}.$$

Theorem 3.3 Let A and B be fuzzy sets.

$$\text{If } A \subseteq B \text{, then } A_\alpha \subseteq B_\alpha \text{ and } A_{\alpha+} \subseteq B_{\alpha+}.$$

Theorem 3.4 For any fuzzy set A and any $\alpha \in [0, 1]$, $A_{\alpha+} \subseteq A_\alpha$,

$$A_{\alpha+} = \bigcup_{\beta>\alpha} A_\beta \text{, and } A_\alpha = \bigcap_{\beta<\alpha} A_{\beta+}.$$

Proof. The first inclusion is obtained from the implication

$$m_A(x) > \alpha \Rightarrow m_A(x) \ge \alpha .$$

Equality $A_{\alpha+} = \bigcup_{\beta>\alpha} A_\beta$ holds since

$$\begin{aligned} x \in A_{\alpha+} &\Leftrightarrow m_A(x) > \alpha \Leftrightarrow \exists \beta > \alpha \text{ s.t. } m_A(x) \ge \beta \Leftrightarrow \exists \beta > \alpha \text{ s.t. } x \in A_\beta \\ &\Leftrightarrow x \in \bigcup_{\beta>\alpha} A_\beta , \end{aligned}$$

where symbol "$\Leftrightarrow$" means "is equivalent to". As for the last equality $A_\alpha = \bigcap_{\beta<\alpha} A_{\beta+}$, it can be obtained from

$$\begin{aligned} x \in A_\alpha &\Leftrightarrow m_A(x) \ge \alpha \Leftrightarrow \forall \beta < \alpha, m_A(x) > \beta \Leftrightarrow \forall \beta < \alpha, x \in A_{\beta+} \\ &\Leftrightarrow x \in \bigcap_{\beta<\alpha} A_{\beta+} . \end{aligned}$$

□

Theorem 3.5 Let $\{A_t \mid t \in T\}$ be a class of fuzzy sets, where T is an index set. Then, for every $\alpha \in [0, 1]$,

(1) $\bigcup_{t\in T}(A_t)_\alpha \subseteq (\bigcup_{t\in T} A_t)_\alpha$;

(2) $\bigcap_{t\in T}(A_t)_\alpha = (\bigcap_{t\in T} A_t)_\alpha$;

(3) $\bigcup_{t\in T}(A_t)_{\alpha+} = (\bigcup_{t\in T} A_t)_{\alpha+}$;

(4) $\bigcap_{t\in T}(A_t)_{\alpha+} \supseteq (\bigcap_{t\in T} A_t)_{\alpha+}$.

Proof. We only prove (1). In fact,

$$x \in \bigcup_{t\in T}(A_t)_\alpha \Rightarrow \exists t \in T \text{ s.t.} x \in (A_t)_\alpha \Rightarrow \exists t \in T \text{ s.t.} m_{A_t}(x) \geq \alpha \Rightarrow \sup_{t\in T} m_{A_t}(x) \geq \alpha$$
$$\Rightarrow m_{\bigcup_{t\in T} A_t}(x) \geq \alpha \Rightarrow x \in (\bigcup_{t\in T} A_t)_\alpha.$$

The proofs of (2), (3), and (4) are similar to the proof of (1). □

It should be noted that the inverse inclusions of inclusions in (1) and (4) of Theorem 3.5 do not hold in general. A counterexample is shown as follows.

Example 3.8 Let X be a singleton $\{a\}$, A_t have membership function $m_{A_t}(a) = t$, and $T = [0, 1)$. Then $(A_t)_1 = \varnothing$ for every $t \in T$ such that $\bigcup_{t\in T}(A_t)_1 = \varnothing$. However, $\bigcup_{t\in T} A_t$ has membership function $m_{\bigcup_{t\in T} A_t}(a) = \sup_{t\in T} t = 1$ such that $\bigcup_{t\in T}(A_t)_1 = X$. This shows that the inclusion in (1) of Theorem 3.5 cannot be replaced by equality generally.

In a special case when T is a finite index set, we have better conclusions.

Theorem 3.6 Let $\{A_t \mid t \in T\}$ be a set of fuzzy sets, where T is a finite index set. Then, for every $\alpha \in [0, 1]$,

(1*) $\bigcup_{t\in T}(A_t)_\alpha = (\bigcup_{t\in T} A_t)_\alpha$;

(4*) $\bigcap_{t\in T}(A_t)_{\alpha+} = (\bigcap_{t\in T} A_t)_{\alpha+}$.

Theorem 3.7 For any fuzzy set A and any $\alpha \in [0, 1]$, $(\overline{A})_\alpha = \overline{A_{(1-\alpha)+}}$.

Proof. The conclusion comes from

$$(\overline{A})_\alpha = \{x \mid 1 - m_A(x) \geq \alpha\} = \{x \mid m_A(x) \leq 1-\alpha\} = \overline{\{x \mid m_A(x) > 1-\alpha\}}$$
$$= \overline{A_{(1-\alpha)+}} \ .$$

□

3.4 Convex Fuzzy Sets

In this section, we consider the fuzzy subsets of n-dimensional Euclidean space (or its convex subset), i.e., $X = R^n$, $n = 1, 2, \cdots$. An important class of fuzzy sets is the class of convex fuzzy sets.

A crisp subset of R^n is *convex* if, for any two points x_1 and x_2 in this subset, point $cx_1 + (1-c)x_2$ is also in this subset, where c may be any real number in [0, 1]. In one-dimensional case, any convex set is an interval (either closed, or open, or left closed right open, or left open right closed) and vice versa.

Definition 3.8 A fuzzy set is *convex* iff its α-cut is convex for every $\alpha \in [0, 1]$.

Theorem 3.8 Fuzzy set A is convex if and only if

$$m_A(cx_1 + (1-c)x_2) \geq \min[m_A(x_1), m_A(x_2)]$$

for any $c \in [0, 1]$.

Proof. *Necessity*: For any $x_1 \in X$ and $x_2 \in X$, taking $\alpha = \min[m_A(x_1), m_A(x_2)]$, we have $x_1 \in A_\alpha$ and $x_2 \in A_\alpha$. From the convexity of A_α, we know that $cx_1 + (1-c)x_2 \in A_\alpha$ for any $c \in [0,1]$. This means that $m_A(cx_1 + (1-c)x_2) \geq \alpha = \min[m_A(x_1), m_A(x_2)]$ for any $c \in [0,1]$.
Sufficiency: We want to show that for any $\alpha \in [0,1]$, A_α is convex if $m_A(cx_1 + (1-c)x_2) \geq \min[m_A(x_1), m_A(x_2)]$ for any $c \in [0,1]$. In fact, for any given $\alpha \in [0,1]$, if $x_1 \in A_\alpha$ and $x_2 \in A_\alpha$, that is, $m_A(x_1) \geq \alpha$ and $m_A(x_2) \geq \alpha$, then for any $c \in [0,1]$,

$$m_A(cx_1 + (1-c)x_2) \geq \min[m_A(x_1), m_A(x_2)] \geq \alpha .$$

This means that $cx_1 + (1-c)x_2 \in A_\alpha$. So, A_α is convex. □

Example 3.9 In Examples 3.1 and 3.3, Fuzzy sets *Y*, *O*, *M*, and $\overline{Y}$ are convex fuzzy sets.

The membership function of a convex fuzzy set is not necessarily convex (also called concave down, in some books) in the meaning discussed in calculus.

Example 3.10 Let *X* be the real line *R*. the fuzzy set with membership function

$$m(x) = e^{-x^2}$$

is convex since all α-cuts are intervals (see Figure 3.7). However, function e^{-x^2} is not convex (concave down) on *R* in calculus.

3.5 Decomposition Theorems

In this section, we discuss how to express a fuzzy set by its α-cuts. Let *A* be a fuzzy subset of universal set *X*. For any crisp set *E* and any real number $\alpha \in [0,1]$, we use αE to denote the fuzzy set having membership function

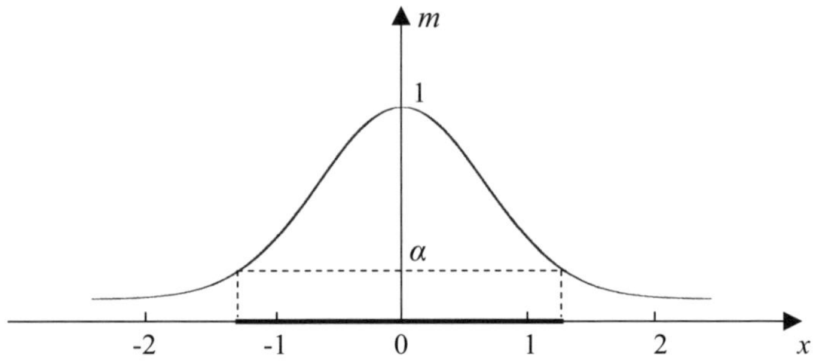

Fig. 3.7 An α-cut of convex fuzzy set with membership function $m(x) = e^{-x^2}$.

$$m_{\alpha E}(x) = \begin{cases} \alpha & \text{if } x \in E \\ 0 & \text{if } x \notin E \end{cases} \quad \forall x \in X .$$

Theorem 3.9 (Decomposition Theorem I)

$$A = \bigcup_{\alpha \in [0,1]} \alpha A_\alpha = \bigcup_{\alpha \in (0,1]} \alpha A_\alpha .$$

Proof. On one hand, $m_A(x) \ge \alpha \chi_{A_\alpha}(x) = m_{\alpha A_\alpha}(x)$ for every $x \in X$ and every $\alpha \in [0, 1]$. So, $m_A(x) \ge \sup_{\alpha \in [0,1]} m_{\alpha A_\alpha}(x)$ for every $x \in X$, i.e., $A \supseteq \bigcup_{\alpha \in [0,1]} \alpha A_\alpha$. On the other hand, for any given $x \in X$, denoting $m_A(x)$ by $\alpha(x)$, we have

$$m_A(x) = \alpha(x) = \alpha(x) \chi_{A_{\alpha(x)}}(x) \le \sup_{\alpha \in [0,1]} \alpha \chi_{A_\alpha}(x) = m_{\bigcup_{\alpha \in [0,1]} \alpha A_\alpha}(x)$$

since $x \in A_{\alpha(x)}$. This means $A \subseteq \bigcup_{\alpha \in [0,1]} \alpha A_\alpha$. Consequently, $A = \bigcup_{\alpha \in [0,1]} \alpha A_\alpha$. The second equality $\bigcup_{\alpha \in [0,1]} \alpha A_\alpha = \bigcup_{\alpha \in (0,1]} \alpha A_\alpha$ is evident. □

Similarly, the decomposition can also be made by using strong α-cuts as shown in the next theorem.

Theorem 3.10 (*Decomposition Theorem II*)

$$A = \bigcup_{\alpha \in [0,1]} \alpha A_{\alpha+} = \bigcup_{\alpha \in (0,1]} \alpha A_{\alpha+} \,.$$

Proof. The second equality is also evident. As for the first equality, based on the result obtained in Theorem 3.9, we only need to show $A \subseteq \bigcup_{\alpha\in[0,1]} \alpha A_{\alpha+}$ since $\bigcup_{\alpha\in[0,1]} \alpha A_{\alpha+} \subseteq \bigcup_{\alpha\in[0,1]} \alpha A_{\alpha}$. For any given $x \in X$ and $\varepsilon > 0$ that is small enough, we have

$$\begin{aligned} m_A(x) &= \sup_{\alpha\in[0,1]} \alpha \chi_{A_\alpha}(x) = \sup_{\alpha\in[-\varepsilon,1-\varepsilon]} [\alpha+\varepsilon]\chi_{A_{\alpha+\varepsilon}}(x) = \sup_{\alpha\in[-\varepsilon,1]} [\alpha+\varepsilon]\chi_{A_{\alpha+\varepsilon}}(x) \\ &\le \sup_{\alpha\in[-\varepsilon,1]} \alpha\chi_{A_{\alpha+\varepsilon}}(x) + \varepsilon \le \sup_{\alpha\in[-\varepsilon,1]} \alpha\chi_{A_{\alpha+}}(x) + \varepsilon = \sup_{\alpha\in[0,1]} \alpha\chi_{A_{\alpha+}}(x) + \varepsilon \\ &= m_{\bigcup_{\alpha\in(0,1]} \alpha A_{\alpha+}}(x) + \varepsilon \,. \end{aligned}$$

Since ε can be arbitrarily close to 0, we obtain that

$$m_A(x) \le m_{\bigcup_{\alpha\in(0,1]} \alpha A_{\alpha+}}(x) \,. \qquad \square$$

We can establish the third decomposition theorem after introducing the concept of level-value set.

Definition 3.9 Set $\{\alpha \mid m_A(x) = \alpha \text{ for some } x \in X\}$ is called the *level-value set* of fuzzy set *A* and denoted by L_A .

Example 3.11 In Example 3.1, $L_Y = L_O = [0, 1]$; while in Example 3.2, $L_D = \{0, 0.3, 0.8, 1\}$.

Based on the concept of level-valued set, we may delete some (may be most) values of α for taking the union in the expression shown in Theorem 3.9 to obtain the expression in the next theorem.

Theorem 3.11 (*Decomposition Theorem III*)

$$A = \bigcup_{\alpha \in L_A} \alpha A_\alpha .$$

Proof. Due to the Decomposition Theorem I, we have

$$A = \bigcup_{\alpha \in [0,1]} \alpha A_\alpha \supseteq \bigcup_{\alpha \in L_A} \alpha A_\alpha .$$

Hence, only inclusion $A \subseteq \bigcup_{\alpha \in L_A} \alpha A_\alpha$ needs to be shown. In fact, for any given $x \in X$, let $m_A(x) = \alpha$. It means that $\alpha \in L_A$ and $x \in A_\alpha$. This yields that

$$m_A(x) = \alpha \chi_{A_\alpha}(x) \le \sup_{\alpha \in L_A} \alpha \chi_{A_\alpha}(x) . \qquad \square$$

There are some examples to show that the α-cuts in the express of Decomposition Theorem III cannot be replaced by strong α-cuts.

3.6 The Extension Principle

The following *extension principle* is a useful tool for fuzzifying classical mathematical concepts such as functions and common binary operators for real numbers.

Extension Principle: Let X_1, X_2, $\cdots$, X_n, and Y be nonempty crisp sets, $U = X_1 \times X_2 \times \cdots \times X_n$ be the product set of X_1, X_2, $\cdots$, and X_n, and let $f : U \to Y$ be a mapping from U to Y. Then, f can be extended to be f: $\mathscr{F}(X_1) \times \mathscr{F}(X_2) \times \cdots \times \mathscr{F}(X_n) \to \mathscr{F}(Y)$ as follows: for any given n fuzzy sets $A_i \in \mathscr{F}(X_i)$, $i = 1, 2, \cdots, n$, fuzzy set $B = f(A_1, A_2, \cdots, A_n) \in \mathscr{F}(Y)$ has membership function

$$m_B(y) = \sup_{x_1, x_2, \ldots, x_n | y = f(x_1, x_2, \cdots, x_n)} \min[m_{A_1}(x_1), m_{A_2}(x_2), \cdots, m_{A_n}(x_n)]$$

with a convention

$$\sup_{\varnothing}\{x \mid x \in [0,1]\} = 0\,.$$

As a special case, if $*$ is a binary operator on universal set X, that is, $*: X \times X \to X$, then, by the extension principle, we can obtain a binary operator on $\mathscr{F}(X)$: for any $A, B \in \mathscr{F}(X)$,

$$m_{A*B}(z) = \sup_{x,y|x*y=z} [m_A(x) \wedge m_B(y)] \qquad \forall z \in X\,.$$

Example 3.12 Let X be the set of all nonnegative integers. The traditional addition for integers can be extended to be a binary operator for fuzzy subsets of X. For instance, assume that fuzzy sets "around 10", denoted by D, and "around 5", denoted by F, have membership functions

$$m_D(x) = \begin{cases} 0.3 & \text{if } x = 10 \pm 2 \\ 0.8 & \text{if } x = 10 \pm 1 \\ 1 & \text{if } x = 10 \\ 0 & \text{otherwise} \end{cases}$$

and

$$m_F(x) = \begin{cases} 0.2 & \text{if } x = 5 \pm 2 \\ 0.7 & \text{if } x = 5 \pm 1 \\ 1 & \text{if } x = 5 \\ 0 & \text{otherwise} \end{cases}$$

$\forall x \in X$, respectively. Then the sum of D and F, denoted by $D + F$, has its membership function

$$m_{D+F}(z) = \sup_{x,y|x+y=z} [m_D(x) \wedge m_F(y)] = \begin{cases} 0.2 & \text{if } z = 15 \pm 4 \\ 0.3 & \text{if } z = 15 \pm 3 \\ 0.7 & \text{if } z = 15 \pm 2 \\ 0.8 & \text{if } z = 15 \pm 1 \\ 1 & \text{if } z = 15 \\ 0 & \text{otherwise} \end{cases}, \quad \forall z \in X .$$

The membership functions of D, F, and $D + F$ are shown in Figure 3.8, where we use the height of a solid small circle to indicate the value of a function at each point.

Fuzzy sets D, F, and $D+F$ are called fuzzy integers that are defined in Section 3.10. Example 3.12 shows the addition of two fuzzy integers obtained by the extension principle.

3.7 Interval Numbers

Let R be the set of all real numbers, i.e., $R = (-\infty, \infty)$.

Definition 3.10 Any closed interval $[l, r]$ is called an *interval number*, where $l \leq r$.

Any crisp real number, a, can be regarded as an interval number $[a, a]$. The set of all interval numbers is denoted by $\mathcal{N}_I$.

Definition 3.11 (*Classical extension*) Let $*$ be a binary operator for real numbers. Operator $*$ can be extended to be a binary operator for interval numbers as follows. Let $[a, b]$ and $[c, d]$ be two interval numbers. Then

$$[a, b] * [c, d] = \{x * y \mid a \leq x \leq b, c \leq y \leq d\}$$

if $x * y$ is well defined for all $x \in [a, b]$ and $y \in [c, d]$.

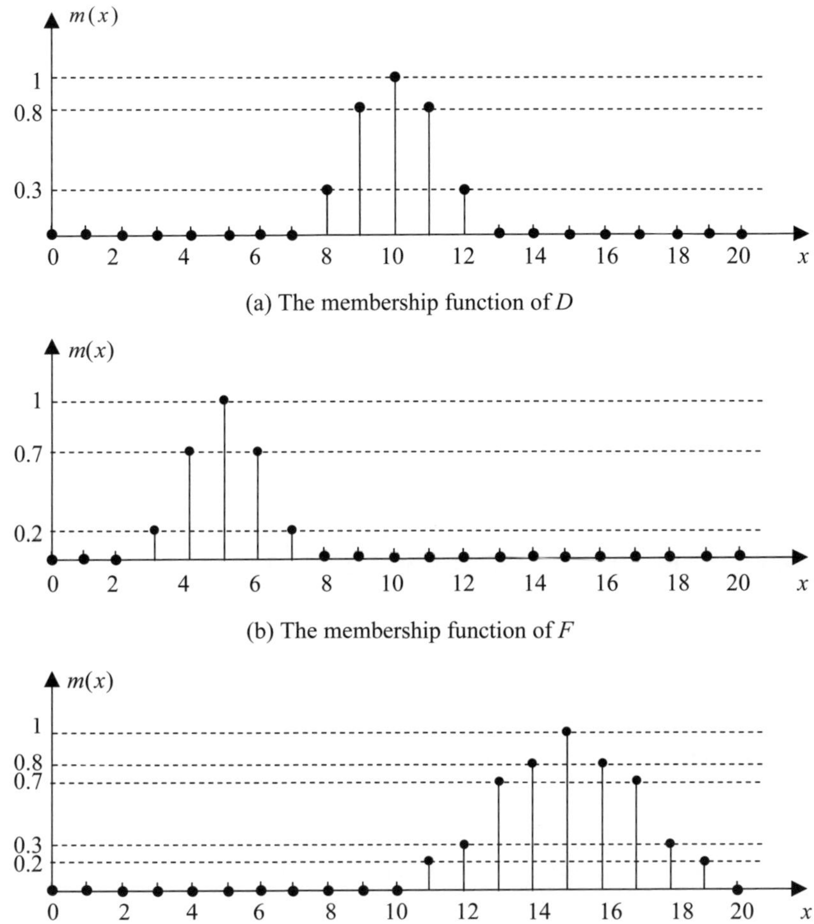

(a) The membership function of D

(b) The membership function of F

(c) The membership function of D+F

Fig. 3.8 The membership function of D+F obtained by the extension principle.

Now, for real numbers, we consider six binary operators: addition +, subtraction −, multiplication ×, division /, maximum ∨, and minimum ∧. According to Definition 3.11, we have

addition: $[a, b]+[c, d]=[a+c, b+d]$;
subtraction: $[a, b]-[c, d]=[a-d, b-c]$;

multiplication:

$$[a, b] \times [c, d] = [\min(ac, ad, bc, bd), \max(ac, ad, bc, bd)];$$

division:

$$[a, b]/[c, d] = [\min(a/c, a/d, b/c, b/d), \max(a/c, a/d, b/c, b/d)]$$
if $0 \notin [c, d]$;

maximum: $[a, b] \vee [c, d] = [a \vee c, b \vee d]$;

minimum: $[a, b] \wedge [c, d] = [a \wedge c, b \wedge d]$.

Example 3.13 $[1, 5] + [-2, 3] = [-1, 8]$, while $[-1, 8] - [-2, 3] = [-4, 10]$. So, the subtraction is not the inverse operation of the addition.

Example 3.14 $[1, 5] \times [-2, 3] = [\min(-2, 3, -10, 15), \max(-2, 3, -10, -15)]$
$= [-10, 15]$, while
$[1, 5] \times [-3, -2] = [\min(-3, -2, -15, -10), \max(-3, -2, -15, -10)]$
$= [-15, -2]$.

Example 3.15
$[-15, -2]/[1, 5] = [\min(-15, -3, -2, -0.4), \max(-15, -3, -2, -0.4)]$
$= [-15, -0.4]$.
So, the division is not the inverse operation of the multiplication.

From Definition 3.11, we have the following property for binary operations of interval numbers.

Property. If $[a_1, b_1] \subseteq [a_2, b_2]$, then $[a_1, b_1] * [c, d] \subseteq [a_2, b_2] * [c, d]$ and $[c, d] * [a_1, b_1] \subseteq [c, d] * [a_2, b_2]$ for any interval numbers $[a_1, b_1]$, $[a_2, b_2]$, and $[c, d]$, provided the involved operations are well defined.

A partial ordering on $\mathcal{N}_I$ can be defined as follows.

Definition 3.12 We say that interval number $[a, b] \in \mathcal{N}_I$ is not larger than (or say, *less than or equal to*) interval number $[c, d] \in \mathcal{N}_I$, denoted by $[a, b] \leq [c, d]$, iff $a \leq c$ and $b \leq d$.

Thus, $(\mathcal{N}_I, \leq)$ is a poset. Unlike the set of all real numbers that is well ordered with respect to the common $\leq$, poset $(\mathcal{N}_I, \leq)$ is not well ordered. For instance, we cannot compare $[0, 3]$ and $[1, 2]$ according to $\leq$ defined above for interval numbers. For any given two interval numbers $[a, b]$ and $[c, d]$, their supremum and infimum exist. In fact, they are interval numbers $[a \vee c, b \vee d]$ and $[a \wedge c, b \wedge d]$ respectively. Hence, poset $(\mathcal{N}_I, \leq)$ with binary operators $\vee$ and $\wedge$ forms a lattice.

3.8 Fuzzy Numbers and Linguistic Attribute

To quantify fuzzy concepts, we may use some types of fuzzy subsets of $R = (-\infty, \infty)$. Fuzzy numbers are most common type of fuzzy subsets of R for this purpose.

Definition 3.13 A *fuzzy number*, denoted by a capital letter with a wave such as $\widetilde{A}$, is a fuzzy subset of R with membership function $m: R \rightarrow [0,1]$ satisfying the following conditions:

(FN1) $\widetilde{A}_\alpha$, the α-cut of $\widetilde{A}$, is a closed interval for any $\alpha \in (0, 1]$;
(FN2) $\widetilde{A}_{0+}$ is bounded.

Condition (FN1) implies the convexity of $\widetilde{A}$, i.e., any fuzzy number is a convex fuzzy subset of R. For any $\alpha \in (0, 1]$, the α-cut of a fuzzy number is an interval number. The set of all fuzzy numbers is denoted by $\mathcal{N}_F$.

Theorem 3.12 Condition (FN1) is equivalent to the following conditions:

(FN 1.1) there exists at least one real number a_0 such that $m(a_0) = 1$;
(FN 1.2) $m(t)$ is nondecreasing on $(-\infty, a_0]$ and nonincreasing on $[a_0, \infty)$;

(FN 1.3) $m(t)$ is upper semi-continuous, or say, $m(t)$ is right-continuous on $(-\infty, a_0)$, i.e., $\lim_{t\to t_0+} m(t) = m(t_0)$ when $t_0 < a_0$, and is left-continuous on (a_0, ∞), i.e., $\lim_{t\to t_0-} m(t) = m(t_0)$) when $t_0 > a_0$.

Proof. Let $\tilde{A}$ be a fuzzy subset of R with membership function $m(t)$.
(FN1) $\Rightarrow$ (FN1.1): Since $\tilde{A}_{\alpha=1}$ is a closed interval, it is nonempty. Taking $a_0 \in \tilde{A}_{\alpha=1}$, we have $m(a_0) = 1$.
(FN1)$\Rightarrow$(FN1.2): A proof by contradiction is used here. Assume that $m(t)$ is not nondecreasing on $(-\infty, a_0]$, that is, there are $x, y \in (-\infty, a_0]$ with $x < y$ such that $m(x) > m(y)$. Taking $\alpha = m(x) > 0$, we have $x \in \tilde{A}_\alpha$ and $a_0 \in \tilde{A}_\alpha$, but $y \notin \tilde{A}_\alpha$. This contradicts the fact that $\tilde{A}_\alpha$ is an interval. Similarly, we can show that $m(t)$ is not nonincreasing on $[a_0, \infty)$ if $\tilde{A}_\alpha$ is an interval for any $\alpha \in (0, 1]$.
(FN1)$\Rightarrow$(FN1.3): We just need to show that $m(t)$ is right-continuous on $(-\infty, a_0]$ and left-continuous on $[a_0, \infty)$. A proof by contradiction is still used. Assume that $m(t)$ is not right-continuous on $(-\infty, a_0]$, that is, there exists a point $x < a_0$ such that $\lim_{t\to x+} m(t) \neq m(x)$ (the limit exists due to the monotonicity of $m(t)$ on $(-\infty, a_0]$). Since $m(t)$ is nondecreasing on $(-\infty, a_0]$, we have $\lim_{t\to x+} m(t) > m(x)$. Thus, taking $\alpha = \lim_{t\to x+} m(t)$, we have $t \in \tilde{A}_\alpha$ for all $t \in (x, a_0]$ but $x \notin \tilde{A}_\alpha$. This contradicts the fact that $\tilde{A}_\alpha$ is a closed interval.
(FN1.1), (FN1.2), and (FN1.3) $\Rightarrow$ (FN1): For any $\alpha \in (0, 1]$, from (FN1.1) we know that $\tilde{A}_\alpha$ is nonempty; from (FN1.2) we know that $\tilde{A}_\alpha$ is an interval; from (FN1.3) we know that $\tilde{A}_\alpha$ is closed.

The proof of the theorem is now complete. □

The boundedness of the support set $\tilde{A}_{0+}$ implies condition $\int_{-\infty}^{\infty} m(t)\,dt < \infty$. In this book, the latter is also used to weaken the requirement for fuzzy numbers sometimes.

For any fuzzy number with membership function $m(t)$, there exists a closed interval $[a_b, a_c]$ such that

$$m(t) = \begin{cases} 1 & \text{if } t \in [a_b, a_c] \\ l(t) & \text{if } t \in (-\infty, a_b) \\ r(t) & \text{if } t \in (a_c, \infty) \end{cases},$$

where $0 \le l(t) < 1$ is nondecreasing and $0 \le r(t) < 1$ is nonincreasing. Functions $l(t)$ and $r(t)$ are called the *left branch* and the *right branch* of $m(t)$, respectively.

Now, we turn to discuss several special types of fuzzy numbers that are commonly used.

Definition 3.14 A *rectangular fuzzy number* is a fuzzy number with membership function having a form as

$$m(t) = \begin{cases} 1 & \text{if } t \in [a_l, a_r] \\ 0 & \text{otherwise} \end{cases},$$

where a_l, $a_r \in R$ with $a_l \le a_r$ (see Figure 3.9).

A fuzzy number is rectangular if and only if the left branch and right branch of its membership function are zero. It is identified with the corresponding vector $[a_l \ a_r]$ and is an interval number essentially. Any crisp real number a can be regarded as a special rectangular fuzzy number with $a_l = a_r = a$.

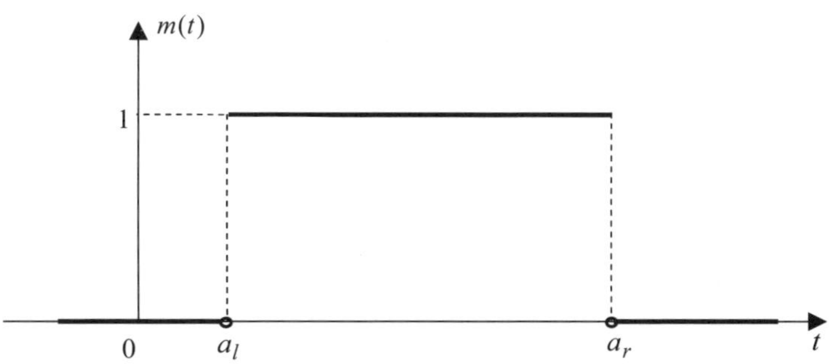

Fig. 3.9 The membership function of a rectangular fuzzy number.

Definition 3.15 A *triangular fuzzy number* is a fuzzy number with membership function having a form as

$$m(t)=\begin{cases}\dfrac{t-a_l}{a_0-a_l} & \text{if } t\in[a_l,a_0)\\ 1 & \text{if } t=a_0\\ \dfrac{t-a_r}{a_0-a_r} & \text{if } t\in(a_0,a_r]\\ 0 & \text{otherwise}\end{cases},$$

where a_l, a_0, $a_r \in R$ with $a_l \le a_0 \le a_r$ (see Figure 3.10).

A triangular fuzzy number is identified with the corresponding vector $[a_l\ a_0\ a_r]$. Any crisp real number a can be regarded as a special triangular fuzzy number with $a_l = a_0 = a_r = a$.

Definition 3.16 A *trapezoidal fuzzy number* is a fuzzy number with membership function having a form as

$$m(t)=\begin{cases}\dfrac{t-a_l}{a_b-a_l} & \text{if } t\in[a_l,a_b)\\ 1 & \text{if } t=[a_b,a_c]\\ \dfrac{t-a_r}{a_c-a_r} & \text{if } t\in(a_c,a_r]\\ 0 & \text{otherwise}\end{cases},$$

where a_l, a_b, a_c, $a_r \in R$ with $a_l \le a_b \le a_c \le a_r$ (see Figure 3.11).

A trapezoidal fuzzy number is identified with the corresponding vector $[a_l\ a_b\ a_c\ a_r]$. Any rectangular fuzzy number $[a_l\ a_r]$ can be regarded as a special trapezoidal fuzzy number with $a_l = a_b$ and $a_c = a_r$. Similarly, any triangular fuzzy number $[a_l\ a_0\ a_r]$ can be

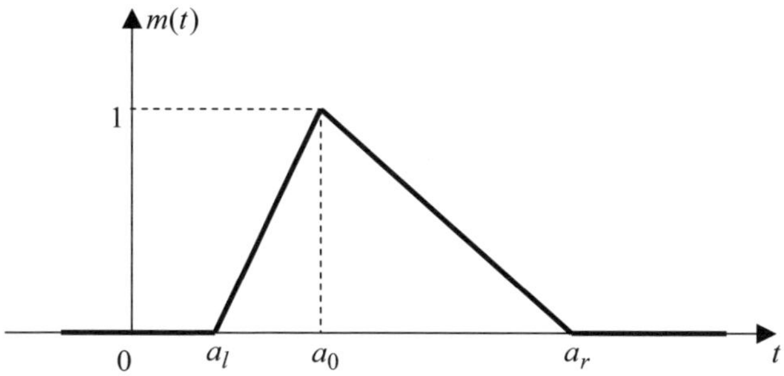

Fig. 3.10 The membership function of a triangular fuzzy number.

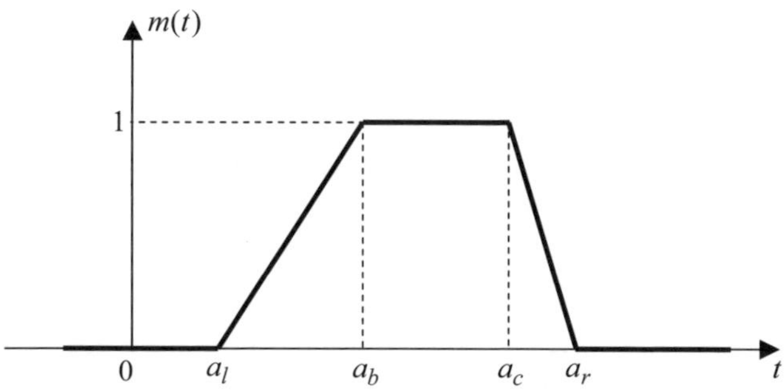

Fig. 3.11 The membership function of a trapezoidal fuzzy number.

regarded as a special trapezoidal fuzzy number with $a_b = a_c = a_0$. Of course, any crisp real number a can be regarded as a special trapezoidal fuzzy number with $a_l = a_b = a_c = a_r = a$. Thus, our discussion and models can be applied to databases involving even both crisp and fuzzy data.

Example 3.16 Fuzzy sets *Y*, *M*, *O*, and $\overline{Y}$ discussed in Examples 3.1 and 3.3 are trapezoidal fuzzy numbers.

Both the left branch and the right branch of the membership function of a trapezoidal fuzzy number are piecewise linear. Hence, it is convenient to calculate the sum and difference of trapezoidal fuzzy numbers. This can be seen in the next section.

Definition 3.17 A fuzzy number is called a *cosine fuzzy number* if its membership function has a form as

$$m(t) = \begin{cases} \frac{1}{2}[1+\cos\frac{2\pi(t-a)}{\theta}] & \text{if } a-\frac{\theta}{2} \le t \le a+\frac{\theta}{2} \\ 0 & \text{otherwise} \end{cases},$$

where real number a is the center and positive number θ is the length of its support set (see Figure 3.12).

According to Definition 3.13, the fuzzy subset of $R=(-\infty,\infty)$ with membership function $m(x)=e^{-x^2}$ (see Figure 3.7) discussed in Example 3.10 is not a fuzzy number since it violates the requirement of boundedness for its support set. However, if we weaken this requirement by $\int_{-\infty}^{\infty} m(x)\,dx < \infty$, such a fuzzy set can also be regarded as a fuzzy number and is called a *normal fuzzy number*. In general, a normal fuzzy number has the membership function with a form $m(x)=e^{-(x-a)^2/2\sigma^2}$, $\forall x \in (-\infty,\infty)$, where a is a real number indicating the center of the fuzzy number and σ is a positive real number indicating its "width". It is easy to know (from either calculus or the probability theory) that $m(a)=1$ and $\int_{-\infty}^{\infty} m(x)\,dx = \sqrt{2\pi}\sigma < \infty$.

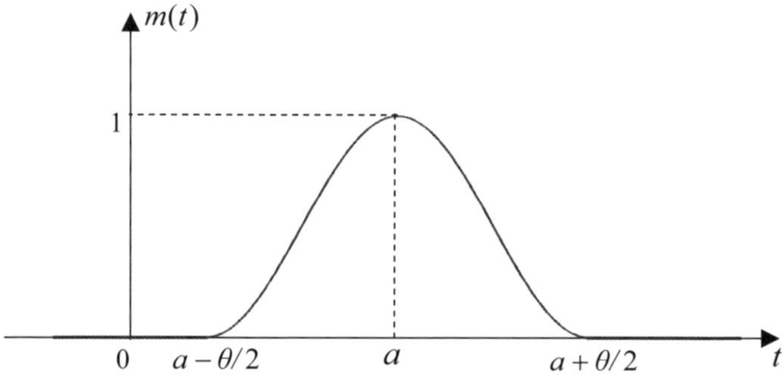

Fig. 3.12 The membership function of a cosine fuzzy number.

Beyond numerical attributes, some categorical, even linguistic attributes, may exist in databases. A linguistic attribute is a variable whose range is a finite set of descriptive words. One way to quantify a linguistic attribute is assigning a fuzzy number to each word. These fuzzy numbers should form a fuzzy partition of an appropriate interval.

Example 3.17 In Example 3.5, each criterion for submitted papers is a linguistic attribute whose range is the finite set {bad, weak, fair, good, excellent}. We assign trapezoidal fuzzy numbers $\widetilde{a}_b$, $\widetilde{a}_w$, $\widetilde{a}_f$, $\widetilde{a}_g$, and $\widetilde{a}_e$ to these five descriptive words respectively. These five fuzzy numbers form a fuzzy partition of interval $I=[0, 5]$ as shown in Figure 3.5.

3.9 Binary Operations for Fuzzy Numbers

In this section, we discuss the fuzzy arithmetic. By using the extension principle, we may extend the six common binary operators for real numbers to be corresponding binary operators for fuzzy numbers. Let $\widetilde{A}$ and $\widetilde{B}$ be two fuzzy numbers with membership functions m_A and m_B respectively and let $*$ be one of six common binary operators (+, −, ×, /, ∨, and ∧) for real numbers. The extensions then are shown as follows.

Definition 3.18 $\widetilde{A} * \widetilde{B}$ is a fuzzy subset of R with membership function

$$m_{A*B}(z) = \sup_{x,y|x*y=z} [m_A(x) \wedge m_B(y)] \qquad \forall z \in R.$$

To develop the methods for calculating the membership function of $\widetilde{A} * \widetilde{B}$, we need an important property of its α-cuts shown in the next theorem.

Theorem 3.13 $(\widetilde{A} * \widetilde{B})_\alpha = \widetilde{A}_\alpha * \widetilde{B}_\alpha$ for any $\alpha \in (0,1]$, and $(\widetilde{A} * \widetilde{B})_{\alpha+} = \widetilde{A}_{\alpha+} * \widetilde{B}_{\alpha+}$ for any $\alpha \in [0,1)$, provided all involved operations are well defined.

Proof. For any given $\alpha \in (0,1]$, on one hand, if $z \in (\widetilde{A} * \widetilde{B})_\alpha$, then $m_{A*B}(z) \geq \alpha$. From Definition 3.18 and by the property of the supremum shown in Theorem 2.5, we know that, for any integer $n > 2/\alpha$, there exist $x_n, y_n \in R$ such that $x_n * y_n = z$ and both $m_A(x_n) \geq \alpha - 1/n \geq \alpha/2$ and $m_B(y_n) \geq \alpha - 1/n \geq \alpha/2$, that is, $x_n \in \widetilde{A}_{\alpha/2}$ and $y_n \in \widetilde{B}_{\alpha/2}$. Since $\widetilde{A}_{\alpha/2}$ is a bounded interval, there exists a convergent subsequence $\{x_{n_i}\}$ of sequence $\{x_n\}$. Let $\lim_{i\to\infty} x_{n_i} = x_0$. We have $m_A(x_0) \geq \alpha - 1/n$ due to the closure property of $\widetilde{A}_{\alpha-1/n}$ and the fact that $m_A(x_m) \geq \alpha - 1/n$ for all $m \geq n$ when n is large enough. Thus, $m_A(x_0) \geq \alpha$. This means $x_0 \in \widetilde{A}_\alpha$. Similarly, from sequence $\{y_{n_i}\}$ we may choose a convergent subsequence $\{y_{n_{i_j}}\}$ with limit y_0 such that $y_0 \in \widetilde{B}_\alpha$. From $x_n * y_n = z$ for all integers n that are large enough, we know that $x_0 * y_0 = z$ due to the continuity of binary operator $*$. Thus, $z \in \widetilde{A}_\alpha * \widetilde{B}_\alpha$. On the other hand, if $z \in \widetilde{A}_\alpha * \widetilde{B}_\alpha$, then there exist $x_0 \in \widetilde{A}_\alpha$ and $y_0 \in \widetilde{B}_\alpha$ such that $z = x_0 * y_0$. From $m_A(x_0) \geq \alpha$ and $m_B(y_0) \geq \alpha$, we know that $\sup_{x*y=z}[m_A(x) \wedge m_B(y)] \geq \alpha$. This means that $z \in (\widetilde{A} * \widetilde{B})_\alpha$. The first equality is now proved.

As for the second equality, we may obtain it through the following equivalences where $\alpha \in [0,1)$.

$$
\begin{aligned}
z \in (\widetilde{A} * \widetilde{B})_{\alpha+} &\Leftrightarrow m_{A*B}(z) > \alpha \\
&\Leftrightarrow \sup_{x*y=z}[m_A(x) \wedge m_B(y)] > \alpha \\
&\Leftrightarrow \exists x_0, y_0 \text{ such that } x_0 * y_0 = z,\ m_A(x_0) > \alpha, \text{and } m_B(y_0) > \alpha \\
&\Leftrightarrow \exists x_0 \in \widetilde{A}_{\alpha+} \text{ and } y_0 \in \widetilde{B}_{\alpha+} \text{ such that } x_0 * y_0 = z \\
&\Leftrightarrow z \in \widetilde{A}_{\alpha+} * \widetilde{B}_{\alpha+}
\end{aligned}
$$

The proof of the theorem is now complete. □

Theorem 3.14 If $\widetilde{A}$ and $\widetilde{B}$ are fuzzy numbers, then so is $\widetilde{A} * \widetilde{B}$ unless $0 \in \widetilde{B}_{0+}$ when operator $*$ is the division.

Proof. From the first equality in Theorem 3.13 and the closure properties of these operations for interval numbers, we know that $\widetilde{A} * \widetilde{B}$ satisfies condition (FN1). Furthermore, from the second equality in Theorem 3.13, we have $(\widetilde{A} * \widetilde{B})_{0+} = \widetilde{A}_{0+} * \widetilde{B}_{0+}$ and, therefore, $(\widetilde{A} * \widetilde{B})_{0+}$ is bounded since both $\widetilde{A}_{0+}$ and $\widetilde{B}_{0+}$ are bounded and $0 \notin \widetilde{B}_{0+}$ when operator $*$ is the division. □

From Theorem 3.13 and Theorem 3.9 (Decomposition Theorem I), we have the following representation theorem directly.

Theorem 3.15 $\widetilde{A} * \widetilde{B} = \bigcup_{\alpha \in (0,1]} \alpha(A_\alpha * B_\alpha)$.

Theorem 3.16 If $\widetilde{A}$ and $\widetilde{B}$ are rectangular fuzzy numbers, then so is $\widetilde{A} * \widetilde{B}$ unless $0 \in \widetilde{B}_{0+}$ when operator $*$ is the division.

Proof. Since any rectangular fuzzy number is just an interval number, the conclusion of the theorem can be obtained from the discussion of section 3.7. □

Theorem 3.17 If $\widetilde{A}$ and $\widetilde{B}$ are triangular (or trapezoidal) fuzzy numbers, then so are $\widetilde{A} + \widetilde{B}$ and $\widetilde{A} - \widetilde{B}$.

Proof. A fuzzy number with membership function $m(t)$ is triangular if and only if there is a unique point a_0 such that $m(a_0) = 1$ and both $l(t)$ and $r(t)$ are linear in their nonzero part. From Theorem 3.15 and then the fact that the addition and the subtraction preserve the linearity of $l(t)$ and $r(t)$ as well as the uniqueness of a_0, we know that $\widetilde{A} + \widetilde{B}$ and $\widetilde{A} - \widetilde{B}$ are triangular fuzzy numbers provided $\widetilde{A}$ and $\widetilde{B}$ are triangular fuzzy numbers. Ignoring the uniqueness of a_0, a similar conclusion for trapezoidal fuzzy numbers can be obtained. □

Example 3.18 Let fuzzy numbers $\widetilde{A}$ and $\widetilde{B}$ have membership functions

$$m_A(t) = \begin{cases} (t+2)/2 & \text{if } t \in [-2, 0] \\ 1 - t & \text{if } t \in (0, 1] \\ 0 & \text{else} \end{cases}$$

and

$$m_B(t) = \begin{cases} t-2 & \text{if } t \in [2, 3] \\ 4-t & \text{if } t \in (3, 4] \\ 0 & \text{else} \end{cases} ,$$

respectively. Setting $m_A(t) = \alpha$ and expressing t in terms of α, we can obtain the α-cuts of A: $A_\alpha = [-2+2\alpha, 1-\alpha]$ for $\alpha \in (0, 1]$. In a similar way, we have $B_\alpha = [2+\alpha, 4-\alpha]$ for $\alpha \in (0, 1]$. By using Theorem 3.13, we obtain

$$(A+B)_\alpha = A_\alpha + B_\alpha = [3\alpha, 5-2\alpha],$$

$$(A-B)_\alpha = A_\alpha - B_\alpha = [-6+3\alpha, -1-2\alpha],$$

$$(A \cdot B)_\alpha = A_\alpha \cdot B_\alpha = [-2\alpha^2 + 10\alpha - 8, \alpha^2 - 5\alpha + 4],$$

$$(A/B)_\alpha = A_\alpha / B_\alpha = [(-2+2\alpha)/(2+\alpha), (1-\alpha)/(2+\alpha)].$$

Thus, composing the α-cuts (in an inverse way of calculating the α-cuts), we get

$$m_{A+B}(t) = \begin{cases} t/3 & \text{if } t \in [0, 3] \\ (5-t)/2 & \text{if } t \in (3, 5] \\ 0 & \text{else} \end{cases} ,$$

$$m_{A-B}(t) = \begin{cases} (t+6)/3 & \text{if } t \in [-6, -3] \\ (-1-t)/2 & \text{if } t \in (-3, -1] \\ 0 & \text{else} \end{cases} ,$$

$$m_{A\cdot B}(t)=\begin{cases}(5-\sqrt{9-2t})/2 & \text{if } t\in[-8,0]\\(5-\sqrt{9+4t})/2 & \text{if } t\in(0,4]\\0 & \text{else}\end{cases},$$

$$m_{A/B}(t)=\begin{cases}(2t+2)/(2-t) & \text{if } t\in[-1,0]\\(1-2t)/(1+t) & \text{if } t\in(0,0.5]\\0 & \text{else}\end{cases}.$$

They are shown in Figure 3.13.

From Example 3.18 and Figure 3.13, we can see that the product and the quotient of two triangular fuzzy numbers may not be triangular. A similar situation may occur for trapezoidal fuzzy numbers.

Example 3.19 Let fuzzy numbers $\widetilde{A}$ and $\widetilde{B}$ have membership functions

$$m_A(t)=\begin{cases}t/5 & \text{if } t\in[0,5]\\6-t & \text{if } t\in(5,6]\\0 & \text{else}\end{cases}$$

and

$$m_B(t)=\begin{cases}t-2 & \text{if } t\in[2,3]\\4-t & \text{if } t\in(3,4]\\0 & \text{else}\end{cases},$$

respectively. Then $A_\alpha=[5\alpha,6-\alpha]$ and $B_\alpha=[2+\alpha,4-\alpha]$ for $\alpha\in(0,1]$. By using Theorem 3.13, we have

$$(A\vee B)_\alpha=A_\alpha\vee B_\alpha=\begin{cases}[2+\alpha,6-\alpha] & \text{if } \alpha\in(0,0.5]\\[5\alpha,6-\alpha] & \text{if } \alpha\in(0.5,1]\end{cases},$$

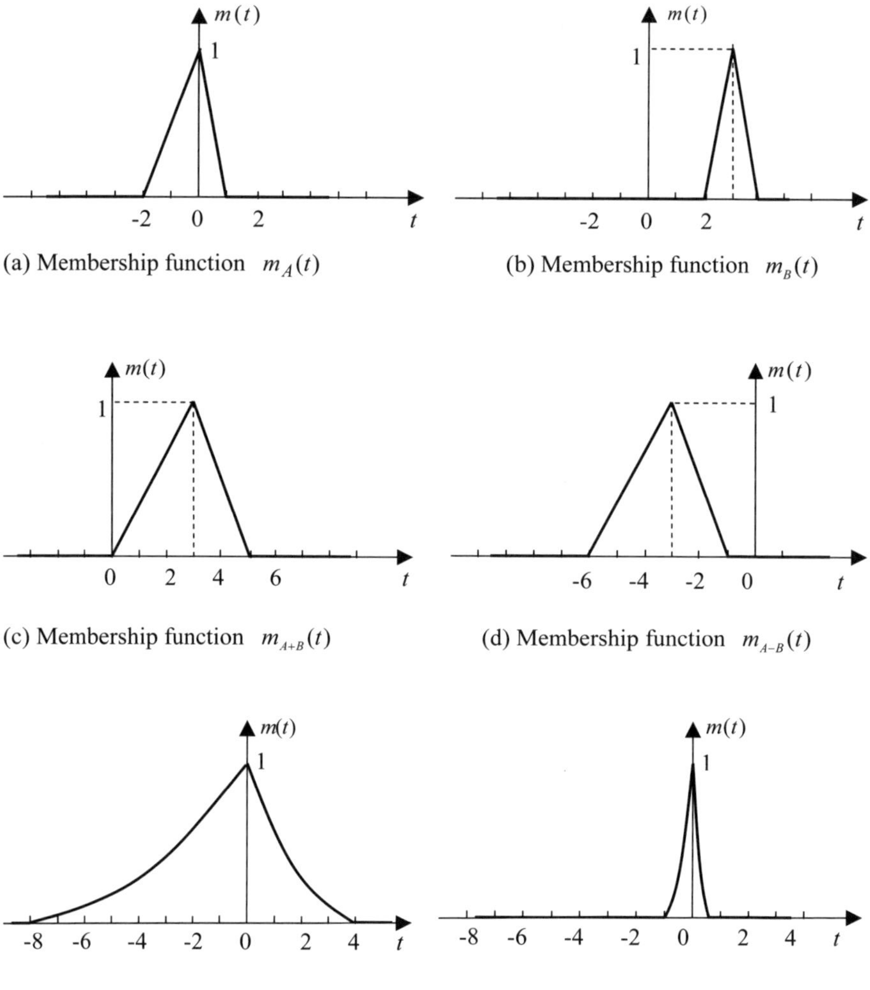

(a) Membership function $m_A(t)$ (b) Membership function $m_B(t)$

(c) Membership function $m_{A+B}(t)$ (d) Membership function $m_{A-B}(t)$

(e) membership function $m_{A\cdot B}(t)$ (f) membership function $m_{A/B}(t)$

Fig. 3.13 Membership functions in Example 3.18.

and

$$(A \wedge B)_\alpha = A_\alpha \wedge B_\alpha = \begin{cases} [5\alpha, 4-\alpha] & \text{if } \alpha \in (0, 0.5] \\ [2+\alpha, 4-\alpha] & \text{if } \alpha \in (0.5, 1] \end{cases}.$$

Thus,

$$m_{A\vee B}(t)=\begin{cases} t-2 & \text{if } t\in(2,2.5] \\ t/5 & \text{if } t\in(2.5,5] \\ 6-t & \text{if } t\in(5,6] \\ 0 & \text{else} \end{cases}, \qquad m_{A\wedge B}(t)=\begin{cases} t/5 & \text{if } t\in(0,2.5] \\ t-2 & \text{if } t\in(2.5,3] \\ 4-t & \text{if } t\in(3,4] \\ 0 & \text{else} \end{cases}.$$

They are shown in Figure 3.14.

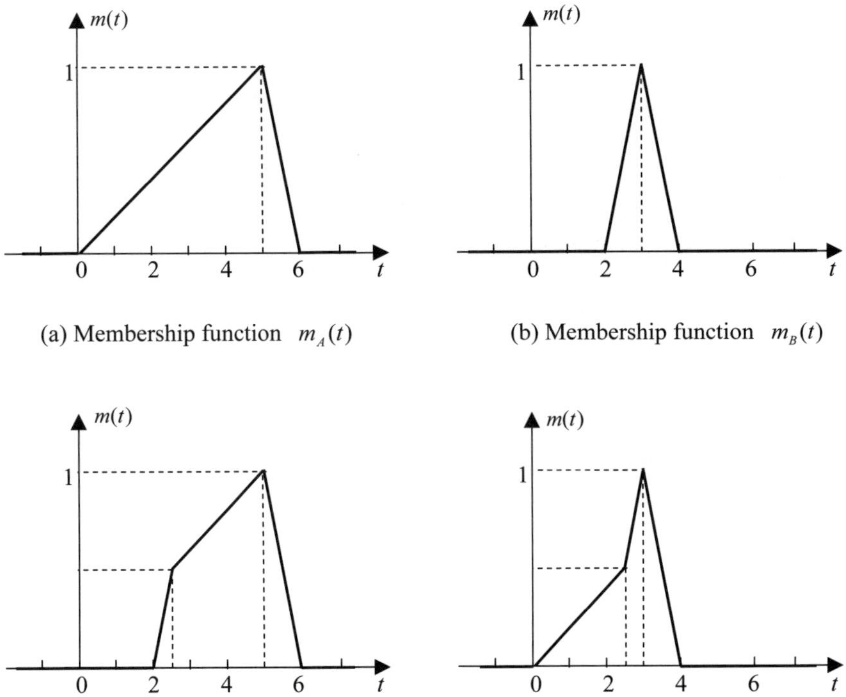

(a) Membership function $m_A(t)$

(b) Membership function $m_B(t)$

(c) Membership function $m_{A\vee B}(t)$

(d) Membership function $m_{A\wedge B}(t)$

Fig. 3.14 Membership functions in Example 3.19.

Here, we also see that the maximum and the minimum of two triangular fuzzy numbers may not be triangular. A similar situation may occur for trapezoidal fuzzy numbers as well.

We should note that, generally, $m_{A\vee B}$ is different from $m_{A\cup B}$, and $m_{A\wedge B}$ is different from $m_{A\cap B}$. Moreover, for fuzzy numbers, the subtraction is not the inverse operation of the addition; and the division is not the inverse operation of the multiplication.

We may also define a partial ordering on $\mathcal{N}_F$.

Definition 3.19 Let $\widetilde{A}$ and $\widetilde{B}$ be two fuzzy numbers. We say $\widetilde{A} \le \widetilde{B}$ iff $\widetilde{A}_\alpha \le \widetilde{B}_\alpha$ for every $\alpha \in (0, 1]$.

Relation $\le$ is a partial ordering on $\mathcal{N}_F$ and, therefore, $(\mathcal{N}_F, \le)$ is a poset. Furthermore, for any two given fuzzy numbers $\widetilde{A}$ and $\widetilde{B}$, we have

$$\sup\{\widetilde{A}, \widetilde{B}\} = \widetilde{A} \vee \widetilde{B} = \bigcup_{\alpha\in(0,1]} \alpha(\widetilde{A}_\alpha \vee \widetilde{B}_\alpha)$$

and

$$\inf\{\widetilde{A}, \widetilde{B}\} = \widetilde{A} \wedge \widetilde{B} = \bigcup_{\alpha\in(0,1]} \alpha(\widetilde{A}_\alpha \wedge \widetilde{B}_\alpha).$$

Thus, $(\mathcal{N}_F, \le)$ is a lattice.

3.10 Fuzzy Integers

Let Z be the set of all integers.

Definition 3.20 A *fuzzy integer* is a fuzzy subset of Z with membership function $m(i)$, $i \in Z$, satisfying the following conditions:

(FI1) there exists at least one integer i_0 such that $m(i_0) = 1$;
(FI2) $m(i_2) \ge \min[m(i_1), m(i_3)]$ whenever $i_1, i_2, i_3 \in Z$ and $i_1 \le i_2 \le i_3$;
(FI3) $\{i \mid m(i) > 0\}$ is a finite subset of Z.

Example 3.20 Fuzzy sets "around 10" and "around 5", denoted by D and F respectively, as well as $D+F$ in Example 3.12 are fuzzy integers.

The set of all fuzzy integers is denoted by $\mathscr{I}_F$. The same as for fuzzy numbers, by the extension principle, we may extend the six common binary operators for integers to be corresponding binary operators for fuzzy integers. The extension for the addition is shown in Example 3.12. It should be emphasized that, unlike the common binary operators for crisp integers, the difference for fuzzy integers is not an inverse operator of the addition and the division for fuzzy integers is not the inverse operator of the multiplication.

Similar to Definition 3.19, we may also define a partial ordering $\leq$ on $\mathscr{I}_F$ such that $(\mathscr{I}_F, \leq)$ is a poset. Poset $(\mathscr{I}_F, \leq)$ with binary operators $\vee$ and $\wedge$ forms a lattice.

Exercises

Exercise 3.1 Let $\{A_t \mid t \in T\}$ be a class of fuzzy sets, where T is an index set. Prove that $\bigcap_{t\in T}(A_t)_{\alpha+} \supseteq (\bigcap_{t\in T} A_t)_{\alpha+}$ for every $\alpha \in [0, 1]$. Cite a counterexample to show that the inverse inclusion may not hold.

Exercise 3.2 Fuzzy set M is given in Example 3.3. Find the membership function of $\overline{M}$. Is it convex? What is $\overline{M}_{0.8}$?

Exercise 3.3 Prove that a fuzzy subset of R^n is convex if and only if its strong α-cut is convex for every $\alpha \in [0, 1]$.

Exercise 3.4 May we establish the fourth decomposition theorem as

$$A = \bigcup_{\alpha \in L_A} \alpha A_{\alpha+} \ ?$$

Why?

Exercise 3.5 Let the universal set X be the set of all real numbers, i.e., $X = (-\infty, \infty)$, and A, B, and C be fuzzy subsets of X. For each of these fuzzy sets with respective membership function given below, show its α-cuts as a function of α on (0, 1].

(1) $m_A(t)=\begin{cases}1 & \text{if } t=0\\ 0 & \text{otherwise}\end{cases}$, for $t\in X$.

(2) $m_B(t)=\begin{cases}t & \text{if } t\in(0,1)\\ 1 & \text{if } t\in[1,3]\\ 1-(t-3)/3 & \text{if } t\in(3,6)\\ 0 & \text{otherwise}\end{cases}$, for $t\in X$.

(3) $m_C(t)=e^{-t^2}$ for $t\in X$.

Exercise 3.6 Let the universal set *X* be the set of all real numbers, i.e., $X=(-\infty,\infty)$, and *A*, *B*, and *C* be fuzzy subsets of *X*. Knowing their α-cuts given below, find their membership function respectively.

(1) $A_\alpha=\begin{cases}[0,1] & \text{if } \alpha\in(0,1/2]\\ \varnothing & \text{if } \alpha\in(1/2,1]\end{cases}$.

(2) $B_\alpha=[\alpha,2-\alpha]$ if $\alpha\in(0,1]$.

(3) $C_\alpha=[2\alpha,6-3\alpha]$ if $\alpha\in(0,1]$.

Exercise 3.7 Find the membership function of $D-F$, where fuzzy sets *D* and *F* are given in Example 3.12

Exercise 3.8 Let fuzzy numbers $\widetilde{A}$ and $\widetilde{B}$ have membership functions

$$m_A(t)=\begin{cases}(t+1)/2 & \text{if } t\in(-1,1]\\ (3-t)/2 & \text{if } t\in(1,3]\\ 0 & \text{else}\end{cases}$$

and

$$m_B(t)=\begin{cases}(t-1)/2 & \text{if } t\in(1,3]\\ (5-t)/2 & \text{if } t\in(3,5]\\ 0 & \text{else}\end{cases}$$

respectively. Find $\widetilde{A}+\widetilde{B}$, $\widetilde{A}-\widetilde{B}$, $\widetilde{A}\cdot\widetilde{B}$, and $\widetilde{A}/\widetilde{B}$.

Exercise 3.9 Let fuzzy numbers $\widetilde{A}$ and $\widetilde{B}$ have membership functions

$$m_A(t) = \begin{cases} (t+2)/3 & \text{if } t \in (-2, 1] \\ (4-t)/3 & \text{if } t \in (1, 4] \\ 0 & \text{else} \end{cases}$$

and

$$m_B(t) = \begin{cases} t-1 & \text{if } t \in (1, 2] \\ 3-t & \text{if } t \in (2, 3] \\ 0 & \text{else} \end{cases}$$

respectively. Find $\widetilde{A} \vee \widetilde{B}$ and $\widetilde{A} \wedge \widetilde{B}$.

Exercise 3.10 Prove that any fuzzy number with membership function $m(t)$ is a crisp number if and only if $\int_{-\infty}^{\infty} m(t)dt = 0$.

Chapter 4

Set Functions

From now on, we use the following conventions:

$$\sup_{x\in\varnothing}\{x \mid x\in[0,a]\}=0 \quad \text{for any} \quad a\in[0,\infty],$$

$$\inf_{x\in\varnothing}\{x \mid x\in[0,a]\}=a \quad \text{for any} \quad a\in[0,\infty],$$

$$0\times\infty=\infty\times 0=0,$$

$$a/\infty=0 \quad \text{for any} \quad a\in(-\infty,\infty),$$

$$\textstyle\sum_{i\in\varnothing}a_i=0 \quad \text{and} \quad \prod_{i\in\varnothing}a_i=1 \quad \text{for any real number sequence} \quad \{a_i\}.$$

In this chapter, starting from the classical additive measures and regarding them as special examples of *nonadditive set functions*, which are not necessarily additive, we introduce the concept of monotone measures in general case where the universal set may be infinite. Such kind of set functions abandons the additivity of the classical measures but keep the monotonicity, sometimes also the continuity (or semi-continuity) if necessary. From Sections 4.4 to 4.9, we discuss several special but common types of monotone measures. In Section 4.10, we abandon the monotonicity to introduce more general nonadditive set functions. The set functions possessing the nonadditivity can be adopted to describe the interaction among contribution rates from a number of attributes towards a certain target and are very useful and powerful in information fusion and data mining.

4.1 Weights and Classical Measures

Consider n attributes $x_1, x_2, \cdots, x_n$ in a database. They may be the information sources in information fusion (see Chapter 6) or some variables in a real problem. These attributes form the universal set, denoted by X, that is, $X = \{x_1, x_2, \cdots, x_n\}$. The weights of these attributes, $w_1, w_2, \cdots, w_n$, can be regarded as a mapping from the class of all singletons, $\{\{x_i\} \mid i = 1, 2, \cdots, n\}$, to interval [0, 1] satisfying $\sum_{i=1}^{n} w_i = 1$. The concept of weights can be generalized to a countable (infinite) university set $X = \{x_1, x_2, \cdots\}$. Corresponding to these infinitely many but countable attributes $x_1, x_2, \cdots$, nonnegative weights $w_1, w_2, \cdots$ should satisfy $\sum_{i=1}^{\infty} w_i = 1$.

Generally, let X be the universal set and $\mathscr{R}_\sigma$ be a σ-ring of subsets of X. Then $(X, \mathscr{R}_\sigma)$ is called a *measurable space*. In most case, using a σ-algebra $\mathscr{F}$ as $\mathscr{R}_\sigma$ is convenient for defining measures.

Example 4.1 $(R, \mathscr{B})$ is a measurable space where $R = (-\infty, \infty)$, the real line, and $\mathscr{B}$ is the Borel field.

As a special case, $(X, \mathscr{P}(X))$ is a measurable space since $\mathscr{P}(X)$ is a σ-ring. With set operations union and complement, it satisfies the laws shown in Theorem 2.1 and forms a *Boolean algebra*.

Let $\mathscr{C}$ be a nonempty class of subsets of X, $\emptyset \in \mathscr{C}$, and μ be a mapping from $\mathscr{C}$ to the extended real line $(-\infty, \infty]$. μ is called an *extended real-valued set function*, or *set function* simply if there is no confusion, and denoted as $\mu : \mathscr{C} \to (-\infty, \infty]$.

Definition 4.1 Set function μ is *additive* on $\mathscr{C}$ iff $\mu(E \cup F) = \mu(E) + \mu(F)$ whenever $E \in \mathscr{C}$, $F \in \mathscr{C}$, $E \cup F \in \mathscr{C}$, and $E \cap F = \emptyset$. μ is *finitely additive* on $\mathscr{C}$ iff

$$\mu(\bigcup_{i=1}^{n} E_i) = \sum_{i=1}^{n} \mu(E_i)$$

for any finite disjoint class $\{E_1, E_2, \cdots, E_n\}$ of sets in $\mathscr{C}$ satisfying $\bigcup_{i=1}^n E_i \in \mathscr{C}$.

Definition 4.2 Set function μ is *σ-additive* (or *countably additive*) on $\mathscr{C}$ iff $\mu(\bigcup_{i=1}^\infty A_i) = \sum_{i=1}^\infty \mu(A_i)$ whenever $\{A_i\}$ is a disjoint sequence of sets in $\mathscr{C}$ and $\bigcup_{i=1}^\infty A_i$ is also in $\mathscr{C}$.

It is evident that, when $\mu(\varnothing) < \infty$, the σ-additivity implies the finite additivity, and the latter implies the additivity and $\mu(\varnothing) = 0$ for set function μ defined on $\mathscr{C}$. Furthermore, if $\mathscr{C}$ is finite, then the additivity of μ is equivalent to its σ-additivity.

Definition 4.3 Nonnegative set function $\mu : \mathscr{C} \to [0, \infty]$ is called a *measure* on $\mathscr{C}$ iff μ is σ-additive on $\mathscr{C}$ and there exists $E \in \mathscr{C}$ such that $\mu(E) < \infty$.

Theorem 4.1 If μ is a measure on $\mathscr{C}$, then $\mu(\varnothing) = 0$.

Proof. Let set $E \in \mathscr{C}$ such that $\mu(E) < \infty$. Take a set sequence $\{A_i\}$ with $A_1 = E$ and $A_i = \varnothing$ for all $i = 2, 3, \cdots$. Set sequence $\{A_i\}$ is disjoint and $\bigcup_{i=1}^\infty A_i = E$. By the σ-additivity of μ, we have

$$\sum_{i=1}^{\infty} \mu(A_i) = \mu(E) + \sum_{i=2}^{\infty} \mu(\varnothing) = \mu(E) .$$

This means that $\sum_{i=2}^\infty \mu(\varnothing) = 0$. Since μ is nonnegative, we conclude that $\mu(\varnothing) = 0$. □

If the universal set X is finite, $\mathscr{C}$ must be also finite and, therefore, any additive set function on $\mathscr{C}$ is σ-additive.

Definition 4.4 Set function μ is said to be *finite* iff it never takes infinite value; μ is said to be *σ-finite* iff for any $E \in \mathscr{C}$, there exists $E_i \in \mathscr{C}$ with $\mu(E_i) < \infty$, $i = 1, 2, \cdots$, such that $\bigcup_{i=1}^\infty E_i \supseteq E$.

Definition 4.5 Let $(X, \mathscr{R}_\sigma)$ be a measurable space and μ be a measure on $\mathscr{R}_\sigma$ (we also say that μ is a measure on $(X, \mathscr{R}_\sigma)$). Then triple $(X, \mathscr{R}_\sigma, \mu)$ is called a *measure space*.

Definition 4.6 Let $(X, \mathscr{F}, \mu)$ be a measure space. If $\mu(X)=1$, then both μ and $(X, \mathscr{F}, \mu)$ are said to be *normalized*. A normalized measure is also called a *probability measure* (or, simply, *probability*). A probability measure is *discrete* iff there exists a countable set of singletons $\{\{x_1\}, \{x_2\}, \cdots\} \subseteq \mathscr{F}$ such that $\sum_{i=1}^{\infty} \mu(\{x_i\})=1$.

The definition of discrete probability measure given above can be generalized by replacing set $\{\{x_1\}, \{x_2\}, \cdots\}$ with disjoint class $\{A_1, A_2, \cdots\} \subseteq \mathscr{F}$ satisfying $\sum_{i=1}^{\infty} \mu(A_i)=1$ and $\mu(B_i)$ equals either $\mu(A_i)$ or zero for any $B_i \subseteq A_i$ with $B_i \in \mathscr{F}$, $i=1,2,\cdots$. However, we do not adopt such a generalization in this book.

Example 4.2 Let $X=\{a,b,c,d\}$, $\mathscr{C}=\mathscr{P}(X)$, and $\mu(E)=|E|$ for every $E \in \mathscr{C}$, where $|E|$ is the cardinality of set E, i.e., the number of points in E. Then μ is a finite measure on $\mathscr{C}$. If we take $P(E)=\mu(E)/4$ for every $E \in \mathscr{C}$, then P is a discrete probability measure on $\mathscr{C}$.

Example 4.3 Let $X=\{x_1, x_2, \cdots\}$ and $\mu(E)=|E|$ for $E \in \mathscr{P}(X)$. Then μ is a σ-finite measure on $\mathscr{P}(X)$.

Example 4.4 Let $X=\{x_1, x_2, \cdots, x_n\}$, and let $\mu(\{x_i\})=w_i \in [0, \infty)$, $i=1, 2, \cdots, n$, and $\mu(\emptyset)=0$. Then μ is a finite measure on class $\mathscr{C}=\{\{x_i\} \mid i=1, 2, \cdots, n\} \cup \{\emptyset\}$.

Example 4.5 Let $X=\{x_1, x_2, \cdots\}$ and $\mu(E)=\sum_{x_i \in E} 2^{-i}$ for every $E \in \mathscr{P}(X)$. Then μ is a discrete probability measure on $\mathscr{P}(X)$. Positive real number 2^{-i} can be regarded as the weight of attribute x_i for each $i=1, 2, \cdots$.

Example 4.6 Let $X=R$ and $\mu(E)=|E|$ for $E \in \mathscr{P}(X)$. Then μ is a measure on $\mathscr{P}(X)$ but it is not σ-finite.

Example 4.7 Let $X = R$ and $\mathscr{S}$ be the class of all bounded left closed right open intervals discussed in Example 2.10. Then $\mathscr{S}$ is a semiring. For each interval $[a, b)$ in $\mathscr{S}$, define $\mu([a,b)) = b - a$. Then μ is a finite measure on $\mathscr{S}$.

Theorem 4.2 Let $(X, \mathscr{R}_\sigma, \mu)$ be a measure space. Then Measure μ has the following properties.

(M1) *Monotonicity*: $\mu(E) \le \mu(F)$ whenever $E \in \mathscr{R}_\sigma$, $F \in \mathscr{R}_\sigma$, and $E \subseteq F$.

(M2) *Continuity from below*: $\lim_{i\to\infty} \mu(E_i) = \mu(\bigcup_{i=1}^{\infty} E_i)$, whenever $E_i \in \mathscr{R}_\sigma$, $i = 1, 2, \cdots$, and $\{E_i\}$ is nondecreasing.

(M3) *Continuity from above*: $\lim_{i\to\infty} \mu(E_i) = \mu(\bigcap_{i=1}^{\infty} E_i)$, whenever $E_i \in \mathscr{R}_\sigma$, $i = 1, 2, \cdots$, $\{E_i\}$ is nonincreasing, and there exists i_0 such that $\mu(E_{i_0}) < \infty$.

Proof. Only (M1) and (M2) are proved here. The third is left to readers as an exercise.

For (M1). Given $E \in \mathscr{R}_\sigma$, $F \in \mathscr{R}_\sigma$, and $E \subseteq F$, let $G = F - E$. Then $G \in \mathscr{R}_\sigma$, $G \cap E = \varnothing$, and $\mu(G) \ge 0$. Thus, from $\mu(G) + \mu(E) = \mu(F)$, we have $\mu(E) \le \mu(F)$.

For (M2). For any given nondecreasing set sequence $E_i \in \mathscr{R}_\sigma$, $i = 1, 2, \cdots$, let $F_1 = E_1$ and $F_i = E_i - E_{i-1}$ for $i = 2, 3, \cdots$. Since $\bigcup_{i=1}^{\infty} E_i = \bigcup_{i=1}^{\infty} F_i$, $E_n = \bigcup_{i=1}^{\infty} F_i$ for $n = 1, 2, \cdots$, and $\{F_i\}$ is disjoint, we have

$$\mu\left(\bigcup_{i=1}^{\infty} E_i\right) = \mu\left(\bigcup_{i=1}^{\infty} F_i\right) = \sum_{i=1}^{\infty} \mu(F_i) = \lim_{n\to\infty} \sum_{i=1}^{n} \mu(F_i) = \lim_{n\to\infty} \mu(E_n).$$

□

4.2 Extension of Measures

Let X be the universal set, $\mathscr{C}$ be a nonempty class of subsets of X with $\varnothing \in \mathscr{C}$ and $\mu : \mathscr{C} \to [0, \infty]$ be a measure.

Definition 4.7 If $\mathscr{D} \supseteq \mathscr{C}$ and there exists a measure μ' on $\mathscr{D}$ such that $\mu'(E) = \mu(E)$ for every $E \in \mathscr{C}$, then μ' is an *extension* of μ from $\mathscr{C}$ to $\mathscr{D}$.

Among the structured classes we have discussed, the semiring has the simplest structure. Now let us consider how to extend a measure from a semiring to the ring (or even σ-ring) generated by this semiring.

Let μ be a measure on a semiring $\mathscr{S}$. It is easy to extend μ to $\mathscr{R}(\mathscr{S})$. In fact, for any $E \in \mathscr{R}(\mathscr{S})$, we have $E = \bigcup_{i=1}^{n} E_i$ for some integer n, where $\{E_i\}$ is a disjoint finite class of sets in $\mathscr{S}$. Thus, we just let $\mu'(E) = \sum_{i=1}^{n} \mu(E_i)$. μ' is then an extension of μ from $\mathscr{S}$ to $\mathscr{R}(\mathscr{S})$.

Example 4.8 In Example 4.4, $X = \{x_1, x_2, \cdots, x_n\}$ .Class $\mathscr{S}=\{\{x_i\} \mid i = 1, 2, \cdots, n\} \cup \{\varnothing\}$ is a semiring. Measure μ on $\mathscr{S}$ can be uniquely extended onto the ring generated by $\mathscr{S}$, that is, onto $\mathscr{R}(\mathscr{S}) = \{E \mid E \text{ is finite}\}$. In fact, $\mathscr{R}(\mathscr{S})$ is a finite class. Taking for any $E \in \mathscr{R}(\mathscr{S})$ $\mu'(E) = \sum_{i:x_i \in E} w_i$, where $w_i = \mu(\{x_i\})$, $i = 1, 2, \cdots, n$, it is not difficult to verify the additivity of μ' on $\mathscr{R}(\mathscr{S})$.

Example 4.9 Similar to Example 4.8, let $X = \{x_1, x_2, \cdots\}$ and $\mathscr{S}$ consist of all singletons and the empty set $\varnothing$. Then $\mathscr{R}(\mathscr{S})$ consists of all finite sets. For any singleton $\{x_i\}$, take $\mu(\{x_i\}) = 2^{-i}$. With $\mu(\varnothing) = 0$, μ is a measure on $\mathscr{S}$. For any set $E \in \mathscr{R}(\mathscr{S})$, it can be expressed as a finite disjoint union of singletons, that is, $E = \bigcup_{j=1}^{k} \{x_{i_j}\}$ for some nonnegative integer k. Thus, defining

$$\mu'(E) = \sum_{j=1}^{k} \mu(E_{i_j}) = \sum_{j=1}^{k} 2^{-i_j} ,$$

we get an extension of μ from $\mathscr{S}$ onto $\mathscr{R}(\mathscr{S})$.

Example 4.10 In Example 4.7, consider the ring generated by $\mathscr{S}$, $\mathscr{R}(\mathscr{S})$. It consists of all finite unions of left closed right open intervals. Each set E in $\mathscr{R}(\mathscr{S})$ can be expressed as a finite disjoint union of sets in $\mathscr{S}$, say $E_i = [a_i, b_i)$, $i = 1, 2, \cdots, k$. We just need to define $\mu'(E) = \sum_{i=1}^{k} \mu(E_i)$

$=\sum_{i=1}^{k}(b_i - a_i)$. Thus, μ' is a finite measure on $\mathscr{R}(\mathscr{S})$ and is the extension of μ on $\mathscr{S}$.

In general, since each set E in $\mathscr{R}(\mathscr{S})$ is a finite disjoint union of sets in $\mathscr{S}$, i.e., $E=\bigcup_{i=1}^{k} E_i$, where $\{E_i\} \subseteq \mathscr{S}$, to extend a measure μ from $\mathscr{S}$ onto $\mathscr{R}(\mathscr{S})$, we just need to define μ' on $\mathscr{R}(\mathscr{S})$ by $\mu'(E)=\sum_{i=1}^{k}\mu(E_i)$.

Without any confusion, we may omit the prime on μ after extending. As for the extension of a measure from a ring to the σ-ring generated by this ring, the situation is rather complex. First, let us continue the example above.

Example 4.11 In Example 4.8, $\mathscr{R}_\sigma(\mathscr{R}(\mathscr{S})) = \mathscr{R}(\mathscr{S}) = \mathscr{P}(X)$ and μ has already been extended onto $\mathscr{R}_\sigma(\mathscr{R}(\mathscr{S}))$ there. In Example 4.9, $\mathscr{R}_\sigma(\mathscr{R}(\mathscr{S})) = \mathscr{P}(X)$. Each set in $\mathscr{P}(X)$ is a countable set and can be expressed as a countable disjoint union of singletons, that is,

$$\forall E \in \mathscr{P}(X), \quad E = \bigcup_{j=1}^{\infty} \{x_{i_j}\},$$

where all x_{i_j} are different. Let $E_k = \bigcup_{j=1}^{k}\{x_{i_j}\}$. Then $E_k \in \mathscr{R}(\mathscr{S})$, $k = 1, 2, \cdots$, and $\{E_k\}$ is nondecreasing. Thus, let

$$\mu'(E) = \lim_k \mu(E_k) = \lim_k \sum_{j=1}^{k} 2^{-i_j} = \sum_{j=1}^{\infty} 2^{-i_j}.$$

This completes the extension of μ from $\mathscr{R}(\mathscr{S})$ to $\mathscr{R}_\sigma(\mathscr{R}(\mathscr{S}))$.

However, not all examples have so simple extension. The complexity of the extension for μ from a ring onto the σ-ring generated by this ring depends on the structure of the ring. The extension procedure of μ given in Example 4.10 from $\mathscr{R}(\mathscr{S})$ to $\mathscr{R}_\sigma(\mathscr{R}(\mathscr{S}))$ is very complex. We omit the discussion here. Interested readers may refer to [Halmos 1950]. In general, we have the following theorem. Its proof is omitted too.

Theorem 4.3 If μ is a finite measure on a semiring $\mathscr{S}$, then μ can be uniquely extended to be a σ-finite measure on $\mathscr{R}_\sigma(\mathscr{S})$.

By using Theorem 4.3, measure μ in Example 4.10 can be uniquely extended to be a σ-finite measure on the Borel field $\mathscr{B}$. Furthermore, let $\mathscr{L} = \{E \cup F \mid E \in \mathscr{B}, \quad F \subseteq D \in \mathscr{B} \text{ with } \mu(D) = 0\}$ and let $\overline{\mu}(E \cup F) = \mu(E)$ for $E \cup F \in \mathscr{L}$. Then $\overline{\mu}$ is the *completion* of μ and is called the *Lebesgue measure* on the real line. It is a generalization of the concept of the length of intervals. Class $\mathscr{L}$ is called the *Lebesgue field.* We should be sure that the Lebesgue field is far smaller than the power set of R.

4.3 Monotone Measures

Let $(X, \mathscr{F})$ be a measurable space and $\mu : \mathscr{F} \to [0, \infty]$ be an extended real-valued set function, where $\mathscr{F}$ is a σ-algebra of subsets of X. When X is finite, usually, we take the power set $\mathscr{P}(X)$ as $\mathscr{F}$.

Definition 4.8 Set function μ: $\mathscr{F} \to [0, \infty]$ is called a *monotone measure* on $(X, \mathscr{F})$ iff it satisfies the following requirements:

(MM1) $\mu(\varnothing) = 0$ (vanishing at the empty set);

(MM2) $\mu(E) \leq \mu(F)$ whenever $E \in \mathscr{F}$, $F \in \mathscr{F}$, and $E \subseteq F$ (monotonicity).

In this case, $(X, \mathscr{F}, \mu)$ is called a *monotone measure space.*

Definition 4.9 Monotone measure $\mu : \mathscr{F} \to [0, \infty]$ is *lower-semi-continuous* (or *continuous from below*) iff it satisfies property (M2) given in Theorem 4.2, that is,

$$\lim_i \mu(E_i) = \mu(\bigcup_{i=1}^{\infty} E_i)$$

whenever $\{E_n\} \subseteq \mathscr{F}$, $E_1 \subseteq E_2 \subseteq \cdots$; μ is *upper-semi-continuous* (or *continuous from above*) iff it satisfies property (M3) given in Theorem 4.2, that is,

$$\lim_i \mu(E_i) = \mu(\bigcap_{i=1}^{\infty} E_i)$$

whenever $\{E_n\} \subseteq \mathscr{F}$, $E_1 \subseteq E_2 \subseteq \cdots$ and there exists positive integer i_0 such that $\mu(E_{i_0}) < \infty$; μ is *continuous* iff it is both lower-semi-continuous and upper-semi-continuous.

From the properties shown in Theorem 4.2, we know that any measure is a continuous monotone measure. So, the concept of monotone measures is a generalization of the concept of measures. However, the requirement of the additivity has been abandoned for the monotone measure, that is, monotone measures are nonadditive generally.

Similar to measures, we may also define the normalization for monotone measures as follows.

Definition 4.10 A monotone measure μ on $(X, \mathscr{F})$ is *normalized* iff $\mu(X) = 1$.

When $\mathscr{F}$ is finite, or more specially, when X is finite, any monotone measure is continuous. In databases, the number of attributes is always finite. So, we just need to consider monotone measures defined on the power set of a finite universal set for describing the individual and joint contribution rates from some attributes towards a certain target.

Definition 4.11 A monotone measure μ is *subadditive* iff $\mu(E \cup F) \le \mu(E) + \mu(F)$ for any $E \in \mathscr{F}$ and $F \in \mathscr{F}$.

Definition 4.12 A monotone measure μ is *superadditive* iff $\mu(E \cup F) \ge \mu(E) + \mu(F)$ for any $E \in \mathscr{F}$ and $F \in \mathscr{F}$ with $E \cap F = \varnothing$.

It is easy to see that a monotone measure μ is additive if and only if it is both subadditive and superadditive.

Example 4.12 Three workers, x_1, x_2, and x_3, are hired for manufacturing a certain kind of wooden toys. Denote $X = \{x_1, x_2, x_3\}$. Working separately, they produce 5, 6, and 7 toys per hour respectively. This can be considered as their individual efficiencies, i.e., the contribution rates towards the target "total amount of manufactured toys". When the three workers (or two of them) work together sometimes, we must consider their joint efficiencies to calculate the total number of the manufactured toys in a given period of time. We may use $\mu(\{x_1, x_2\})$ to denote the joint efficiency of x_1 and x_2. Similarly, $\mu(\{x_1, x_3\})$, $\mu(\{x_2, x_3\})$, and $\mu(X)$ are joint efficiencies of x_1 and x_3, x_2 and x_3, and all of them respectively. Assume that $\mu(\{x_1, x_2\}) = 14$, $\mu(\{x_1, x_3\}) = 13$, $\mu(\{x_2, x_3\}) = 9$, and $\mu(X) = 17$. Then, with $\mu(\emptyset)=0$, $\mu : \mathscr{P}(X) \to [0, \infty)$ is a monotone measure. It is nonadditive. For instance, $\mu(\{x_1, x_2\}) > \mu(\{x_1\}) + \mu(\{x_2\})$. This inequality means that workers x_1 and x_2 cooperate well. The nonadditivity of μ describes the interaction among the contribution rates from these three workers towards the total amount of their manufactured toys.

Example 4.13 In diagnosis of common cold, a doctor usually uses three symptoms, namely, running nose, soar throat and coughing represented by x_1, x_2, and x_3 respectively. Denote $X = \{x_1, x_2, x_3\}$. If each symptom occurs separately, we can diagnose cold respectively with certainties 0.6, 0.5, and 0.4. We may consider them as single-symptom diagnosis certainties, i.e., the contribution rates towards the target "overall certainty of the diagnosis". When the three symptoms (or two of them) occur together sometimes, we must consider their joint certainties to calculate the overall certainty for given periods of occurrence of the symptoms. Similar to Example 4.12, $\mu(\{x_1, x_2\})$, $\mu(\{x_1, x_3\})$, $\mu(\{x_2, x_3\})$, and $\mu(X)$ are used to denote the joint certainty of x_1 and x_2, of x_1 and x_3, of x_2 and x_3, and for all of them respectively. Assume that $\mu(\{x_1, x_2\}) = 0.8$, $\mu(\{x_1, x_3\}) = 0.76$, $\mu(\{x_2, x_3\}) = 0.7$, $\mu(X) = 0.88$, and $\mu(\emptyset) = 0$. Then, $\mu : \mathscr{P}(X) \to [0, \infty)$ is a monotone measure. It is subadditive. For instance, $\mu(\{x_1, x_2\}) < \mu(\{x_1\}) + \mu(\{x_2\})$. These joint certainties reflect the evidence combination (see Chapter 6) of symptoms in medical diagnoses.

Example 4.14 To evaluate used TV sets, we roughly adopt two quality factors: "picture" and "sound". These are denoted by x_1 and x_2, respectively, and the corresponding weights may be taken as $w_1 = 0.7$ and $w_2 = 0.3$. Let $X = \{x_1, x_2\}$. In this case, if we take $w_{12} = 1$ as the weight of X and $w_0 = 0$ as the weight of $\varnothing$, an additive measure w: $\mathscr{P}(X) \to [0, 1]$ is obtained. However, such a measure is not reasonable for evaluating the global quality of a TV set. We prefer to assign an importance 0.3 to "picture" and importance 0.1 to "sound". With assigning 1 to X and 0 to $\varnothing$, a superadditive monotone measure $v : \mathscr{P}(X) \to [0, 1]$ is formed as

$$v(E) = \begin{cases} 0 & \text{if } E = \varnothing \\ 0.3 & \text{if } E = \{x_1\} \\ 0.1 & \text{if } E = \{x_2\} \\ 1 & \text{if } E = X \end{cases} .$$

The details for using this monotone measure in synthetic evaluation can be found in Chapter 6.

In the following, we show more mathematical examples for monotone measures.

Example 4.15 Let $X = \{1, 2, \cdots, n\}$. Given a positive real number k, if we define set function $\mu : \mathscr{P}(X) \to [0, 1]$ by

$$\mu(E) = \left(\frac{|E|}{n} \right)^k \quad \text{for } E \in \mathscr{P}(X),$$

where $|E|$ is the cardinality of E, then μ is a normalized monotone measure. It is superadditive when $k > 1$, subadditive when $k < 1$, and additive when $k = 1$.

Example 4.16 Let X be the real line R, $\mathscr{F}$ be the Lebesgue field $\mathscr{L}$, and μ be the Lebesgue measure on $\mathscr{L}$. For any set $E \in \mathscr{P}(X)$, define

$$\overline{\mu}(E) = \inf\{\mu(F) \mid F \supseteq E, F \in \mathscr{L}\}$$

and

$$\underline{\mu}(E) = \sup\{\mu(F) \mid F \subseteq E, F \in \mathscr{L}\}.$$

Then $\overline{\mu}$ is a subadditive monotone measure and $\underline{\mu}$ is a superadditive monotone measure on $\mathscr{P}(X)$. In real analysis, $\overline{\mu}$ and $\underline{\mu}$ are called *outer measure* and *inner measure* generated by the Lebesgue measure, respectively.

Example 4.17 Let $X_0 = \{1, 2, \cdots\}$, $X = X_0 \times X_0$, and $\mathscr{F} = \mathscr{P}(X)$. For any $E \in \mathscr{F}$, define $\mu(E) = |\operatorname{Proj}_x E|$, where $\operatorname{Proj}_x E = \{x \mid (x, y) \in E\}$. Then μ is a lower-semi-continuous monotone measure on $\mathscr{F}$. It is not upper-semi-continuous. In fact, if $E_n = \{1\} \times \{n, n+1, \cdots\}$, then $E_1 \supseteq E_2 \supseteq \cdots$, and $\mu(E_n) = 1$ for every $n = 1, 2, \cdots$, but $\bigcap_{i=1}^{\infty} E_n = \varnothing$ such that $\mu(\bigcap_{i=1}^{\infty} E_n) = 0$. This violates the upper-semi-continuity.

Example 4.18 Let $f(x)$ be a nonnegative, real-valued function defined on $X = (-\infty, \infty)$. If we define

$$\mu(E) = \sup_{x \in E} f(x) \quad \text{for every set } E \text{ of real numbers,}$$

then μ is a lower-semi-continuous monotone measure on measurable space $(X, \mathscr{P}(X))$. It is not upper-semi-continuous in general. For example, taking $f(x) = 1$, $\forall x \in X$, $E_i = [i, \infty)$, $i = 1, 2, \cdots$, and $E = \bigcap_{i=1}^{\infty} E_i = \varnothing$, we have $\mu(E) = 0$ but $\mu(E_i) = \sup_{x \in E_i} f(x) = 1 < \infty$ for every $i = 1, 2, \cdots$. This violates the continuity from above.

Example 4.19 The measurable space is the same as used in Example 4.18. Taking a function $f: X \to [0, 1]$ that satisfies $\inf_{x \in X} f(x) = 0$, set function μ defined for every $E \in \mathscr{P}(X)$ by

$$\mu(E) = \inf_{x \notin E} f(x)$$

is a normalized upper-semi-continuous monotone measure on $(X, \mathscr{P}(X))$. It is not lower-semi-continuous in general. For example, take $f(x) = 1/(1+x^2)$ for every real number x and $E_i = (-\infty, i]$ for $i = 1, 2, \cdots$. Set sequence $\{E_i\}$ is nondecreasing and $\bigcup_{i=1}^{\infty} E_i = (-\infty, \infty) = X$. But we have $\mu(X) = \inf_{x \notin X} f(x) = \inf_{x \in \emptyset} f(x) = 1$, according to the conventions given at the beginning of this chapter, and

$$\mu(E_i) = \inf_{x \notin E_i} f(x) = \inf_{x \in (i, \infty)} f(x) = 0 \quad \text{for} \quad i = 1, 2, \cdots.$$

This violates the lower-semi-continuity.

Definition 4.13 Let $\mu : \mathscr{F} \to [0, \infty)$ is a monotone measure on measurable space $(X, \mathscr{F})$. Denote $\mu(X) = c$. Set function ν defined on $(X, \mathscr{F})$ by

$$\nu(E) = c - \mu(\overline{E}) \quad \text{for every} \quad E \in \mathscr{F}$$

is called the *dual* of μ.

If ν is the dual of μ, then μ is the dual of ν. It is also easy to know that μ is normalized if and only if its dual ν is normalized. If μ is upper-semi-continuous, then its dual ν is lower-semi-continuous and vice versa.

4.4 λ-Measures

The most common type of monotone measures in literature is the λ-measure (also called Sugeno's λ-fuzzy measure). In comparison with the classical measures, they have only one more parameter, λ, that indicates the magnitude of the interaction mentioned in the last section in a special and simple way. In the following, we discuss λ-measures in a more general aspect than the Sugeno's original model of the λ-fuzzy measure.

Let (X, $\mathscr{F}$) be a measurable space, μ be a nonnegative extended real-valued set function on $\mathscr{F}$, and $\lambda \in (-1/\sup\mu, \infty) \cup \{0\}$ be a constant, where $\sup\mu = \sup\{\mu(E) \mid E \in \mathscr{F}\}$. When μ is monotone and normalized, it is obviously that $\sup\mu = 1$.

Definition 4.14 Nonnegative set function μ is said to satisfy the *λ-rule* iff

$$\mu(E \cup F) = \mu(E) + \mu(F) + \lambda\mu(E)\mu(F) \tag{4.1}$$

whenever $E \in \mathscr{F}$, $F \in \mathscr{F}$, and $E \cap F = \emptyset$.

Theorem 4.4 If nonnegative set function μ satisfies the λ-rule, then

$$\mu(\bigcup_{i=1}^{n} E_i) = \begin{cases} \frac{1}{\lambda}(\prod_{i=1}^{n}[1 + \lambda \cdot \mu(E_i)] - 1) & \text{as } \lambda \neq 0 \\ \sum_{i=1}^{n} \mu(E_i) & \text{as } \lambda = 0, \end{cases} \tag{4.2}$$

for any finite disjoint class $\{E_1, E_2, \cdots, E_n\}$ of sets in $\mathscr{F}$.

Proof. When $\lambda = 0$, the conclusion is just that the additivity implies the finite additivity. When $\lambda \neq 0$, equation (4.1) can be rewritten as

$$\mu(E_1 \cup E_2) = \frac{1}{\lambda}([1 + \lambda \cdot \mu(E_1)][1 + \lambda \cdot \mu(E_2)] - 1).$$

For any given positive integer n>2, assuming that

$$\mu(\bigcup_{i=1}^{n-1} E_i) = \frac{1}{\lambda}(\prod_{i=1}^{n-1}[1 + \lambda \cdot \mu(E_i)] - 1),$$

we have

$$\begin{aligned}\mu(\bigcup_{i=1}^{n} E_i) &= \mu((\bigcup_{i=1}^{n-1} E_i) \cup E_n) \\ &= \frac{1}{\lambda}([1+\lambda \cdot \frac{1}{\lambda}(\prod_{i=1}^{n-1}[1+\lambda \cdot \mu(E_i)]-1)\ ][1+\lambda \cdot \mu(E_n)]-1) \\ &= \frac{1}{\lambda}(\prod_{i=1}^{n-1}[1+\lambda \cdot \mu(E_i)][1+\lambda \cdot \mu(E_n)]-1) \\ &= \frac{1}{\lambda}(\prod_{i=1}^{n}[1+\lambda \cdot \mu(E_i)]-1) \end{aligned}$$

The proof of the theorem is now completed by using the mathematical induction. □

Definition 4.15 Nonnegative set function μ is said to satisfy the *σ-λ-rule* iff

$$\mu(\bigcup_{i=1}^{\infty} E_i) = \begin{cases} \frac{1}{\lambda}(\prod_{i=1}^{\infty}[1+\lambda \cdot \mu(E_i)]-1) & \text{as } \lambda \neq 0 \\ \sum_{i=1}^{\infty} \mu(E_i) & \text{as } \lambda = 0 \end{cases} \tag{4.3}$$

for any disjoint sequence $\{E_i\}$ of sets in $\mathscr{F}$.

Similar to measures, the σ-λ-rule of nonnegative set functions that vanishes at the empty set is equivalent to the λ-rule when $\mathscr{F}$ is finite, especially, when X is finite.

Definition 4.16 Nonnegative set function $\mu : \mathscr{F} \to [0, \infty]$ is called a *λ-measure* on $\mathscr{F}$ iff μ satisfies the σ-λ-rule on $\mathscr{F}$ for some $\lambda \in (-1/\sup \mu, \infty) \cup \{0\}$ and there exists $E \in \mathscr{F}$ such that $\mu(E) < \infty$.

A λ-measure is usually denoted as g_λ. A normalized λ-measure is called a *Sugeno measure* (or, Sugeno's *λ-fuzzy measure*). It is easy to see that any λ-measure is superadditive if $\lambda > 0$ and is subadditive if $\lambda < 0$. Any λ-measure is a classical measure if and only if $\lambda = 0$.

Example 4.20 Let $X = \{x_1, x_2\}$ and $\mathscr{F}=\mathscr{P}(X)$. If

$$\mu(E) = \begin{cases} 0 & \text{when } E = \emptyset \\ 0.2 & \text{when } E = \{x_1\} \\ 0.5 & \text{when } E = \{x_2\} \\ 1 & \text{when } E = X \end{cases},$$

then μ is a Sugeno measure with λ = 3. For it, we just need to verify that

$$\mu(X) = \mu(\{x_1\}) + \mu(\{x_2\}) + 3\mu(\{x_1\}) \cdot \mu(\{x_2\}).$$

In fact,

$$\mu(\{x_1\}) + \mu(\{x_2\}) + 3\mu(\{x_1\}) \cdot \mu(\{x_2\}) = 0.2 + 0.5 + 3 \times 0.2 \times 0.5 = 1 = \mu(X).$$

Theorem 4.5 If g_λ is a λ-measure on $\mathscr{F}$, then $g_\lambda(\emptyset) = 0$ and g_λ satisfies the λ-rule.

Proof. We only need to prove the theorem when $\lambda \neq 0$ since a similar result for measures has already been proved. From Definition 4.16, there exists set $E \in \mathscr{F}$ with $g_\lambda(E) < \infty$. Let $E_1 = E$ and $E_i = \emptyset$ for $i = 2, 3, \cdots$. Then set sequence $\{E_1, E_2, \cdots\}$ is disjoint and $E_1 = \bigcup_{i=1}^{\infty} E_i$. By using the σ-λ-rule (4.3) of g_λ, we have

$$g_\lambda(E_1) = \frac{1}{\lambda}\{\prod_{i=2}^{\infty}[1 + \lambda \cdot g_\lambda(E_i)] \cdot [1 + \lambda \cdot g_\lambda(E_1)] - 1\},$$

that is,

$$1+\lambda\cdot g_\lambda(E_1)=\prod_{i=2}^{\infty}[1+\lambda\cdot g_\lambda(E_i)]\cdot[1+\lambda\cdot g_\lambda(E_1)].$$

Since $\lambda\in(-1/\sup g_\lambda,\infty)\cup\{0\}$ and $g_\lambda(E_1)<\infty$, we know that $0<1+\lambda\cdot g_\lambda(E_1)<\infty$. Thus,

$$\prod_{i=2}^{\infty}[1+\lambda\cdot g_\lambda(E_i)]=1$$

and, therefore,

$$1+\lambda\cdot g_\lambda(\emptyset)=1.$$

Consequently, $g_\lambda(\emptyset)=0$ since $\lambda\neq 0$.

By using $g_\lambda(\emptyset)=0$, the second result of the theorem is clear. The theorem is now complete. □

When $X=\{x_1,x_2,\cdots\}$ and $\mathscr{F}=\mathscr{P}(X)$, knowing the values of λ-measure g_λ at every singleton and the value of λ, equation (4.3) can be used to calculate the value of g_λ at any set in $\mathscr{F}$. Conversely, restricting the universal set being finite and based on equation (4.2), we can prove the following theorem, by which the value of λ may be uniquely determined if the values of λ-measure g_λ at every singleton and at X are known. The proof of the theorem is omitted here. The interested readers may refer to [Wang and Klir, 1992 or Wang and Klir 2008].

Theorem 4.6 Let $X=\{x_1,x_2,\cdots,x_n\}$, where $n\geq 2$, and g_λ be a λ-measure on $\mathscr{P}(X)$. Knowing $g_\lambda(\{x_i\})=a_i\geq 0$ (with at least two of them being non-zero) and $g_\lambda(X)=b>a_i$, $i=1,2,\cdots,n$, the value of λ can be uniquely determined by equation

$$1 + b\lambda = \prod_{i=1}^{n} (1 + a_i\lambda):$$

(1) $\lambda > 0$ when $\sum_{i=1}^{n} a_i < b$;
(2) $\lambda = 0$ when $\sum_{i=1}^{n} a_i = b$;
(3) $-1/b < \lambda < 0$ when $\sum_{i=1}^{n} a_i > b$.

Example 4.21 Let $X = \{x_1, x_2, x_3\}$ and g_λ be a λ-measure on $\mathscr{P}(X)$. Knowing $g_\lambda(\{x_1\}) = 0.1$, $g_\lambda(\{x_2\}) = g_\lambda(\{x_3\}) = 0.2$, and $g_\lambda(X) = 1$, we want to find the value of parameter λ and the values of g_λ at the other sets in $\mathscr{P}(X)$.

From Theorem 4.6, we know that $\lambda > 0$ and

$$1 + \lambda = (1 + 0.1\lambda)(1 + 0.2\lambda)(1 + 0.2\lambda),$$

that is,

$$0.004\lambda^2 + 0.08\lambda - 0.5 = 0.$$

Solving this quadratic equation, we obtain

$$\begin{aligned}\lambda &= \frac{-0.08 \pm \sqrt{0.08^2 - 4 \times 0.004 \times (-0.5)}}{2 \times 0.004} \\ &= \frac{-0.08 \pm 0.12}{0.008} \\ &= 5 \text{ or } -25\end{aligned}$$

Since $\lambda = -25$ violates $\lambda > 0$, we obtain the unique feasible solution $\lambda = 5$.

Furthermore, using equation (4.1), we have

$$g_\lambda(\{x_1, x_2\}) = 0.1 + 0.2 + 5 \times 0.1 \times 0.2 = 0.4,$$

$$g_\lambda(\{x_1, x_3\}) = 0.1 + 0.2 + 5 \times 0.1 \times 0.2 = 0.4,$$

$$g_\lambda(\{x_2, x_3\}) = 0.2 + 0.2 + 5 \times 0.2 \times 0.2 = 0.6 .$$

Theorem 4.7 Let $X = \{x_1, x_2, \cdots, x_n\}$ and g_λ be a λ-measure on $\mathscr{P}(X)$ with $g_\lambda(X) = c > 0$ and parameter $\lambda \in (-1/c, \infty)$. For any sets $E \in \mathscr{P}(X)$ and $F \in \mathscr{P}(X)$,

(1) $g_\lambda(E - F) = \dfrac{g_\lambda(E) - g_\lambda(E \cap F)}{1 + \lambda \cdot g_\lambda(E \cap F)}$,

(2) $g_\lambda(E \cup F) = \dfrac{g_\lambda(E) + g_\lambda(F) - g_\lambda(E \cap F) + \lambda \cdot g_\lambda(E) \cdot g_\lambda(F)}{1 + \lambda \cdot g_\lambda(E \cap F)}$,

(3) $g_\lambda(\overline{E}) = \dfrac{c - g_\lambda(E)}{1 + \lambda \cdot g_\lambda(E)}$.

Proof. Set E can be expressed as a disjoint union $(E \cap F) \cup (E - F)$. By (4.1), we have

$$\begin{aligned} g_\lambda(E) &= g_\lambda(E \cap F) + g_\lambda(E - F) + \lambda \cdot g_\lambda(E \cap F) \cdot g_\lambda(E - F) \\ &= g_\lambda(E \cap F) + g_\lambda(E - F)[1 + \lambda \cdot g_\lambda(E \cap F)] . \end{aligned}$$

Since $\lambda \in (-1/c, \infty)$, we know that $1 + \lambda \cdot g_\lambda(E \cap F) > 0$ and, therefore, obtain (1). As for (2), using a similar strategy and (1), we get

$$\begin{aligned} g_\lambda(E \cup F) &= g_\lambda(E \cup [F - (E \cap F)]) \\ &= g_\lambda(E) + g_\lambda(F - (E \cap F)) \cdot [1 + \lambda \cdot g_\lambda(E)] \\ &= g_\lambda(E) + \frac{g_\lambda(F) - g_\lambda(F \cap E \cap F)}{1 + \lambda \cdot g_\lambda(F \cap E \cap F)} \cdot [1 + \lambda \cdot g_\lambda(E)] \\ &= g_\lambda(E) + \frac{g_\lambda(F) - g_\lambda(E \cap F)}{1 + \lambda \cdot g_\lambda(E \cap F)} \cdot [1 + \lambda \cdot g_\lambda(E)] \\ &= \frac{g_\lambda(E) + g_\lambda(F) - g_\lambda(E \cap F) + \lambda \cdot g_\lambda(E) \cdot g_\lambda(F)}{1 + \lambda \cdot g_\lambda(E \cap F)} . \end{aligned}$$

Finally, conclusion (3) is obtained by substituting X for E and E for F in (1). □

Theorem 4.8 Let $X=\{x_1, x_2, \cdots, x_n\}$ and g_λ be a normalized λ-measure on $\mathscr{P}(X)$ and parameter $\lambda = a \in (-1, \infty)$. The dual of g_λ, denoted by μ, is a normalized λ-measure on $\mathscr{P}(X)$ as well and its parameter is $\lambda' = -a/(a+1)$.

Proof. From the definition of the dual, we have $\mu(X) = 1 - g_\lambda(\overline{X}) = 1 - g_\lambda(\varnothing) = 1 - 0 = 1$. To show that μ is also a λ-measure on $\mathscr{P}(X)$ with parameter $\lambda' = -a/(a+1)$, we just need to verify the corresponding λ-rule for μ. In fact, for any given $E \in \mathscr{P}(X)$ and $F \in \mathscr{P}(X)$ with $E \cap F = \varnothing$, from (3) of Theorem 4.7, we have

$$\begin{aligned}
\mu(E)+\mu(F)+\lambda'\mu(E)\cdot\mu(F) &= \mu(E)+\mu(F)-\frac{a}{a+1}\mu(E)\cdot\mu(F) \\
&= 1-g_\lambda(\overline{E})+1-g_\lambda(\overline{F})-\frac{a}{a+1}[1-g_\lambda(\overline{E})]\cdot[1-g_\lambda(\overline{F})] \\
&= \frac{(1+a)g_\lambda(E)}{1+ag_\lambda(E)}+\frac{(1+a)g_\lambda(F)}{1+ag_\lambda(F)}-a\frac{(1+a)g_\lambda(E)g_\lambda(F)}{[1+ag_\lambda(E)][1+ag_\lambda(F)]} \\
&= \frac{(1+a)[g_\lambda(E)+g_\lambda(F)+ag_\lambda(E)g_\lambda(F)]}{[1+ag_\lambda(E)][1+ag_\lambda(F)]} \\
&= \frac{(1+a)g_\lambda(E\cup F)}{[1+ag_\lambda(E\cup F)]} \\
&= 1-g_\lambda(\overline{E\cup F}) \\
&= \mu(E\cup F)\ .
\end{aligned}$$

The proof is now complete. □

The above theorem also shows that the dual of any superadditive normalized λ-measure must be subadditive and vice versa. This conclusion is also true for any λ-measure.

4.5 Quasi-Measures

Given a λ-measure μ on $\mathscr{F}$ with $\lambda \neq 0$, from (4.3), we have

$$1+\lambda \cdot \mu(\bigcup_{i=1}^{\infty} E_i) = \prod_{i=1}^{\infty}[1+\lambda \cdot \mu(E_i)] \tag{4.4}$$

for any disjoint sequence $\{E_i\}$ of sets in $\mathscr{F}$. Taking logarithm in both sides of (4.4), we obtain

$$\ln(1+\lambda \cdot \mu(\bigcup_{i=1}^{\infty} E_i)) = \sum_{i-1}^{\infty} \ln[1+\lambda \cdot \mu(E_i)] \ . \tag{4.5}$$

If we define a new set function ν on $\mathscr{F}$ by

$$\nu(E) = \ln[1+\lambda \cdot \mu(E)] \quad \text{for every} \quad E \in \mathscr{F},$$

Equation (4.5) becomes

$$\nu(\bigcup_{i=1}^{\infty} E_i) = \sum_{i=1}^{\infty} \nu(E_i) \, .$$

This means that the new set function ν possesses the σ-additivity and, therefore, with the fact that

$$\nu(\varnothing) = \ln(1+\lambda \cdot \mu(\varnothing) = \ln 1 = 0 \, ,$$

is a classical measure. Such a result provides a new approach for constructing λ-measures and suggest us to introduce a wider type of monotone measures that includes λ-measures as special examples.

Definition 4.17 Let $a \in (0, \infty]$. An extended real-valued function $\theta : [0, a] \to [0, \infty]$ is called a *T-function* iff it is continuous, strictly increasing, and such that $\theta^{-1}(0) = 0$ and

$$\theta^{-1}(\{\infty\}) = \begin{cases} \varnothing & \text{when } a < \infty \\ \{\infty\} & \text{when } a = \infty. \end{cases}$$

In the above definition, expression "$\theta^{-1}(\{\infty\}) = \varnothing$ or $\{\infty\}$" means that the image of any finite value by mapping θ must be finite.

Definition 4.18 Let μ be a set function on $\mathscr{F}$. μ is *quasi-additive* iff there exists a T-function θ, whose domain contains the range of μ, such that the set function $\theta \circ \mu$ defined by $(\theta \circ \mu)(E) = \theta(\mu(E))$ for every $E \in \mathscr{F}$ is additive; μ is *quasi-σ-additive* iff there exists a T-function θ, whose domain contains the range of μ, such that the set function $\theta \circ \mu$ defined by $(\theta \circ \mu)(E) = \theta(\mu(E))$ for every $E \in \mathscr{F}$ is σ-additive; μ is called a *quasi-measure* iff there exists a T-function θ such that $\theta \circ \mu$ is a classical measure on $\mathscr{F}$. In this case, T-function θ is called the *proper T-function* of μ. A normalized quasi-measure is called a *quasi-probability*.

Obviously, for any given measure μ on $\mathscr{F}$ and any T-function θ whose range covers the range of μ, set function $\theta^{-1} \circ \mu$ is a quasi-measure on $\mathscr{F}$. It should be noted that, for a given quasi-measure, its proper T-functions are not unique. If μ is a finite quasi-measure, a T-function θ such that $\theta \circ \mu$ is a normalized measure (i.e., a probability measure) is called its *standard T-function*. Any classical measure is a special case of quasi-measure with the identity function as one of its proper T-functions.

Example 4.22 The monotone measure given in Example 4.15 is a quasi-measure. Its standard T-function is $\theta(y) = y^{1/k}$, $y \in [0, 1]$.

The following theorem shows that any λ-measure is a special example of quasi-measure.

Theorem 4.9 Any λ-measure g_λ on $\mathscr{F}$ with parameter $\lambda \neq 0$ is a quasi-measure having

$$\theta_\lambda(y) = \frac{\ln(1+\lambda y)}{k\lambda}, \quad y \in [0, \sup g_\lambda]$$

as its proper T-function, where k may be any finite positive real number. Conversely, if μ is a classical measure on $\mathscr{F}$, then $\theta_\lambda^{-1} \circ \mu$ is a λ-measure, where

$$\theta_\lambda^{-1}(x) = \frac{e^{k\lambda x} - 1}{\lambda}, \quad x \in [0, \infty]$$

and k may be any finite positive real number.

Proof. Since θ_λ is continuous and strictly increasing with $\theta_\lambda(0) = (\ln 1)/(k\lambda) = 0$, it is a T-function. Set function g_λ as a λ-measure has at least one set $E_0 \in \mathscr{F}$ such that $0 \le g_\lambda(E_0) < \infty$. Hence, by the behavior of θ_λ that it does not map any finite value to the infinite, we know that

$$(\theta_\lambda \circ g_\lambda)(E_0) = \theta_\lambda(g_\lambda(E_0)) < \infty .$$

So, we only need to verify the σ-additivity of $\theta_\lambda \circ g_\lambda$. In fact, for any disjoint set sequence $\{E_i\}$ in $\mathscr{F}$,

$$\begin{aligned}(\theta_\lambda \circ g_\lambda)(\bigcup_{i=1}^{\infty} E_i) &= \theta_\lambda(g_\lambda(\bigcup_{i=1}^{\infty} E_i)) \\ &= \frac{1}{k\lambda}\ln(1+\lambda \cdot g_\lambda(\bigcup_{i=1}^{\infty} E_i)) \\ &= \frac{1}{k\lambda}\ln(1+(\prod_{i=1}^{\infty}[1+\lambda \cdot g_\lambda(E_i)]) - 1)\end{aligned}$$

$$= \frac{1}{k\lambda} \ln \prod_{i=1}^{\infty} [1 + \lambda \cdot g_\lambda(E_i)]$$

$$= \frac{1}{k\lambda} \sum_{i=1}^{\infty} \ln[1 + \lambda \cdot g_\lambda(E_i)]$$

$$= \sum_{i=1}^{\infty} \frac{\ln[1 + \lambda \cdot g_\lambda(E_i)]}{k\lambda}$$

$$= \sum_{i=1}^{\infty} (\theta_\lambda \circ g_\lambda)(E_i) \quad .$$

Conversely, if μ is a classical measure on $\mathscr{F}$, then it is σ-additive and $\mu(\varnothing) = 0$. Thus,

$$(\theta_\lambda^{-1} \circ \mu)(\varnothing) = \theta_\lambda^{-1}(\mu(\varnothing)) = \theta_\lambda^{-1}(0) = 0$$

and

$$(\theta_\lambda^{-1} \circ \mu)(\bigcup_{i=1}^{\infty} E_i) = \theta_\lambda^{-1}(\mu(\bigcup_{i=1}^{\infty} E_i)$$

$$= \theta_\lambda^{-1}(\sum_{i=1}^{\infty} \mu(E_i))$$

$$= \frac{\exp(k\lambda \sum_{i=1}^{\infty} \mu(E_i)) - 1}{\lambda}$$

$$= \frac{\prod_{i=1}^{\infty} e^{k\lambda \cdot \mu(E_i)} - 1}{\lambda}$$

$$= \frac{1}{\lambda} (\prod_{i=1}^{\infty} [1 + \lambda \cdot \theta_\lambda^{-1}(\mu(E_i))] - 1)$$

$$= \frac{1}{\lambda} (\prod_{i=1}^{\infty} [1 + \lambda \cdot (\theta_\lambda^{-1} \circ \mu)(E_i)] - 1),$$

that is, $\theta_\lambda^{-1} \circ \mu$ satisfies the σ-λ-rule. So, it is a λ-measure. □

Example 4.23 In Example 20, $X = \{x_1, x_2\}$, $\mathscr{F} = \mathscr{P}(X)$ and

$$g_\lambda(E) = \begin{cases} 0 & \text{if } E = \varnothing \\ 0.2 & \text{if } E = \{x_1\} \\ 0.4 & \text{if } E = \{x_2\} \\ 1 & \text{if } E = X \end{cases} .$$

Set function g_λ is a λ-measure with parameter $\lambda = 5$. If we take

$$\theta_\lambda(y) = \frac{\ln(1+\lambda y)}{\ln(1+\lambda)} = \frac{\ln(1+5y)}{\ln 6},$$

then

$$(\theta_\lambda \circ g_\lambda)(E) = \begin{cases} 0 & \text{if } E = \varnothing \\ 0.386\cdots & \text{if } E = \{x_1\} \\ 0.613\cdots & \text{if } E = \{x_2\} \\ 1 & \text{if } E = X \end{cases} .$$

Set function $\theta_\lambda \circ g_\lambda$ is a probability measure on $\mathscr{P}(X)$. The above θ_λ is the standard T-function of g_λ.

The following example shows how to construct a λ-measure from an existing probability measure with a given value of parameter λ.

Example 4.24 Let $X = \{x_1, x_2\}$, $\mathscr{F} = \mathscr{P}(X)$ and

$$\mu(E) = \begin{cases} 0 & \text{if } E = \varnothing \\ 0.34 & \text{if } E = \{x_1\} \\ 0.66 & \text{if } E = \{x_2\} \\ 1 & \text{if } E = X \end{cases} .$$

For given $\lambda = -0.75$, taking T-function

$$\theta_\lambda(y) = \frac{\ln(1-0.75y)}{\ln 0.25},$$

we have

$$\theta_\lambda^{-1}(x) = \frac{1-0.25^x}{0.75}.$$

Thus

$$g_\lambda(E) = (\theta_\lambda^{-1} \circ \mu)(E) = \theta_\lambda^{-1}(\mu(E)) = \begin{cases} 0 & \text{if } E = \varnothing \\ 0.50\cdots & \text{if } E = \{x_1\} \\ 0.79\cdots & \text{if } E = \{x_2\} \\ 1 & \text{if } E = X \end{cases}$$

is a λ-measure with parameter $\lambda = -0.75$. As a quasi-measure, its standard T-function is $\theta_\lambda(y)$ shown above.

4.6 Möbius and Zeta Transformations

We have seen that the nonadditivity of a monotone measure describes the interaction among the contribution rates of considered attributes towards a certain target. Now the question is what the amounts of the mentioned interaction are. The following example shows the idea for introducing the Möbius and zeta transformations.

Example 4.25 Let $X = \{x_1, x_2\}$ and set function $\mu : \mathscr{P}(X) \to [0, \infty)$ be a monotone measure. Set function μ describes the individual as well as the joint contribution rates from attributes towards a certain target. Let $\nu(\{x_1, x_2\}) = \mu(\{x_1, x_2\}) - (\mu(\{x_1\}) + \mu(\{x_2\}))$. Due to the nonadditivity of μ, $\nu(\{x_1, x_2\})$ may not be zero. The amount of $\nu(\{x_1, x_2\})$ can be understood as the "pure" interaction between the contribution rates from attributes x_1 and x_2.

When the number of attributes is larger than 2, the expression describing the "pure" interaction among considered attributes is not so simple. The following definition gives a general expression that can describe the "pure" interaction among the contribution rates from attributes towards the target.

Definition 4.19 Let $X = \{x_1, x_2, \cdots, x_n\}$ and μ be a real-valued set function on $\mathscr{P}(X)$. Define set function ν by

$$\nu(E) = \sum_{F \subseteq E} (-1)^{|E-F|} \mu(F) \tag{4.6}$$

for every $E \in \mathscr{P}(X)$. Set function ν on $\mathscr{P}(X)$ is called the *Möbius representation* of μ. Equation(4.6) is also called the *Möbius transformation*.

Example 4.26 Let $X = \{x_1, x_2, x_3\}$ and $\mu : \mathscr{P}(X) \to [0, \infty)$ be defined as

$$\mu(E) = \begin{cases} 0 & \text{if } E = \emptyset \\ 0.1 & \text{if } |E| = 1 \\ 0.35 & \text{if } |E| = 2 \\ 1 & \text{if } E = X \end{cases}$$

for $E \in \mathscr{P}(X)$. Then its Möbius representation ν has values

$$\nu(E) = \begin{cases} 0 & \text{if } E = \emptyset \\ 0.1 & \text{if } |E| = 1 \\ 0.15 & \text{if } |E| = 2 \\ 0.25 & \text{if } E = X \end{cases}$$

for $E \in \mathscr{P}(X)$. Besides the contribution rates from individual attributes, set function ν also describes the "pure" interaction amounts, that is, the amount of "pure" interaction between any two attributes is 0.15, while the amount of "pure" interaction among all three attributes is 0.25. Here, we can see that the total sum of "pure" contribution rate (including the interactions) from sets in $\mathscr{P}(X)$ is

$$\sum_{E \subseteq X} \nu(E) = 0 + 0.1 + 0.1 + 0.1 + 0.15 + 0.15 + 0.15 + 0.25 = 1 .$$

Definition 4.20 For given set function ν on $\mathscr{P}(X)$, transformation

$$\mu(E) = \sum_{F \subseteq E} \nu(F) \tag{4.7}$$

for every $E \in \mathscr{P}(X)$ is called the *zeta transformation.*

We should note that, in Definitions 4.19 and 4.20, both set functions μ and ν are not necessarily nonnegative. However, if ν is nonnegative, then so is μ.

Lemma 4.1 For any given finite sets G and E satisfying $G \subseteq E$,

$$\sum_{F | G \subseteq F \subseteq E} (-1)^{|F-G|} = \begin{cases} 0 & \text{if } G \subset E \\ 1 & \text{if } G = E . \end{cases}$$

Proof. When $G \subset E$, denoting $n = |E - G|$, we have

$$\begin{aligned} \sum_{F | G \subseteq F \subseteq E} (-1)^{|F-G|} &= \sum_{i=0}^{n} C(n,i)(-1)^i \\ &= (1-1)^n \\ &= 0 \quad . \end{aligned}$$

When $G = E$, it is obvious that

$$\sum_{F | G \subseteq F \subseteq E} (-1)^{|F-G|} = (-1)^0 = 1 .$$

□

Taking $G = \varnothing$, we have the following Corollary.

Corollary 4.1 For any given nonempty finite sets *E*,

$$\sum_{F \subseteq E} (-1)^{|F|} = 0 \ .$$

Theorem 4.10 The Möbius transformation and the zeta transformation are the inverse to each other.

Proof. We verify (4.7) from (4.6). In fact, by using Lemma 4.1,

$$\sum_{F \subseteq E} \nu(F) = \sum_{F \subseteq E} \sum_{G \subseteq F} (-1)^{|F-G|} \mu(G) = \sum_{G \subseteq E} \sum_{F | G \subseteq F \subseteq E} (-1)^{|F-G|} \mu(G)$$
$$= \sum_{G = E} \mu(G) = \mu(E) \ .$$

It is similar for verifying (4.6) from (4.7). □

Example 4.27 We use set functions μ and ν shown in Example 4.26 to confirm the conclusion of Theorem 4.10. When $|E| = 2$,

$$\begin{aligned} \mu(E) &= \sum_{F \subseteq E} \nu(F) \\ &= 0.1 + 0.1 + 0.15 \\ &= 0.35 ; \end{aligned}$$

while $|E| = 3$, that is, $E = X$,

$$\begin{aligned} \mu(E) &= \sum_{F \subseteq E} \nu(F) \\ &= 0.1 + 0.1 + 0.1 + 0.15 + 0.15 + 0.15 + 0.25 \\ &= 1 . \end{aligned}$$

The case of $|E|=0$ or $|E|=1$ is trivial.

4.7 Belief Measures and Plausibility Measures

In this section, two common types of monotone measures and the relation to the probability are discussed. We still use X to denote the universal set.

Definition 4.21 Set function $m: \mathscr{P}(X) \to [0,1]$ is called a *basic probability assignment* if there exists a countable class of sets $\{A_i \mid i=1,2,\cdots\} \subseteq \mathscr{P}(X)-\{\varnothing\}$ such that $\sum_{i=1}^{\infty} m(A_i)=1$ and $m(E)=0$ for any $E \notin \{A_i \mid i=1,2,\cdots\}$.

From Definition 4.21, we know that $m(\varnothing)=0$. Defining $p(\varnothing)=0$ and $p(\{E\})=m(E)$ for every $E \in \mathscr{P}(X)$, p is a normalized measure on the semiring consisting of the empty set and all singletons in $\mathscr{P}(\mathscr{P}(X))$. This normalized measure can be uniquely extended to be a discrete probability measure on $\mathscr{P}(\mathscr{P}(X))$.

Example 4.28 Let $X=\{x_1, x_2, x_3\}$ and $m: \mathscr{P}(X) \to [0,\infty)$ be defined as

$$m(E)=\begin{cases} 0.1 & \text{if } E=\{x_1\} \\ 0.3 & \text{if } E=\{x_3\} \\ 0.6 & \text{if } E=\{x_1, x_2\} \\ 0 & \text{else}\ . \end{cases}$$

Then m is a basic probability assignment on $\mathscr{P}(X)$. Denote $E_0=\varnothing$, $E_1=\{x_1\}$, $E_2=\{x_2\}$, $E_3=\{x_1,x_2\}$, $E_4=\{x_3\}$, $E_5=\{x_1,x_3\}$, $E_6=\{x_2,x_3\}$, and $E_7=\{x_1,x_2,x_3\}$. Then $\mathscr{P}(X) = \{E_i \mid i=0,1,\cdots,7\}$ and, regarding $\mathscr{P}(X)$ as the universal set and each set E_i as an element, class $\mathscr{S}=\{\varnothing, \{E_1\}, \{E_2\}, \{E_3\}, \{E_4\}, \{E_5\}, \{E_6\}, \{E_7\}\}$ is a semiring. If we define $p(\{E_i\})=m(E_i)$ for $i=0,1,\cdots,7$, then p is a probability measure on $\mathscr{S}$. It can be uniquely extended to a discrete probability distribution on $\mathscr{P}(\mathscr{P}(X))$, the σ-algebra generated by $\mathscr{S}$, as follows:

$$p(\hat{E}) = \sum_{E_i \in \hat{E}} m(E_i) \tag{4.8}$$

for every $\hat{E} \in \mathscr{P}(\mathscr{P}(X))$. For example,

$$\begin{aligned} p(\{\{x_2\},\{x_1,x_3\}\}) &= p(\{E_2,E_5\}) \\ &= p(\{E_2\}) + p(\{E_5\}) \\ &= 0 \quad , \end{aligned}$$

while

$$p(\{\{x_3\},\{x_1,x_2\}\}) = p(\{E_4,E_3\}) = p(\{E_4\}) + p(\{E_3\}) = 0.9 \quad .$$

Definition 4.22 Let set function m be a basic probability assignment on $\mathscr{P}(X)$. The set function $\mu : \mathscr{P}(X) \to [0,1]$, determined by

$$\mu(E) = \sum_{F \subseteq E} m(F)$$

for every $E \in \mathscr{P}(X)$ is called a *belief measure* on measurable space $(X, \mathscr{P}(X))$ and we say that μ is induced from m. A belief measure is usually indicated as *Bel*.

When the universal set X is finite, we may see that any belief measure is just the zeta transformation of a basic probability assignment; conversely, the Möbius representation of any belief measure is a basic probability assignment.

Example 4.29 Based on the basic probability assignment shown in Example 4.28, using (4.8) we obtain a belief measure *Bel* on $\mathscr{P}(X)$ as

$$Bel(E) = \begin{cases} 0 & \text{if } E = \emptyset \text{ or } \{x_2\} \\ 0.1 & \text{if } E = \{x_1\} \\ 0.3 & \text{if } E = \{x_3\} \text{ or } \{x_2, x_3\} \\ 0.4 & \text{if } E = \{x_1, x_3\} \\ 0.7 & \text{if } E = \{x_1, x_2\} \\ 1 & \text{if } E = \{x_1, x_2, x_3\} \ . \end{cases}$$

The following theorem shows that any belief measure is a superadditive normalized monotone measure, which is continuous from above. The theorem also shows an important inequality for belief measures.

Theorem 4.11 If μ is a belief measure on $\mathscr{P}(X)$, then it has the following properties:

(BM1) $\mu(\emptyset) = 0$ and $\mu(X) = 1$;
(BM2) $\mu(\bigcup_{i=1}^{n} E_i) \geq \sum_{I \subseteq \{1,\cdots,n\}, I \neq \emptyset} (-1)^{|I|-1} \mu(\bigcap_{i \in I} E_i)$ for any finite subclass $\{E_1, \cdots, E_n\}$ of $\mathscr{P}(X)$;
(BM3) μ is superadditive;
(BM4) μ is monotone;
(BM5) μ is upper-continuous (continuous from above).

Proof. Property (BM1) is obtained from the definitions of basic probability assignment m and relative belief measure μ. To property (BM2), consider any given finite subclass $\{E_1, \cdots, E_n\}$. Denoting $I(F) = \{i \mid 1 \leq i \leq n, F \subseteq E_i\}$ for any given set F and using Corollary (4.1), we have

$$\sum_{I \subseteq \{1,\cdots,n\}, I \neq \emptyset} (-1)^{|I|-1} \mu(\bigcap_{i \in I} E_i) = \sum_{I \subseteq \{1,\cdots,n\}, I \neq \emptyset} [(-1)^{|I|-1} \sum_{F \subseteq \bigcap_{i \in I} E_i} m(F)]$$
$$= \sum_{F \mid I(F) \neq \emptyset} [m(F) \sum_{I \subseteq I(F), I \neq \emptyset} (-1)^{|I|-1}]$$

$$
\begin{aligned}
&= \sum_{F|I(F)\neq\varnothing} [m(F)(1-\sum_{I\subseteq I(F)}(-1)^{|I|})] \\
&= \sum_{F|I(F)\neq\varnothing} m(F) \\
&= \sum_{F\subseteq E_i \text{ for some } i} m(F) \\
&\le \sum_{F\subseteq \bigcup_{i=1}^{n} E_i} m(F) \\
&= \mu(\bigcup_{i=1}^{n} E_i) \quad .
\end{aligned}
$$

As for property (BM3), considering any given sets E and F with $E \cap F = \varnothing$ and using (BM2), we have

$$
\begin{aligned}
\mu(E\cup F) &\ge \mu(E)+\mu(F)-\mu(E\cap F) \\
&= \mu(E)+\mu(F) \quad .
\end{aligned}
$$

Property (BM4) is a direct result of (BM3). In fact, let $E \subseteq F$. Since $E\cap(F-E)=\varnothing$ and μ is nonnegative, we have

$$
\begin{aligned}
\mu(F) &= \mu(E\cup(F-E)) \\
&\ge \mu(E)+\mu(F-E) \\
&\ge \mu(E) \quad .
\end{aligned}
$$

Finally, we show property (BM5). For any given belief measure μ, let m be the corresponding basic probability assignment. From Definition 4.21, we know that there exists a countable class of sets $\{A_i \mid i=1,2,\cdots\} \subseteq \mathscr{P}(X)-\{\varnothing\}$ such that $\sum_{i=1}^{\infty} m(A_i)=1$ and $m(E)=0$ for any $E \notin \{A_i \mid i=1,2,\cdots\}$. Hence, for any given $\varepsilon>0$, there exists positive integer n_0, such that $\sum_{n>n_0} m(A_i)<\varepsilon$. Consider any given nonincreasing set sequence $\{E_i\}$ with $\bigcap_{i=1}^{\infty} E_i = E$. For each A_n with $n \le n_0$, if $A_n - E \neq \varnothing$, then there exists $i(n)$ such that $A_n - E_{i(n)} \neq \varnothing$.

Write $i_0 = \max(i(1),\cdots,i(n_0))$. If $A_n - E \neq \varnothing$, then $A_n - E_{i(n)} \neq \varnothing$ for every $n \leq n_0$. So, we have

$$\begin{aligned}
\mu(E) &= \sum_{F \subseteq E} m(F) \\
&= \sum_{A_n \subseteq E} m(A_n) \\
&\geq \sum_{A_n \subseteq E, n \leq n_0} m(A_n) \\
&\geq \sum_{A_n \subseteq E_{i_0}, n \leq n_0} m(A_n) \\
&\geq \sum_{A_n \subseteq E_{i_0}} m(A_n) - \sum_{n > n_0} m(A_n) \\
&> \sum_{F \subseteq E_{i_0}} m(F) - \varepsilon \\
&= \mu(E_{i_0}) - \varepsilon .
\end{aligned}$$

Thus, by the monotonicity of μ, we know $\mu(E) = \lim_i \mu(E_i)$. The proof of the theorem is now complete. □

Properties (BM1) and (BM2) shown in Theorem 4.11 are essential to belief measures. This can be seen in the next theorem.

Theorem 4.12 Let X be finite. If a set function $\mu : \mathscr{P}(X) \to [0,1]$ satisfies conditions (BM1) and (BM2), then its Möbius representation $m : \mathscr{P}(X) \to [0,1]$ is a basic probability assignment and, furthermore, μ is the belief measure induced from m.

Proof. Since m is the Möbius representation of μ, using (4.6) we have

$$m(\varnothing) = \sum_{F \subseteq \varnothing} (-1)^{|\varnothing - F|} \mu(F) = (-1)^0 \mu(\varnothing) = 0 .$$

We know that the zeta transformation is the inverse of the Möbius transformation. So,

$$\sum_{E\subseteq X} m(E)=\mu(X)=1\,.$$

Thus, we only need to show $m(E)\geq 0$ for every subset E of X. In fact, since X is finite, any subset of X must also be finite. For any given subset E, say, $E=\{x_1,x_2,\cdots,x_n\}$, denoting $E_i = E-\{x_i\}$, we have $E=\bigcup_{i=1}^{n} E_i$ and

$$\begin{aligned} m(E) &= \sum_{F\subseteq E}(-1)^{|E-F|}\mu(F) \\ &= \mu(E)-\sum_{I\subseteq\{1,\cdots,n\},I\neq\varnothing}(-1)^{|I|-1}\mu(\bigcap_{i\in I}E_i) \\ &= \mu(\bigcup_{i=1}^{n}E_i)-\sum_{I\subseteq\{1,\cdots,n\},I\neq\varnothing}(-1)^{|I|-1}\mu(\bigcap_{i\in I}E_i) \\ &\geq 0\ , \end{aligned}$$

due to (BM2). Finally, we know that μ is the belief measure induced from m since it is the zeta transformation of m. □

Corollary 4.2 Let X be finite. Set function $\mu:\mathscr{P}(X)\to[0,1]$ is a belief measure if and only if (BM1) and (BM2) hold.

Definition 4.23 Let set function m be a basic probability assignment on $\mathscr{P}(X)$. The set function, $\mu:\mathscr{P}(X)\to[0,1]$, determined by

$$\mu(E)=\sum_{F\cap E\neq\varnothing} m(F)$$

for every $E\in\mathscr{P}(X)$ is called a *plausibility measure* on measurable space $(X, \mathscr{P}(X))$ and we say that μ is induced from m. A plausibility measure is usually indicated as *Pl*.

Theorem 4.13 If *Bel* and *Pl* are the belief measure and plausibility measure on $\mathscr{P}(X)$ induced from a basic probability assignment m, then they are dual to each other, that is, $Pl(E) = 1 - Bel(\overline{E})$ for every $E \in \mathscr{P}(X)$, and $Bel \le Pl$.

Proof. From Definitions 4.22 and 4.23, for any $E \in \mathscr{P}(X)$, we have

$$\begin{aligned} Pl(E) &= \sum_{F \cap E \neq \emptyset} m(F) \\ &= \sum_{F \subseteq X} m(F) - \sum_{F \cap E = \emptyset} m(F) \\ &= 1 - \sum_{F \subseteq \overline{E}} m(F) \\ &= 1 - Bel(E) \quad . \end{aligned}$$

Furthermore, since $\{F \mid F \subseteq E\} \subseteq \{F \mid F \cap E \neq \emptyset\}$, we have

$$Bel(E) = \sum_{F \subseteq E} m(F) \le \sum_{F \cap F \neq \emptyset} m(F) = Pl(E)$$

for every $E \in \mathscr{P}(X)$. □

Theorem 4.14 If μ is a plausibility measure on $\mathscr{P}(X)$, then it has the following properties:

(PM1) $\mu(\emptyset) = 0$ and $\mu(X) = 1$;
(PM2) $\mu(\bigcap_{i=1}^{n} E_i) \le \sum_{I \subseteq \{1,\cdots,n\}, I \neq \emptyset} (-1)^{|I|-1} \mu(\bigcup_{i \in I} E_i)$ for any finite subclass $\{E_1, \cdots, E_n\}$ of $\mathscr{P}(X)$;
(PM3) μ is subadditive;
(PM4) μ is monotone;
(PM5) μ is lower-continuous (continuous from below).

Proof. Properties (PM1), (PM3), (PM4), and (PM5) are direct result of Theorems 4.11 and 4.13. As for (PM2), we use *Pl* for the plausibility measure and *Bel* for its dual. By using Corollary 4.1, property (BM2), and De Morgan's laws, we have

$$
\begin{aligned}
Pl(\bigcap_{i=1}^{n} E_i) &= 1 - Bel(\overline{\bigcap_{i=1}^{n} E_i}) \\
&= 1 - Bel(\bigcup_{i=1}^{n} \overline{E}_i) \\
&\le 1 - \sum_{I \subseteq \{1,\cdots,n\}, I \neq \varnothing} (-1)^{|I|-1} Bel(\bigcap_{i \in I} \overline{E}_i) \\
&= \sum_{I \subseteq \{1,\cdots,n\}} (-1)^{|I|-1} - (-1)^{|\varnothing|-1} - \sum_{I \subseteq \{1,\cdots,n\}, I \neq \varnothing} (-1)^{|I|-1} Bel(\bigcap_{i \in I} \overline{E}_i) \\
&= \sum_{I \subseteq \{1,\cdots,n\}, I \neq \varnothing} (-1)^{|I|-1} [1 - Bel(\bigcap_{i \in I} \overline{E}_i)] \\
&= \sum_{I \subseteq \{1,\cdots,n\}, I \neq \varnothing} (-1)^{|I|-1} [1 - Bel(\overline{\bigcup_{i \in I} E_i})] \\
&= \sum_{I \subseteq \{1,\cdots,n\}, I \neq \varnothing} (-1)^{|I|-1} Pl(\bigcup_{i \in I} E_i) \quad .
\end{aligned}
$$

The proof is now complete. □

The following three theorems establish the relation among belief measures, plausibility measures, and discrete probability measures.

Theorem 4.15 Let p: $\mathscr{P}(X) \to [0, 1]$ be a discrete probability measure. Then p is both a belief measure and a plausibility measure. The corresponding basic probability assignment m focuses on the singletons in $\mathscr{P}(X)$.

Proof. Since p is a discrete probability measure, there exists countable set $\{x_1, x_2, \cdots\} \subseteq X$ such that $\sum_{i=1}^{\infty} \mu(\{x_i\}) = 1$. Let

$$
m(E) = \begin{cases} p(E) & \text{if } E = \{x_i\} \text{ for some } x_i \\ 0 & \text{otherwise} \end{cases}
$$

for every $E \in \mathscr{P}(X)$. Then, m is a basic probability assignment focusing on countably many singletons, and

$$p(E) = \sum_{x_i \in E} p(x_i) = \sum_{F \subseteq E} m(F) = \sum_{F \cap E \neq \varnothing} m(F),$$

that is, p is both a belief measure and a plausibility measure. □

Theorem 4.16 If m is a basic probability assignment focusing only on some singletons in $\mathscr{P}(X)$, then the induced belief measure and plausibility measure coincide, resulting in a discrete probability.

Proof. Let m be a basic probability assignment focusing on singletons $\{x_1\}, \{x_2\}, \cdots$, and Bel and Pl be the induced belief measure and plausibility measure respectively. Then, for any $E \in \mathscr{P}(X)$,

$$Bel(E) = \sum_{F \subseteq E} m(F) = \sum_{x_i \in E} p(x_i) = \sum_{F \cap E \neq \varnothing} m(F) = Pl(E).$$

Furthermore, considering any disjoint set sequence $\{E_j\}$ with $\bigcup_{j=1}^{\infty} E_j = E$, we have

$$Bel(E) = \sum_{x_i \in E} p(x_i) = \sum_{x_i \in \bigcup_{j=1}^{\infty} E_j} p(x_i) = \sum_{j=1}^{\infty} \sum_{x_i \in E_j} p(x_i) = \sum_{j=1}^{\infty} Bel(E_j).$$

This means that the induced belief measure (plausibility measure) is σ-additive and, therefore, is a discrete probability measure. □

The above two theorems tell us that any discrete probability is a special case of both belief measures and plausibility measures. The set function shown in Example 4.5 is both a belief measure and a plausibility measure.

Example 4.30 Continue from Example 4.5 where $X = \{x_1, x_2, \cdots\}$ and $\mu(E) = \sum_{x_i \in E} 2^{-i}$ for $E \in \mathscr{P}(X)$. Set function μ is a discrete probability measure on $\mathscr{P}(X)$. The corresponding basic probability assignment m focuses on only singletons $\{x_i\}$, $i = 1, 2, \cdots$, and is expressed as $m(\{x_i\}) = 2^{-i}$ for every $i = 1, 2, \cdots$.

Theorem 4.17 Let m be a basic probability assignment on $(X, \mathscr{P}(X))$. If the induced belief measure *Bel* and plausibility measure *Pl* coincide, then m focuses only on singletons.

Proof. A proof by contradiction is used. Assume that there exists a nonempty set $E \in \mathscr{P}(X)$, which is not a singleton, such that $m(E) > 0$. Then, for any $x \in E$,

$$Bel(\{x\}) = m(\{x\}) < m(\{x\}) + m(E) \leq \sum_{F \cap \{x\} \neq \varnothing} m(F) = Pl(\{x\}).$$

This contradicts the fact that $Bel = Pl$. □

Finally, we show that, when X is countable (including finite), any Sugeno measure is a special case of either belief measures or plausibility measures according to the sign of parameter λ.

Theorem 4.18 Let $X = \{x_1, x_2, \cdots\}$ be a countable universal set and g_λ with $\lambda \neq 0$ be a Sugono measure on $(X, \mathscr{P}(X))$. Then g_λ is a belief measure when $\lambda > 0$ and is a plausibility measure when $\lambda < 0$.

Proof. Since the dual of a plausibility measure is a belief measure and the dual of a Sugeno measure with parameter $\lambda < 0$ is a Sugeno measure with parameter $\lambda' = -\lambda/(\lambda + 1) > 0$, we just need to show the conclusion of the theorem in the case of $\lambda > 0$.

Let g_λ be a Sugeno measure with parameter $\lambda > 0$. Define

$$m(E) = \begin{cases} \lambda^{|E|-1} \prod_{x_i \in E} g_\lambda(\{x_i\}) & \text{if } E \neq \varnothing \\ 0 & \text{otherwise} \end{cases} \tag{4.9}$$

for every $E \in \mathscr{P}(X)$. Set function m is nonnegative. From Definition 4.15, we have

$$
\begin{aligned}
g_\lambda(E) &= \frac{1}{\lambda}[\prod_{x_i \in E}(1+\lambda \cdot g_\lambda(\{x_i\}))-1] \\
&= \frac{1}{\lambda}\sum_{F \subseteq E, F \neq \emptyset}[\lambda^{|F|} \cdot \prod_{x_i \in F} g_\lambda(\{x_i\})] \\
&= \sum_{F \subseteq E, F \neq \emptyset}[\lambda^{|F|-1} \cdot \prod_{x_i \in F} g_\lambda(\{x_i\})] \\
&= \sum_{F \subseteq E, F \neq \emptyset} m(F) \\
&= \sum_{F \subseteq E} m(F)
\end{aligned}
$$

for every $E \in \mathscr{P}(X)$. Moreover, since $g_\lambda(X) = 1$, we have

$$\sum_{F \subseteq X} m(F) = g_\lambda(X) = 1 .$$

Thus, m is a basic probability assignment on $(X, \mathscr{P}(X))$ and, therefore, g_λ is the belief measure induced from m. □

Example 4.31 We use the Sugeno measure discussed in Example 4.20 where $X = \{x_1, x_2\}$ and

$$
g_\lambda(E) = \begin{cases} 0 & \text{when } E = \emptyset \\ 0.2 & \text{when } E = \{x_1\} \\ 0.5 & \text{when } E = \{x_2\} \\ 1 & \text{when } E = X \end{cases} ,
$$

for $E \in \mathscr{F} = \mathscr{P}(X)$. The parameter of Sugeno measure g_λ is $\lambda = 3$. From expression (4.9), the corresponding basic probability assignment $m: \mathscr{P}(X) \to [0, 1]$ can be obtained as follows.

$$m(\{x_1\}) = 3^{1-1} g_\lambda(\{x_1\}) = 1 \times 0.2 = 0.2\,,$$

$$m(\{x_2\}) = 3^{1-1} g_\lambda(\{x_2\}) = 1 \times 0.5 = 0.5\,,$$

$$m(X) = 3^{2-1} g_\lambda(\{x_1\}) \cdot g_\lambda(\{x_2\}) = 3 \times 0.2 \times 0.5 = 0.3\,.$$

Thus,

$$m(E) = \begin{cases} 0.2 & \text{if } E = \{x_1\} \\ 0.5 & \text{if } E = \{x_2\} \\ 0.3 & \text{if } E = X \\ 0 & \text{otherwise} \end{cases}$$

and the induced belief measure *bel* is just the above Sugeno measure g_λ. From this basic probability assignment, the induced plausibility measure is

$$Pl(E) = \begin{cases} 0 & \text{when } E = \varnothing \\ 0.5 & \text{when } E = \{x_1\} \\ 0.8 & \text{when } E = \{x_2\} \\ 1 & \text{when } E = X \end{cases}\,.$$

It is a Sugeno measure with parameter $\lambda' = -\lambda / \lambda + 1 = -0.75$.

4.8 Necessity Measures and Possibility Measures

In this section, we discuss a special type of basic probability assignments and the induced belief measures and plausibility measures.

Let X be a finite universal set and m be a basic probability assignment on $\mathscr{P}(X)$.

Definition 4.24 Basic probability assignment m is *consonant* iff it focuses on a class of nonempty sets that are well ordered by set inclusion, that is, there exists a strictly increasing finite set sequence $\mathscr{C} = \{A_1, A_2, \cdots, A_n\}$ such that $\sum_{i=1}^{n} m(A_i) = 1$ and $m(A) = 0$ for every $A \notin \mathscr{C}$.

The above-mentioned strictly increasing set sequence is called a *nest*.

Example 4.32 Let $X = \{x_1, x_2, x_3, x_4\}$ and basic probability assignment m be given as

$$m(A) = \begin{cases} 0.6 & \text{if } A = \{x_3\} \\ 0.1 & \text{if } A = \{x_1, x_3\} \\ 0.3 & \text{if } A = X \\ 0 & \text{otherwise} . \end{cases}$$

Then m is consonant since $\{x_3\} \subset \{x_1, x_3\} \subset X$.

Definition 4.25 The belief measure induced from a consonant basic probability assignment is called a *necessity measure*; the plausibility measure induced from a consonant basic probability assignment is called a *possibility measure*.

A necessity measure is usually denoted by ν, while a possibility measure is denoted by π.

Example 4.33 The necessity measure $\nu : \mathscr{P}(X) \to [0,1]$ and the possibility measure $\pi : \mathscr{P}(X) \to [0,1]$ induced from basic probability assignment m presented in Example 4.32 are

$$\nu(E) = \begin{cases} 0.6 & \text{if } E = \{x_3\}, \{x_2, x_3\}, \{x_3, x_4\}, \text{or } \{x_2, x_3, x_4\} \\ 0.7 & \text{if } E = \{x_1, x_3\}, \{x_1, x_2, x_3\}, \text{or } \{x_1, x_3, x_4\} \\ 1 & \text{if } E = X \\ 0 & \text{otherwise} \end{cases}$$

and

$$\pi(E)=\begin{cases}0.3 & \text{if } E=\{x_2\},\{x_4\},\text{ or }\{x_2,x_4\}\\ 0.4 & \text{if } E=\{x_1\},\{x_1,x_2\},\{x_1,x_4\},\text{ or }\{x_1,x_2,x_4\}\\ 1 & \text{if } x_3\in E\\ 0 & E=\varnothing\end{cases}$$

respectively.

The following two theorems show some interesting properties of necessity measures and possibility measures. The second theorem can be proved through a similar way as the first one, or by using the duality based on the conclusion of the first one.

Theorem 4.19 Let ν be a necessity measure. For any class of set $\{E_1, E_2, \cdots, E_l\}$,

$$\nu(\bigcap_{j=1}^{l} E_j) = \min[\nu(E_1), \nu(E_2), \cdots, \nu(E_l)].$$

Proof. Let m be the corresponding basic probability assignment focusing on $\{A_1, A_2, \cdots, A_n\}$ that satisfies $A_1 \subset A_2 \subset \cdots \subset A_n$. Then,

$$\begin{aligned}\nu(\bigcap_{j=1}^{l} E_j) &= \sum_{F\subseteq \bigcap_{j=1}^{l} E_j} m(F)\\ &= \sum_{i|A_i\subseteq \bigcap_{j=1}^{l} E_j} m(A_i)\quad .\end{aligned}$$

For each $i = 1, 2, \cdots, n$ and $j = 1, 2, \cdots, l$, using i_j to denote the largest i such that $A_i \subseteq E_j$, we know that $A_i \subseteq \bigcap_{j=1}^{l} E_j$ means $i \le \min_j i_j$. Due to the strict increasingness of $\{A_1, A_2, \cdots, A_n\}$,

$$\sum_{i|A_i \subseteq \bigcap_{j=1}^{l} E_j} m(A_i) = \min_j [\sum_{i|A_i \subseteq E_j} m(A_i)]$$

$$= \min[\nu(E_1), \nu(E_2), \cdots, \nu(E_l)] \quad .$$

So,

$$\nu(\bigcap_{j=1}^{l} E_j) = \min[\nu(E_1), \nu(E_2), \cdots, \nu(E_l)] . \qquad \square$$

Theorem 4.20 Let π be a possibility measure. For any class of set $\{E_1, E_2, \cdots, E_l\}$,

$$\pi(\bigcup_{j=1}^{l} E_j) = \max[\pi(E_1), \pi(E_2), \cdots, \pi(E_l)] .$$

Proof. Since the belief measure and the plausibility measure induced by a basic probability assignment are dual to each other, as a special case, so are the necessity measure and the possibility measure. Thus, by the result obtained in Theorem 4.19,

$$\begin{aligned}
\pi(\bigcup_{j=1}^{l} E_j) &= 1 - \nu(\overline{\bigcup_{j=1}^{l} E_j}) \\
&= 1 - \nu(\bigcap_{j=1}^{l} \overline{E_j}) \\
&= 1 - \min[\nu(\overline{E_1}), \nu(\overline{E_2}), \cdots, \nu(\overline{E_l})] \\
&= \max[1 - \nu(\overline{E_1}), 1 - \nu(\overline{E_2}), \cdots, 1 - \nu(\overline{E_l})] \\
&= \max[\pi(E_1), \pi(E_2), \cdots, \pi(E_l)] ,
\end{aligned}$$

where De Morgan's rule is used. $\square$

To generalize necessity measures and possibility measures to a universal set that is not finite, we need the following concepts of minitivity and maxitivity. Now let X be the universal set that may not be finite.

Definition 4.26 A monotone measure μ on $(X, \mathscr{P}(X))$ is *minitive* (or *maxitive*) iff

$$\mu(\bigcap_{t\in T} E_t) = \inf_{t\in T} \mu(E_t) \quad \text{(or } \mu(\bigcup_{t\in T} E_t) = \sup_{t\in T} \mu(E_t)\text{, respectively)}$$

for any class $\{E_t \mid t \in T\}$, where T is an arbitrary index set.

From definition 4.26, we may say that any necessity measure is minitive, while any possibility measure is maxitive.

Definition 4.27 Let μ be a monotone measure on $(X, \mathscr{P}(X))$. μ is called a *generalized necessity measure* iff it is minitive; μ is called a *generalized possibility measure* iff it is maxitive.

Example 4.34 Let $f : X \to [0, a]$ be a nonnegative real valued function, where a is a nonnegative real number. Define set function μ on $\mathscr{P}(X)$ by

$$\mu(E) = \sup_{x\in E} f(x)$$

for every $E \in \mathscr{P}(X)$. Then μ is a monotone measure on $(X, \mathscr{P}(X))$. Furthermore, μ satisfies the maxitivity and, therefore, it is a generalized possibility measure.

It should be note that a normalized generalized possibility measure may not be a plausibility measure. We can see it from the next example.

Example 4.35 Let the universal set $X = \{r \mid r \text{ is rational}\} \cap [0, 1]$ and $f(x) = x$ for $x \in X$. X is not finite, but countable. Define

$$\pi(E) = \sup_{x \in E} f(x)$$

for every $E \in \mathscr{P}(X)$. Then π is a generalized possibility measure and is normalized. However, it is not a plausibility measure. In fact, since π has infinitely many different values, it is impossible to find a class consists of only finitely many subsets of X, on which a basic probability assignment focuses, such that π is induced from this basic probability assignment.

4.9 *k*-Interactive Measures

Let $X = \{x_1, x_2, \cdots, x_n\}$. As we have seen from Section 1 of this chapter, for identifying an additive measure on $\mathscr{P}(X)$, we need to determine the value of the measure at each singleton. So, there are n unknown parameters. As one of the extreme cases, to determine such an additive measure from data, though the complexity is low, but it cannot capture the interactions among the contribution rates from $x_1, x_2, \cdots, x_n$ towards the given target. In another extreme case, we use a monotone measure to describe all possible interaction among the contribution rates from $x_1, x_2, \cdots, x_n$ towards the given target. It is powerful. However, for identifying a monotone measure, there are $2^n - 1$ unknown parameters. In data mining, when the number of attributes, n, increases, the number of unknown parameters increases exponentially. The complexity of computation is too high and, therefore, is not acceptable. Thus, we face a contradiction of powerfulness and the complexity. A compromised way with this contradiction is to consider only a relatively small number of most common and interesting lower-order interactions but omit the higher-order interactions to reduce the complexity. The following concept of k-interactive measure is one of the proper compromised ways.

Definition 4.28 Let μ be a monotone measure on $(X, \mathscr{P}(X))$ and ν be its Möbius representation. μ is called a *k-interactive measure*, where k is an integer satisfying $2 \le k \le n$, iff $\nu(E) = 0$ for all sets E with $|E| > k$.

From Definition 4.28, we may see that any k-interactive measure is a special case of k'-interactive measures, where k' is an integer lager than k.

As mentioned above, if $|X|=n$, an unknown monotone measure may have up to 2^n-1 unknown parameters, while a k-interactive measure have at most $\sum_{i=1}^{k} C(n,k)$ unknown parameters. When n is large, the computational complexity can be significantly reduced provided the monotone measure is restricted to be a k-interactive measure with a small integer k.

Example 4.36 When $|X|=n=3$ and $k=2$, the difference of the numbers of parameters in above-mentioned two set function models is $(2^n-1)-\sum_{i=1}^{k} C(n,k)=7-6=1$. However, when $n=10$ and $k=2$, a monotone measure may have up to $2^{10}-1=1023$ unknown parameters, but a 2-interactive measure only have at most $C(10,1)+C(10,2)$ $=10+45=55$ unknown parameters.

4.10 Efficiency Measures and Signed Efficiency Measures

From Example 4.12, we have seen that monotone measure μ is used to represent the individual and joint efficiencies of workers. Set function μ satisfies requirement $\mu(\emptyset)$ = 0, which means that there is nothing produced if no worker. The nonadditivity of μ means that there are some interactions among the contribution rates from these workers towards the total number of products. For example, $\mu(\{x_1, x_2\}) > \mu(\{x_1\}) + \mu(\{x_2\})$ means that workers x_1 and x_2 cooperate well such that their joint efficiency is greater than the sum of their individual efficiency; while workers x_2 and x_3 cooperate badly such that their joint efficiency is less than the sum of their individual efficiency. However, set function μ still holds the monotonicity. In some extreme cases, for example, x_2 and x_3 cooperate very badly and they quarrel all the time such that their joint efficiency is very low, say, 4 toys per hour, even lower than their individual efficiencies 6 toys per hour and 7 toys per hour. Thus, the monotonicity of set function μ is violated. Similar situation may occur in many real problems. So, it is necessary to generalize the concept of monotone measure as follows.

Definition 4.29 Let $(X, \mathscr{F})$ be a measurable space. Set function $\mu : \mathscr{F} \to [0, \infty]$ is called an *efficiency measure* iff $\mu(\varnothing) = 0$.

Example 4.37 Similar to Example 4.12, let $X = \{x_1, x_2, x_3\}$, where x_1, x_2, and x_3 are three workers hired for manufacturing a certain kind of wooden toys. Their individual and joint efficiencies can be represented by an efficiency measure μ. For example,

$$\mu(E) = \begin{cases} 0 & \text{if } E = \varnothing \\ 5 & \text{if } E = \{x_1\} \\ 6 & \text{if } E = \{x_2\} \\ 14 & \text{if } E = \{x_1, x_2\} \\ 7 & \text{if } E = \{x_3\} \\ 13 & \text{if } E = \{x_1, x_3\} \\ 4 & \text{if } E = \{x_2, x_3\} \\ 17 & \text{if } E = X \end{cases},$$

where $\mu(\{x_2, x_3\}) < \mu(\{x_2\})$ and $\mu(\{x_2, x_3\}) < \mu(\{x_3\})$ violate the monotonicity.

Now, as an efficiency measure, it is required vanishing at the empty set and to be nonnegative. The first requirement is very natural and is proper in most real problems. However, the second requirement is not satisfied in some real problems. For example, even in the classical linear multiregression, the regression coefficient may be negative. In Chapter 9, the model of multiregression is discussed and a set function serves as the regression coefficients. From there, we can see that even the nonnegativity of the set function should also be dismissed. The classical linear multiregression can be generalized to a nonlinear multiregression based on nonlinear integrals only when a signed set function is adopted. Thus, we further generalize the concept of efficiency measure as follows.

Definition 4.30 Let $(X, \mathscr{F})$ be a measurable space. Set function $\mu : \mathscr{F} \to (-\infty, \infty]$ is called an *signed efficiency measure* iff $\mu(\varnothing) = 0$.

Example 4.38 Let $X=\{x_1, x_2, x_3\}$ and $\mathscr{F} = \mathscr{P}(X)$. Set function $\mu: \mathscr{P}(X) \to (-\infty, \infty]$ is given as

$$\mu(E)=\begin{cases} 0 & \text{if } E=\varnothing \\ 2 & \text{if } E=\{x_1\} \\ -3 & \text{if } E=\{x_2\} \\ -1 & \text{if } E=\{x_1, x_2\} \\ 5 & \text{if } E=\{x_3\} \\ 4 & \text{if } E=\{x_1, x_3\} \\ -2 & \text{if } E=\{x_2, x_3\} \\ 3 & \text{if } E=X \end{cases}.$$

Then μ is a signed efficiency measure on $\mathscr{P}(X)$.

Any signed efficiency measure can be decomposed as the difference of two efficiency measures. In the next chapter, this decomposition is used to define the integral with respect to a signed efficiency measure.

Definition 4.31 Let $(X, \mathscr{F})$ be a measurable space and $\mu: \mathscr{F} \to (-\infty, \infty]$ be a signed efficiency measure. $\mu = \mu^+ - \mu^-$ is called the *reduced decomposition* of μ if both μ^+ and μ^- are efficiency measures on $\mathscr{F}$ and $\mu^+(E)\cdot\mu^-(E)=0$ for every $E \in \mathscr{F}$.

The pair of μ^+ and μ^- is also simply called the reduced decomposition of μ, where μ^+ is called the *positive part* of μ, while μ^- is called the *negative part* of μ. For any given signed efficiency measure, the reduced decomposition is unique. Equality $\mu^+(E)\cdot\mu^-(E)=0$ means that at least one of $\mu^+(E)$ and $\mu^-(E)$ must be zero. In fact,

$$\mu^+(E)=\begin{cases} \mu(E) & \text{if } \mu(E)\ge 0 \\ 0 & \text{otherwise} \end{cases}$$

and

$$\mu^{-}(E)=\begin{cases}-\mu(E) & \text{if } \mu(E)<0\\ 0 & \text{otherwise}\end{cases}.$$

Example 4.39 Consider signed efficiency measure shown in Example 4.38, the reduced decomposition of μ is

$$\mu^{+}(E)=\begin{cases}0 & \text{if } E=\varnothing\\ 2 & \text{if } E=\{x_1\}\\ 0 & \text{if } E=\{x_2\}\\ 0 & \text{if } E=\{x_1,x_2\}\\ 5 & \text{if } E=\{x_3\}\\ 4 & \text{if } E=\{x_1,x_3\}\\ 0 & \text{if } E=\{x_2,x_3\}\\ 3 & \text{if } E=X\end{cases}$$

and

$$\mu^{-}(E)=\begin{cases}0 & \text{if } E=\varnothing\\ 0 & \text{if } E=\{x_1\}\\ 3 & \text{if } E=\{x_2\}\\ 1 & \text{if } E=\{x_1,x_2\}\\ 0 & \text{if } E=\{x_3\}\\ 0 & \text{if } E=\{x_1,x_3\}\\ 2 & \text{if } E=\{x_2,x_3\}\\ 0 & \text{if } E=X\end{cases}.$$

With nonlinear integrals, the signed efficiency measures play a major role in information fusion and data mining. In Chapters 6 and 8-11, we may see the applications of signed efficiency measures.

The following definition gives a concept of boundedness for signed efficiency measures. It is used for discussing the properties of nonlinear integrals in Chapter 5.

Definition 4.32 A signed efficiency measure μ defined on measurable space $(X, \mathscr{F})$ is *bounded* iff there exists a real number M such that $|\mu(A)| \le M$ for every $A \in \mathscr{F}$. In this case, M is called the *bound* of signed efficiency measure μ.

When X is finite, any signed efficiency measure $\mu : \mathscr{F} \to (-\infty, \infty)$ is bounded.

Exercises

Exercise 4.1 Let $X = \{x_1, x_2, x_3\}$. Knowing $\mu(\{x_1\}) = 0.1$ and $\mu(\{x_2\}) = 0.2$, determine probability measure μ on $\mathscr{P}(X)$.

Exercise 4.2 Let $(X, \mathscr{R}_\sigma, \mu)$ be a measure space. Prove Property (M3) given in Theorem 4.2 by using Property (M2).

Exercise 4.3 Show that the Lebesgue measure of any singleton included in the real line is zero. Furthermore, show that the Lebesgue measure of the set consisting of all rational numbers is zero.

Exercise 4.4 Let $X = \{x_1, x_2, x_3\}$ and g_λ be a normalized λ-measure on $\mathscr{P}(X)$. Knowing $g_\lambda(\{x_1\}) = 0.3$, $g_\lambda(\{x_2\}) = 0.4$, and $g_\lambda(\{x_3\}) = 0.5$, determine the value of parameter λ and then the values of g_λ at the other sets in $\mathscr{P}(X)$.

.

Exercise 4.5 Let $X = \{x_1, x_2, x_3\}$. Knowing $g_\lambda(\{x_1\}) = 0.1$, $g_\lambda(\{x_2\}) = 0.2$, and $\lambda = 2$, determine normalized λ-measure g_λ on $\mathscr{P}(X)$.

Exercise 4.6 Let $X = \{x_1, x_2, \cdots, x_n\}$ and $\mu(E) = |E|/(1+|E|)$ for every $E \in \mathscr{P}(X)$. Is μ a quasi-measure on $\mathscr{P}(X)$? If yes, find its standard T-function; if no, simply show your reason.

Exercise 4.7 Let g_λ be a λ-measure with parameter λ on measurable space $(X, \mathscr{F})$ and $c > 0$ be a constant. Is set function $\mu = c \cdot g_\lambda$ also a λ-measure? If yes, what is the relation between its parameter λ' and the original parameter λ? If no, construct a counterexample.

Exercise 4.8 Let $X = \{x_1, x_2, x_3\}$. Set function $\mu : \mathscr{P}(X) \to [0, \infty)$ is given as

$$\mu(E) = \begin{cases} 0 & \text{if } E = \varnothing \\ 2 & \text{if } E = \{x_1\} \\ 3 & \text{if } E = \{x_2\} \\ 1 & \text{if } E = \{x_1, x_2\} \\ 5 & \text{if } E = \{x_3\} \\ 4 & \text{if } E = \{x_1, x_3\} \\ 2 & \text{if } E = \{x_2, x_3\} \\ 3 & \text{if } E = X \end{cases} .$$

Find its Möbius representation ν .

Exercise 4.9 Let $X = \{x_1, x_2, x_3, x_4\}$. Set function $\mu : \mathscr{P}(X) \to [0, \infty)$ is given as

$$\mu(E) = \begin{cases} 0 & \text{if } E = \varnothing \\ 4 & \text{if } E = \{x_1\} \\ 6 & \text{if } E = \{x_2\} \\ 12 & \text{if } E = \{x_1, x_2\} \\ 5 & \text{if } E = \{x_3\} \\ 9 & \text{if } E = \{x_1, x_3\} \\ 11 & \text{if } E = \{x_2, x_3\} \\ 17 & \text{if } E = \{x_1, x_2, x_3\} \\ 3 & \text{if } E = \{x_4\} \\ 7 & \text{if } E = \{x_1, x_4\} \\ 9 & \text{if } E = \{x_2, x_4\} \\ 15 & \text{if } E = \{x_1, x_2, x_4\} \\ 8 & \text{if } E = \{x_3, x_4\} \\ 12 & \text{if } E = \{x_1, x_3, x_4\} \\ 14 & \text{if } E = \{x_2, x_3, x_4\} \\ 20 & \text{if } E = \{x_1, x_2, x_3, x_4\}. \end{cases}$$

Is it a *k*-interactive measure? If yes, show your reason and find the value of *k*. If no, show your reason as well.

Exercise 4.10 Let $X = \{x_1, x_2, x_3\}$ and *m* be a basic probability assignment on $\mathscr{P}(X)$, where

$$m(E)=\begin{cases}0.7 & \text{if } E=\{x_1\}\\ 0.2 & \text{if } E=\{x_1,x_2\}\\ 0.1 & \text{if } E=\{x_1,x_3\}\\ 0 & \text{otherwise}\end{cases}.$$

Find the induced belief measure *Bel* and the induced plausibility measure *Pl* by *m*.

Exercise 4.11 Let $X=\{x_1,x_2,x_3\}$ and monotone measure μ be given as

$$\mu(E)=\begin{cases}0 & \text{if } E=\varnothing\\ 1 & \text{if } E=\{x_1\}\\ 0.3 & \text{if } E=\{x_2\}\\ 1 & \text{if } E=\{x_1,x_2\}\\ 0.5 & \text{if } E=\{x_3\}\\ 1 & \text{if } E=\{x_1,x_3\}\\ 0.5 & \text{if } E=\{x_2,x_3\}\\ 1 & \text{if } E=X\end{cases}.$$

Is it a possibility measure? Why? If yes, find the corresponding basic probability assignment.

Chapter 5

Integrations

In this chapter, the functions and integrals are discussed. A function is a mapping from a measurable space, $(X, \mathscr{F})$, to another measurable space, $(Y, \mathscr{G})$. In most cases, the real line with the Borel field $(R, \mathscr{B})$ is taken as $(Y, \mathscr{G})$. Sometimes, both of them are $(R, \mathscr{B})$ such that the continuity and monotonicity of functions can be considered. Several different types of integrals, including the Riemann integral, the Lebesgue-like integral, the Choquet integral, and the upper and the lower integrals are investigated in this chapter. The first two types of integrals are linear, while the others are generally nonlinear. Any one of them can be chosen as an aggregation tool in information fusion and data mining, which are discussed in Chapters 6 and 8-11.

In this book, only the pair of common addition and common multiplication of real numbers are used as binary operators to define various integrals. Some types of nonlinear integrals involving the other binary operators (the pair of maximum and minimum, or the pan addition and the pan multiplication) of real numbers, such as the Sugeno integral and the pan integrals, are not discussed. The readers interested in those types of nonlinear integrals may refer to [Wang and Klir 1992 or Wang and Klir 2008].

5.1 Measurable Functions

The concept of function has already been accepted and applied in scientific and engineering areas by most readers. In this section, based on the concept of relation discussed in Section 2.5, a general description of functions is given.

Definition 5.1 Let X and Y be two nonempty sets. A relation f from X to Y, denoted as $f: X \to Y$, is called a *function* (or a *mapping*) if each point in X relates to only one point in Y. If $x \in X$ relates to $y \in Y$, we say that y is the *value* of f at x (or the *image* of x under mapping f) and denote it as $y = f(x)$. In this case, x is called the *pre-image* of y. Set X is called the *domain* of function f, and Y is called the *co-domain* of f. Set $\{y \mid y = f(x) \text{ for some } x \in X\}$ is called the *range* of f.

Example 5.1 In a database, there are n attributes, $x_1, x_2, \cdots, x_n$, which form the universal set X, that is, $X = \{x_1, x_2, \cdots, x_n\}$. The data set consists of l real-valued observations (records) to all of these attributes. Denoting the j-th observation of $x_1, x_2, \cdots, x_n$ by x_{j1}, x_{j2}, $\cdots$, x_{jn}, $j = 1, 2, \cdots, l$, respectively, the data set has the following form:

x_1	x_2	$\cdots$	x_n
x_{11}	x_{12}	$\cdots$	x_{1n}
x_{21}	x_{22}	$\cdots$	x_{2n}
$\vdots$	$\vdots$		$\vdots$
x_{l1}	x_{l2}	$\cdots$	x_{ln}

For each $j = 1, 2, \cdots, l$, let $f_j(x_i) = x_{ji}$, $i = 1, 2, \cdots, n$. Then, each row is a function from (X, $\mathscr{P}(X)$) to (R, $\mathscr{B}$), that is, $f_j : X \to (-\infty, \infty)$, $j = 1, 2, \cdots, l$. So, the database can be regarded as a set of l functions on X, where l is called the *size* of the data.

Definition 5.2 Let f be a function from X to Y. For any $A \subseteq X$, set $\{f(x) \mid x \in A\}$ is called the *image* of A under f, denoted by $f(A)$. Conversely, for any $B \subseteq Y$, set $\{x \mid f(x) \in B\}$ is called the *inverse-image* of B, denoted by $f^{-1}(B)$.

In this book, only real-valued (or fuzzy-valued) functions, whose co-domain is a set of real numbers (or fuzzy numbers, respectively), are considered. The characteristic functions discussed in Section 2.1 are of the simplest type of functions, beyond the constant.

Definition 5.3 Any function $f : X \to (-\infty, \infty)$ having a form

$$f = \sum_{i=1}^{m} a_i \chi_{A_i}$$

is called a *simple function*, where m is a positive integer, a_i is a real constant, and $A_i \in \mathscr{F}$ for $i = 1, 2, \cdots, m$.

In the above definition, without any loss of generality, we may assume that sets $A_1, A_2, \cdots, A_m$ are disjoint.

Definition 5.4 Any function $f : X \to (-\infty, \infty)$ having a form

$$f = \sum_{i=1}^{\infty} a_i \chi_{A_i}$$

is called an *elementary function*, where a_i is a real constant, $A_i \in \mathscr{F}$ for $i = 1, 2, \cdots$, and $\{A_i \mid i = 1, 2, \cdots\}$ is a class of disjoint sets.

It is clear that any characteristic function is a simple function and any simple function is an elementary function.

To discuss the real-valued function, more basic knowledge on sets of real numbers is needed. The reader may refer to some textbook on real analysis. One of the important conclusions in real analysis is shown in the following proposition.

Proposition 5.1 Any open subset of the real line $R = (-\infty, \infty)$ can be expressed as a countable union of disjoint open intervals.

From now on, an open subset of the real line is simply called an open set provided there is no confusion.

One of the most common types of real-valued functions defined on the real line, that is, $(X, \mathscr{F}) = (Y, \mathscr{G}) = (R, \mathscr{B})$, is the continuous functions. It plays an important role in calculus.

Definition 5.5 Let $(a, b) = \{x \mid a < x < b \text{ with } -\infty \le a < b \le \infty\}$ be a generalized open interval. Function $f:(a, b) \to (-\infty, \infty)$ is *continuous* on (a, b) iff the inverse-image of any open set is an open set.

The continuity of functions from (a, b) to $(-\infty, \infty)$ described in Definition 5.5 coincides with its common description in calculus. In calculus and real analysis, monotone functions and functions with bounded variation also appear frequently.

Definition 5.6 Function $f:(a, b) \to (-\infty, \infty)$ is *nondecreasing* iff $x_1, x_2 \in (a, b)$ and $x_1 \le x_2$ imply $f(x_1) \le f(x_2)$; f is *nonincreasing* iff $x_1, x_2 \in (a, b)$ and $x_1 \le x_2$ imply $f(x_1) \ge f(x_2)$. Both nondecreasing functions and nonincreasing functions are called *monotone* functions.

Definition 5.7 Function $f:(a, b) \to (-\infty, \infty)$ is *bounded* on (a, b) iff there exists a positive number M such that $|f(x)| \le M$ for every $x \in (a, b)$.

Definition 5.8 Function $f:(a, b) \to (-\infty, \infty)$ is said to have a *bounded variation* iff it can be expressed as the difference of two bounded nondecreasing functions on (a, b).

In Definitions 5.5-5.8, there is no essential difficulty to generalize these concepts on functions by allowing the interval to be closed at finite values (or to be half open half closed at a finite value).

Example 5.2 Function $f(x) = \sin x$ is of bounded variation on $[0, 4\pi]$. In fact, let

$$g(x)=\begin{cases}\sin x & \text{if } x\in[0,\frac{\pi}{2})\\ 1 & \text{if } x\in[\frac{\pi}{2},\frac{3\pi}{2})\\ 2+\sin x & \text{if } x\in[\frac{3\pi}{2},\frac{5\pi}{2})\\ 3 & \text{if } x\in[\frac{5\pi}{2},\frac{7\pi}{2})\\ 4+\sin x & \text{if } x\in[\frac{7\pi}{2},4\pi]\end{cases}$$

and

$$h(x)=\begin{cases}0 & \text{if } x\in[0,\frac{\pi}{2})\\ 1-\sin x & \text{if } x\in[\frac{\pi}{2},\frac{3\pi}{2})\\ 2 & \text{if } x\in[\frac{3\pi}{2},\frac{5\pi}{2})\\ 3-\sin x & \text{if } x\in[\frac{5\pi}{2},\frac{7\pi}{2})\\ 4 & \text{if } x\in[\frac{7\pi}{2},4\pi]\end{cases}.$$

Both h and g are nondecreasing, and they satisfy $f=g-h$.

Definition 5.9 Function $f: X\to(-\infty,\infty)$ is $\mathscr{B}$-$\mathscr{F}$ *measurable* iff $f^{-1}(B)\in\mathscr{F}$ for any $B\in\mathscr{B}$, where "$\mathscr{B}$-$\mathscr{F}$" may be omitted if there is no confusion.

In case $\mathscr{F}=\mathscr{P}(X)$, any function on X is measurable. When X is finite, taking $\mathscr{P}(X)$ (it is finite too) as the σ-algebra $\mathscr{F}$ in the measurable space $(X, \mathscr{F})$ is convenient.

Theorem 5.1 If $f: X\to(-\infty,\infty)$ is a real-valued function, then the following statements are equivalent:

(1) f is measurable;

(2) $\{x \mid f(x) > \alpha\} \in \mathscr{F}$ for any $\alpha \in (-\infty, \infty)$;

(3) $\{x \mid f(x) \le \alpha\} \in \mathscr{F}$ for any $\alpha \in (-\infty, \infty)$;

(4) $\{x \mid f(x) < \alpha\} \in \mathscr{F}$ for any $\alpha \in (-\infty, \infty)$;

(5) $\{x \mid f(x) \ge \alpha\} \in \in \mathscr{F}$ for any $\alpha \in (-\infty, \infty)$.

Proof.

(1)⇒(2): For any $\alpha \in (-\infty, \infty)$, we have $\{x \mid f(x) > \alpha\} = f^{-1}(\alpha, \infty) \in \mathscr{F}$ since interval $(\alpha, \infty) \in \mathscr{B}$.

(2)⇒(3): For any $\alpha \in (-\infty, \infty)$,

$$\{x \mid f(x) \le \alpha\} = \overline{\{x \mid f(x) > \alpha\}} = \overline{f^{-1}((\alpha, \infty))} \in \mathscr{F}.$$

(3)⇒(4): For any $\alpha \in (-\infty, \infty)$,

$$\{x \mid f(x) < \alpha\} = \cup_{i=1}^{\infty} \{x \mid f(x) \le \alpha - 1/i\} \in \mathscr{F}.$$

(4)⇒(5): For any $\alpha \in (-\infty, \infty)$,

$$\{x \mid f(x) \ge \alpha\} = \overline{\{x \mid f(x) < \alpha\}} \in \mathscr{F}.$$

(5)⇒(1): For any left closed right open interval $[a, b)$,

$$f^{-1}([a, b)) = f^{-1}([a, \infty) - [b, \infty)) = f^{-1}([a, \infty)) - f^{-1}([b, \infty)) \in \mathscr{F}. \quad (5.1)$$

Let $\mathscr{A} = \{B \mid f^{-1}(B) \in \mathscr{F}\}$. Given any $E \in \mathscr{A}$, it follows that $\overline{E} \in \mathscr{A}$, since $f^{-1}(\overline{E}) = \overline{f^{-1}(E)} \in \mathscr{F}$, that is, $\mathscr{A}$ is closed under the formation of complements. Similarly, given any sequence $\{E_n\} \in \mathscr{A}$, it follows that $\cup_{i=1}^{\infty} E_n \in \mathscr{A}$ since $f^{-1}(\cup_{i=1}^{\infty} E_n) = \cup_{n=1}^{\infty} f^{-1}(E_n) \in \mathscr{F}$, that is, $\mathscr{A}$ is closed under the formation of countable unions. Hence, $\mathscr{A}$ is a σ-algebra. Denoting the semiring consists of all left closed right open intervals by $\mathscr{S}$, expression (5.1) means that $\mathscr{A} \supseteq \mathscr{S}$. Consequently, according to the definition of $\mathscr{F}(\mathscr{S})$, we have $\mathscr{A} \supseteq \mathscr{F}(\mathscr{S}) = \mathscr{B}$. So, $f^{-1}(B) \in \mathscr{F}$ for any $B \in \mathscr{B}$, that is, f is measurable.

The proof is now complete. □

It is easy to see that any continuous function (even piecewise continuous) or monotone function on an open interval is $\mathscr{B}$-$\mathscr{B}$ measurable and, therefore, any constant, regarded as a function on X, is measurable. The concept of measurable function can also be used for functions defined on a nonempty measurable set.

Theorem 5.2 For any measurable function f on X, there exists a nondecreasing sequence of elementary functions $\{f_n\}$ on X such that $\lim_{n\to\infty} f_n = f$; similarly, there exists a nonincreasing sequence of elementary functions $\{f_n\}$ such that $\lim_{n\to\infty} f_n = f$.

Proof. Only the first conclusion is proved here. The second is similar to the first. Let $f_n(x) = k/n$ when $f(x) \in [k/n, (k+1)/n)$, where $k = \cdots, -2, -1, 0, 1, 2, \cdots$. Then, $\{f_n\}$ is a nondecreasing sequence of elementary functions. Furthermore, $0 \le f - f_n \le 1/n$. Hence, $\lim_{n\to\infty} f_n = f$. □

Restricting a function f to be nonnegative, we may obtain a stronger result as follows.

Theorem 5.3 For any nonnegative measurable function f on X, there exists a nondecreasing sequence of nonnegative simple functions $\{f_n\}$ on X such that $\lim_{n\to\infty} f_n = f$.

Proof. Let

$$f_n(x) = \begin{cases} n & \text{if } f(x) \ge n \\ \dfrac{k-1}{n} & \text{if } f(x) \in [\dfrac{k-1}{n}, \dfrac{k}{n}),\ k = 1, 2, \cdots, n^2 \end{cases}$$

for $n = 1, 2, \cdots$. Then $\{f_n\}$ is a nondecreasing sequence of nonnegative simple functions, and $\lim_{n\to\infty} f_n = f$. □

Theorem 5.4 Let f and g be measurable functions on X and c be a constant.

(1) $c \cdot f$ is measurable;
(2) $f \pm g$ is measurable;
(3) $|f|$ is measurable;
(4) f^2 is measurable;
(5) $f \cdot g$ is measurable;
(6) $1/f$ is measurable if $f(x) \neq 0$ for all $x \in X$;
(7) $f \vee g$ and $f \wedge g$ are measurable.

Proof. Let α be an arbitrarily given constant.

(1) When $c > 0$, we have $\{x \mid (c \cdot f)(x) > \alpha\} = \{x \mid f(x) > \alpha / c\} \in \mathscr{F}$. while $c < 0$, we have $\{x \mid (c \cdot f)(x) > \alpha\} = \{x \mid f(x) < \alpha / c\} \in \mathscr{F}$. As for the case of $c = 0$, $0 \times f = 0$ is a constant function and, therefore, is measurable.

(2) First, we show that $f - g$ is measurable. Inequality $f - g > \alpha$ is equivalent to $f > \alpha + g$. For each $x \in X$, there exists a rational number r_x such that $f(x) > r_x > \alpha + g(x)$. Since there are only countably many rational numbers, we may write them as a sequence $\{r_n\}$. Thus,

$$\{x \mid (f - g)(x) > \alpha\} = \bigcup_{n=1}^{\infty}[\{x \mid f(x) > r_n\} \cap \{x \mid g(x) < r_n - \alpha\}] \in \mathscr{F}.$$

As for $f + g$, regarding -1 as the constant c, the conclusion comes from $f + g = f - (-1 \cdot g)$.

(3) We only need to consider the case of $\alpha > 0$. In this case, $\{x \mid |f(x)| > \alpha\} = \{x \mid f(x) > \alpha\} \cup \{x \mid f(x) < -\alpha\} \in \mathscr{F}$.

(4) Similar to (3), we only need to consider the case of $\alpha > 0$. In this case, $\{x \mid f^2(x) > \alpha\} = \{x \mid f(x) > \sqrt{\alpha}\} \in \mathscr{F}$.

(5) This conclusion can be obtained from

$$f \cdot g = [(f + g)^2 - f^2 - g^2]/2$$

and the above proved conclusions.

(6) $$\{x \mid (1/f)(x) > \alpha\} = [\{x \mid (\alpha f)(x) < 1\} \cap \{x \mid f(x) > 0\}] \cup [\{x \mid (\alpha f)(x) > 1\} \cap \{x \mid f(x) < 0\}] \in \mathscr{F}.$$

(7) $$\{x \mid (f \vee g)(x) > \alpha\} = \{x \mid f(x) \vee g(x) > \alpha\} = \{x \mid f(x) > \alpha\} \cup \{x \mid g(x) > \alpha\} \in \mathscr{F}.$$

$$\{x \mid (f \wedge g)(x) > \alpha\} = \{x \mid f(x) \wedge g(x) > \alpha\} = \{x \mid f(x) > \alpha\} \cap \{x \mid g(x) > \alpha\} \in \mathscr{F}.$$ □

Regarding $f - g$ as a function, the following conclusion is a direct result of Theorems 5.1 and 5.4(2).

Corollary 5.1 Let f and g be measurable functions. Then $\{x \mid f(x) = g(x)\}$, $\{x \mid f(x) > g(x)\}$, and $\{x \mid f(x) \geq g(x)\}$ are measurable sets.

Since the characteristic function of any measurable set is measurable, from Theorem 5.1, we know that all elementary functions (including any simple function) are measurable.

5.2 The Riemann Integral

In this section, we recall the definite integral of function $f : [a, b] \to (-\infty, \infty)$, where $[a, b]$ is a given closed interval, with respect to the Lebesgue measure.

Definition 5.10 A *partition* of $[a, b]$ is a finite sequence $\{t_i \mid i = 0, 1, \cdots, k\}$ satisfying $a = t_0 \leq t_1 \leq \cdots \leq t_k = b$. Number $\max_{1 \leq i \leq k}(t_i - t_{i-1})$ is called the *mesh size* of the partition. A *tagged partition* of $[a, b]$ is a partition $\{t_i \mid i = 0, 1, \cdots, k\}$ with a finite sequence $\{s_i \mid i = 1, \cdots, k\}$ satisfying $s_i \in [t_{i-1}, t_i]$ for $i = 1, \cdots, k$. A refinement of partition $\{t_i \mid i = 0, 1, \cdots, k\}$ is a partition

$$\{t'_j \mid j = 0, 1, \cdots, k'\}$$

such that

$$\{t_i \mid i = 0, 1, \cdots, k\} \subseteq \{t'_j \mid j = 0, 1, \cdots, k'\}.$$

Definition 5.11 Given function f on $[a, b]$ and tagged partition $\{t_i \mid i = 0, 1, \cdots, k\}$ with $\{s_i \mid i = 1, \cdots, k\}$ of $[a, b]$, sum

$$\sum_{i=1}^{k} f(s_i)(t_i - t_{i-1})$$

is called a *Riemann sum* (corresponding to the given tagged partition) of f on $[a, b]$.

Definition 5.12 Let f be a function on $[a, b]$. If there is a real number I_R, for any given $\varepsilon > 0$, there exists $\delta > 0$, such that

$$\left| \sum_{i=1}^{k} f(s_i)(t_i - t_{i-1}) - I_R \right| < \varepsilon$$

whenever the mesh size of the tagged partition $\{t_i \mid i = 0, 1, \cdots, k\}$ with $\{s_i \mid i = 1, \cdots, k\}$ is less than δ, then we say that f is *Riemann integrable* on $[a, b]$ (or say, the Riemann integral of f on $[a, b]$ exists), and I_R is the *Riemann integral* of f on $[a, b]$.

The Riemann integral is also called a *definite integral* in calculus. The definite integral of f on interval $[a, b]$ is denoted as

$$I_R = \int_a^b f(t)\, dt.$$

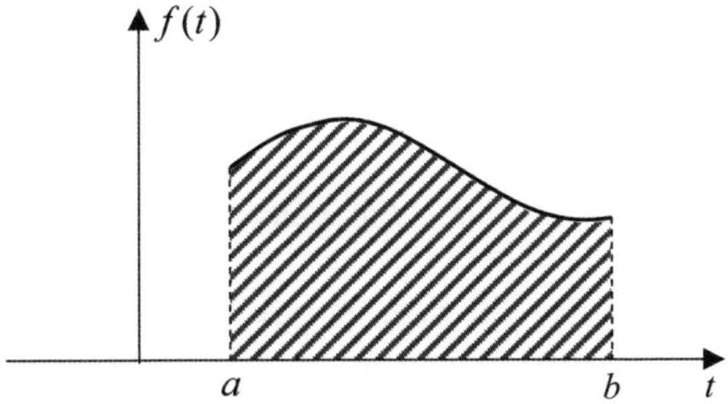

Fig. 5.1 The geometric meaning of a definite integral.

Its geometric meaning, when $f \geq 0$, is the area of the region between the graph of *f* and the *x*-axis from a to b (see Figure 5.1).

Definition 5.13 Let *f* be a function on $[a, b]$ and $P = \{t_i \mid i = 0, 1, \cdots, k\}$ be a partition of $[a, b]$. Denoting

$$M_i = \sup_{t \in [t_{i-1}, t_i]} f(t)$$

and

$$m_i = \inf_{t \in [t_{i-1}, t_i]} f(t),$$

the *upper Darboux sum* of function *f* with partition *P* is

$$\overline{S}(f, P) = \sum_{i=1}^{k} M_i (t_i - t_{i-1}),$$

and the *lower Darboux sum* of function *f* with partition *P* is

$$\underline{S}(f, P) = \sum_{i=1}^{k} m_i (t_i - t_{i-1}).$$

Then the *upper Darporx integral* of *f* on $[a, b]$ is defined as

$$I_{UD} = \inf\{\overline{S}(f, P) \mid P \text{ is a partition of } [a, b]\},$$

and the *lower Darporx integral* of *f* on $[a, b]$ is defined as

$$I_{LD} = \sup\{\underline{S}(f, P) \mid P \text{ is a partition of } [a, b]\}.$$

If $I_{UD} = I_{LD}$, denoted by I_D, then we say that *f* is *Darboux integrable* on $[a, b]$, and I_D is the *Darboux integral* of *f* on $[a, b]$.

The Darboux integral, in fact, is equivalent to the Riemann integral shown in Definitions 5.11 and 5.12, that is, $I_D = I_R$ for very Riemann integrable (or Darboux integrable) function f defined on any given interval $[a, b]$. Thus, from now on, we omit the subscript and simply use *I* to denote the Riemann integral or Darboux integral.

From calculus, we know that any continuous (even piece-wise continuous) function on given interval $[a, b]$ is Riemann integrable. Furthermore, any monotone function and, therefore, any function of bounded variation on $[a, b]$ is Riemann integrable. However, it is easy to cite some examples of measurable functions that are not Riemann integrable defined on a closed interval.

Example 5.3 Consider function $f:[0, 1] \to [0, 1]$ defined by

$$f(x) = \begin{cases} 0 & \text{if } x \in Q_0 \\ 1 & \text{otherwise} \end{cases}$$

for $x \in [0, 1]$, where Q_0 is the set of all rational numbers in $[0, 1]$. Function *f* is discontinuous everywhere in $[0, 1]$, but is measurable on $[0, 1]$. For any partition $P = \{t_i \mid i = 0, 1, \cdots, k\}$ of $[0, 1]$ with $0 = t_0 < t_1 < \cdots < t_k = 1$,

$$M_i = \sup_{t \in [t_{i-1}, t_i]} f(t) = 1$$

and

$$m_i = \inf_{t \in [t_{i-1}, t_i]} f(t) = 0$$

for every $i = 1, \cdots, k$. Hence, the upper Darboux sum of function *f* with partition *P* is

$$\overline{S}(f, P) = \sum_{i=1}^{k} 1 \cdot (t_i - t_{i-1}) = t_k - t_0 = 1,$$

while its lower Darboux sum is

$$\underline{S}(f, P) = \sum_{i=1}^{k} 0 \cdot (t_i - t_{i-1}) = 0.$$

Thus, the upper Darporx integral of f on $[0, 1]$ is 1, but its lower Darporx integral is 0. They are not equal to each other. This shows that the function *f* is not Riemann integrable.

The most important property of the Riemann integral is the *linearity*, that is,

$$\int_a^b [c_1 f(t) + c_2 g(t)]\, dt = c_1 \int_a^b f(t)\, dt + c_2 \int_a^b g(t)\, dt$$

for any real numbers c_1 and c_2 whenever f and g are Riemann integrable on $[a, b]$. Using this linearity, it is not difficult to know that

$$\int_a^c f(t)\, dt = \int_a^b f(t)\, dt + \int_b^c f(t)\, dt$$

if all involved Riemann integrals exist.

Another interesting property of the Riemann integral is $\int_a^b 1\, dt = b - a$ for any interval $[a, b]$. As a special case, $\int_a^a f(t)\, dt = 0$ for any function *f*. Thus, the Riemann integral $\int_a^b f(t)\,dt$ can also be understood as an integral on open interval (a, b) or even on a half open half closed interval.

5.3 The Lebesgue-Like Integral

Let us consider measure space $([0, 1], \mathscr{B}_{[0,1]}, m)$, where $\mathscr{B}_{[0,1]}$ is the class of all Borel sets in $[0,1]$ and *m* is the Lebesgue measure, and check the function shown in Example 5.3. Since there are only countably many rational numbers and the Lebesgue measure of each singleton is zero, by using the countable additivity of the Lebesgue measure *m*, we know that the Lebesgue measure of Q_0, the set consisting of all rational numbers in $[0,1]$, is zero. Therefore, by the additivity of *m*, the Lebesgue measure of $[0,1]-Q_0$, the set consisting of all irrational numbers, is 1. Thus the graph of function *f* given in Example 5.3 almost coincides to the horizontal line with height 1 on $[0,1]$. Intuitively, the area of the region between the graph of function *f* and the *x*-axis should be 1, the same as the constant function 1 has. Unfortunately, Example 5.3 tells us that the above-mentioned area is "unmeasurable", or say, the information carried by such a measurable function is not "aggregatable" by the Riemann integral, though function *f* is measurable. This fact shows that the Riemann integral is not powerful enough as an aggregation tool. Hence, people need to look for another integration approach as a generalization of the Riemann integral such that any measurable function is integrable. The Lebesgue integral is just such an expected tool, which can be established based on Theorem 5.3 step by step as follows.

Definition 5.14 Let $X = (-\infty, \infty)$, E is a Borel set, and g be a nonnegative simple function with expression

$$g(x) = \sum_{j=1}^{n} a_j \chi_{E_j} \ .$$

Then *Lebesgue integral of g* on *E* with respect to the Lebesgue measure *m* is

$$\int_E g\, dm = \sum_{j=1}^{n} a_j \cdot m(E_j \cap E),$$

where $a_j \geq 0$ and $E_j \in \mathscr{B}$ for $j = 1, 2, \cdots, n$. Furthermore, let f be a nonnegative measurable function and $\{g_i\}$ be a nondecreasing sequence of nonnegative simple functions such that $\lim_{i\to\infty} g_i = f$ on *E*. Then the *Lebesgue integral of f* on *E* with respect to the Lebesgue measure *m* is

$$\int_E f\, dm = \lim_{i\to\infty} \int_E g_i dm.$$

The above definition of the Lebesgue integral is unambiguous due to the σ-additivity of *m*. That is, for any two sequences of nondecreasing nonnegative simple functions, $\{g_i\}$ and $\{g_i'\}$, with $\lim_{i\to\infty} g_i = \lim_{i\to\infty} g_i' = f$ on *E*, we have

$$\lim_{i\to\infty} \int_E g_i\, dm = \lim_{i\to\infty} \int_E g_i'\, dm.$$

So, the Lebesgue integral is well defined for any nonnegative measurable function on any given Borel set *E*. When $E = (-\infty, \infty)$, $\int_{(-\infty,\infty)} f\, dm$ is simply written as $\int f\, dm$.

In Definition 5.14, function *f* may be any nonnegative measurable function, including nonnegative piecewise continuous functions, monotone functions, and functions with bounded variation. When a function is Riemann integrable, it is also Lebesgue integrable, and the values of its Lebesgue integral and Riemann integral are the same. Hence, the Lebesgue integral is a generalization of the Riemann integral. The former is more powerful than the latter. This can be seen in the following example.

Example 5.4 We continue the discussion in Example 5.3, where considered function $f:[0,1]\to[0,1]$ is defined as

$$f(x)=\begin{cases}0 & \text{if } x\in Q_0\\ 1 & \text{otherwise}\end{cases}$$

for $x\in[0,1]$, in which $Q_0=\{x\mid x \text{ is rational}\}\cap[0,1]$. Function f is a simple function with $m(Q_0)=0$ and $m([0,1]-Q_0)=1$. Hence,

$$\int_{[0,1]} f\,dm = 0\cdot m(Q_0)+1\cdot m([0,1]-Q_0)=0\times 0+1\times 1=1.$$

That is, f is Lebesgue integrable and the value of the integral coincides with the intuition.

The Lebesgue integral can be immediately generalized to nonnegative measurable functions defined on a general measure space $(X, \mathscr{F}, \mu)$.

Definition 5.15 Given a measure space $(X, \mathscr{F}, \mu)$, let $E\in\mathscr{F}$, f be a nonnegative measurable function on E, and $\{g_i\}$ be a nondecreasing sequence of nonnegative simple functions such that $\lim_{i\to\infty} g_i = f$ on E, where simple function g_i, $i=1,2,\cdots$, has a form

$$g_i(x)=\sum_{j=1}^{n_i} a_{ij}\chi_{E_{ij}},$$

in which $a_{ij}\ge 0$ and $E_{ij}\in\mathscr{F}$ for $j=1,2,\cdots,n_i$. The *Lebesgue-like integral* of f on E with respect to measure μ is

$$\int_E f\,d\mu=\lim_{i\to\infty}\sum_{j=1}^{n_i} a_{ij}\cdot\mu(E_{ij}\cap E). \tag{5.2}$$

In the integral, function f is called the *integrand.*

Similar to the Lebesgue integral, this definition is unambiguous due to the σ-additivity of μ. When $E = X$, we may omit the subscript E from the symbol of the integral as well. Since the Lebesgue integral defined on the Euclidian space is just a special case of the Lebesgue-like integral on general measure space, from now on, we simply call the latter the Lebesgue integral provided there is no confusion. In case we want to emphasize an integral being in Lebesgue's meaning shown in Definition 5.15 to distinct from other types of integrals (they are discussed in Sections 5.5-5.9), a symbol (Leb) $\int f \, d\mu$ is adopted.

As for measurable functions that are not necessarily nonnegative, the following approach can be adopted to define their Lebesgue integral.

Definition 5.16 Given a measurable function f on set $E \in \mathscr{F}$, functions

$$f^+(x) = \begin{cases} f(x) & \text{if } x \in E \text{ and } f(x) \geq 0 \\ 0 & \text{otherwise} \end{cases}$$

and

$$f^-(x) = \begin{cases} -f(x) & \text{if } x \in E \text{ and } f(x) < 0 \\ 0 & \text{otherwise} \end{cases}$$

are called the *positive part* and the *negative part* of f, respectively.

Given measurable function f on measure space $(X, \mathscr{F}, \mu)$, both f^+ and f^- are nonnegative measurable functions on $(X, \mathscr{F}, \mu)$, and $f = f^+ - f^-$ holds. Thus, we may use them to define the Lebesgue integral for any given measurable function f on measurable set E with respect to measure μ as follows.

Definition 5.17 Given a measurable function f on measurable set E, the Lebesgue integral of f on E with respect to measure μ is defined as

$$\int_E f\,d\mu = \int_E f^+\,d\mu - \int_E f^-\,d\mu$$

provided the two terms on the right hand side are not both infinite.

Similar to the Riemann integral, the Lebesgue integral has the following basic properties, where we assume that all involved functions and sets are measurable:

(LIP1) $\int_E f\,d\mu \geq 0$ if $f \geq 0$;

(LIP2) $\int_E 1\,d\mu = \int \chi_E\,d\mu = \mu(E)$;

(LIP3) $\int_E f\,d\mu = \int \chi_E \cdot f\,d\mu$;

(LIP4) $\int_E (c_1 f + c_2 g)\,d\mu = c_1 \int_E f\,d\mu + c_2 \int_E g\,d\mu$.

Property (LIP4) is the linearity. From these basic properties, we may obtain more properties. Some of them are left to the readers as exercises.

Example 5.5 Let $X = \{x_1, x_2, \cdots, x_n\}$ be a set of n attributes (or, information sources) and $w_1, w_2, \cdots, w_n$ are corresponding weights. If $f(x_1), f(x_2), \cdots, f(x_n)$ are an observation (or, received numerical information amounts) of these attributes respectively, then the *weighted sum*

$$\sum_{i=1}^{n} w_i f(x_i) = w_1 f(x_1) + w_2 f(x_2) + \cdots + w_n f(x_n)$$

can be regarded as the Lebesgue integral of function f on X with respect to a certain measure μ on $\mathscr{P}(X)$. In fact, we may define a measure on the semiring, $\mathscr{S}$, that consists of all singletons of X and the empty set by $\mu(\{x_i\}) = w_i$ for $i = 1, 2, \cdots, n$ and $\mu(\emptyset) = 0$. It can be extended to a measure on $\mathscr{P}(X)$, the σ-algebra generated by $\mathscr{S}$, by the additivity uniquely. The observation $f(x_1), f(x_2), \cdots, f(x_n)$ can be regarded as a function $f: X \to (-\infty, \infty)$. Since we now adopt the power set $\mathscr{P}(X)$ as the σ-algebra to form the measure space $(X, \mathscr{P}(X), \mu)$, function f is measurable and, moreover, is a simple function. Thus, the Lebesgue integral of f with respect to measure μ is

$$\int f \, d\mu = \sum_{i=1}^{n} f(x_i) \cdot \mu(\{x_i\}) = \sum_{i=1}^{n} w_i f(x_i).$$

When weights $w_1, w_2, \cdots, w_n$ satisfy the conditions $w_i \geq 0$ for $i = 1, 2, \cdots, n$ and $\sum_{i=1}^{n} w_i = 1$, the weighted sum $\sum_{i=1}^{n} w_i f(x_i)$ is called a *weighted average* of $f(x_1), f(x_2), \cdots, f(x_n)$.

Explaining as well as expressing the weighted sum as a Lebesgue integral is used for the linear multiregression reviewed in Section 9.1. By such point of view, we introduce the nonlinear multiregression in Chapter 9 based on nonlinear integrals, which are discussed in the following several sections.

5.4 The Choquet Integral

Based on the discussion on linear integrals with respect to additive measures in Sections 5.2 and 5.3, beginning from this section, we consider some types of nonlinear integrals with respect to monotone measures.

Let $(X, \mathscr{F}, \mu)$ be a monotone measure space and f be a measurable function on $(X, \mathscr{F})$. Generally, the universal set X is not necessarily finite and σ-algebra $\mathscr{F}$ may not be the power set of X.

To define an integral of f with respect to a monotone measure μ, if the approach shown in Definition 5.15 is still used, we will face a difficulty that the value of the limit in expression (5.2) depends on the choice of sequence $\{g_i\}$ as well as on the expression of each g_i due to the nonadditivity of μ, that is, the unambiguousness of the definition will not be guaranteed. So, the definition of Lebesgue integral with respect to a nonadditive monotone measure fails. Thus, we have to look for another approach to define an integral for measurable function *f* with respect to monotone measures. One of the successful ways is the Choquet integral discussed in this section.

Definition 5.18 Let *f* be a nonnegative measurable function on $(X, \mathscr{F})$ and $E \in \mathscr{F}$. The *Choquet integral* of *f* on *E* with respect to a monotone measure μ, denoted by $(\mathrm{C})\int_E f\, d\mu$, is defined as

$$(\mathrm{C})\int_E f\, d\mu = \int_0^\infty \mu(F_\alpha \cap E)\, d\alpha\,, \tag{5.3}$$

where $F_\alpha = \{x \mid f(x) \geq \alpha\}$, called the α-level set of *f*, for $\alpha \in [0, \infty)$. When $E = X$, $(\mathrm{C})\int_X f\, d\mu$ is simply written as $(\mathrm{C})\int f\, d\mu$.

Since function f in Definition 5.18 is measurable, we know that $F_\alpha = \{x \mid f(x) \geq \alpha\} \in \mathscr{F}$ for every $\alpha \in [0, \infty)$ and, therefore, $F_\alpha \cap E \in \mathscr{F}$. So, $\mu(F_\alpha \cap E)$ is well defined for every $\alpha \in [0, \infty)$. Furthermore, $\{F_\alpha \mid \alpha \in [0, \infty)\}$ is a class of sets that are nonincreasing with respect to α and so are sets in $\{F_\alpha \cap E \mid \alpha \in [0, \infty)\}$. Noting that monotone measure μ is nondecreasing, we know that $\mu(F_\alpha \cap E)$ is a nonincreasing function of α and, therefore, is Riemann integrable. Thus, the Choquet integral of a nonnegative measurable function with respect to a monotone measure on a measurable set is then well defined.

When set function μ is σ-additive, expression (5.3) in Definition 5.18 is just an equivalent definition of the Lebesgue integral of *f* with respect to μ. In literature, this equivalence is called the transformation theorem for the Lebesgue integral. That is to say, when monotone measure μ, as a special case, is a classical measure, the Choquet integral of any given measurable function *f* with respect to μ coincides with the corresponding Lebesgue integral. So, the Choquet integral is a real generalization of the Lebesgue integral. Just by this reason, sometimes people omit "(C)" from the symbol of the Choquet integral and then use the same symbol, $\int f\, d\mu$, as the Lebesgue integral uses if there is no confusion.

Example 5.6 Let $X = [0, 1]$, $f(x) = 2x$ for $x \in [0, 1]$, $\mathscr{F} = \mathscr{B}_{[0,1]}$, the class of all Borel sets in $[0, 1]$, and $\mu(B) = [m(B)]^2$ for $B \in \mathscr{B}_{[0,1]}$, where m is the Lebesgue measure on the real line. Thus, f is a nonnegative measurable function on monotone measure space $(X, \mathscr{B}_{[0,1]}, \mu)$. According to Definition 5.18, the Choquet integral of f with respect to μ is

$$\begin{aligned}
(\mathrm{C})\int f\, d\mu &= \int_0^\infty \mu(\{x \mid f(x) \geq \alpha\} \cap [0, 1])\, d\alpha \\
&= \int_0^\infty \mu(\{x \mid 2x \geq \alpha\} \cap [0, 1])\, d\alpha \\
&= \int_0^2 \mu([\frac{\alpha}{2}, 1])\, d\alpha + \int_2^\infty \mu(\varnothing)\, d\alpha \\
&= \int_0^2 [m([\frac{\alpha}{2}, 1])]^2\, d\alpha + 0 \\
&= \int_0^2 (1 - \frac{\alpha}{2})^2\, d\alpha \\
&= \int_0^2 (1 - \alpha + \frac{\alpha^2}{4})\, d\alpha \\
&= \alpha \,\Big|_0^2 - \frac{1}{2}\alpha^2 \,\Big|_0^2 + \frac{1}{12}\alpha^3 \,\Big|_0^2 \\
&= 2 - 2 + \frac{8}{12} \\
&= \frac{2}{3} \quad .
\end{aligned}$$

When the integrand of the above Riemann integral, $\mu(F_\alpha)$, cannot be expressed as an explicit elementary expression of α, or the expression is too complex, the value of the Choquet integral has to be approximately calculated by using some numerical method (e.g., the Simpson method). However, if the universal set X is finite, such as the set of attributes in a database, we have a simple calculation formula as follows.

Let $X = \{x_1, x_2, \cdots, x_n\}$. In this case, usually, we take the power set of X as the σ-algebra. Thus, $(X, \mathscr{P}(X))$ is a measurable space. Given a monotone measure μ and a nonnegative function f on $(X, \mathscr{P}(X))$, $\mu(F_\alpha) = \mu(\{x \mid f(x) \geq \alpha\})$ is a simple function of α, that is,

$$\mu(F_\alpha) = \sum_{i=1}^{n} \mu(\{x_i^*, x_{i+1}^*, \cdots, x_n^*\}) \cdot \chi_{(f(x_{i-1}^*),\, f(x_i^*)]}(\alpha),$$

for $\alpha \in [0, \infty)$ and, therefore, the Choquet integral of f with respect to μ can be calculated by

$$(\mathrm{C})\int f\, d\mu = \sum_{i=1}^{n} [f(x_i^*) - f(x_{i-1}^*)] \cdot \mu(\{x_i^*, x_{i+1}^*, \cdots, x_n^*\}) \tag{5.4}$$

or, equivalently,

$$(\mathrm{C})\int f\, d\mu = \sum_{i=1}^{n} [\mu(\{x_i^*, x_{i+1}^*, \cdots, x_n^*\}) - \mu(\{x_{i+1}^*, \cdots, x_n^*\})] \cdot f(x_i^*), \tag{5.4*}$$

where $f(x_0^*) = 0$, $\{x_{n+1}^*, \cdots, x_n^*\} = \varnothing$, and $(x_1^*, x_2^*, \cdots, x_n^*)$ is a permutation of $\{x_1, x_2, \cdots, x_n\}$ such that $f(x_1^*) \leq f(x_2^*) \leq \cdots \leq f(x_n^*)$.

Example 5.7 Let $X = \{x_1, x_2, x_3\}$, $\mathscr{F} = \mathscr{P}(X)$, and monotone measure μ on $\mathscr{P}(X)$ be given as

$$\mu(E)=\begin{cases}0 & \text{if } E=\emptyset \\ 0.5 & \text{if } E=\{x_1\} \\ 0.2 & \text{if } E=\{x_2\} \\ 0.6 & \text{if } E=\{x_1,x_2\} \\ 0.4 & \text{if } E=\{x_3\} \\ 0.7 & \text{if } E=\{x_1,x_3\} \\ 0.9 & \text{if } E=\{x_2,x_3\} \\ 1 & \text{if } E=X \quad . \end{cases}$$

Then (X, $\mathscr{P}(X)$, μ) is a monotone measure space. Let function $f: X \to [0, \infty)$ be given as

$$f(x)=\begin{cases}8 & \text{if } x=x_1 \\ 10 & \text{if } x=x_2 \\ 5 & \text{if } x=x_3 \ . \end{cases}$$

Thus, $x_1^*=x_3$, $x_2^*=x_1$, and $x_3^*=x_2$. By using formula (5.4), the Choquet integral of f with respect to μ (on X) can be calculated (see Figure 5.2, the area of the shaded region in the right part is the value of the Choquet integral) as

$$\begin{aligned}(C)\int f\,d\mu &= (f(x_1^*)-0)\cdot\mu(\{x_1^*,x_2^*,x_3^*\})+(f(x_2^*)-f(x_1^*))\cdot\mu(\{x_2^*,x_3^*\}) \\ &\quad +(f(x_3^*)-f(x_2^*))\cdot\mu(\{x_3^*\}) \\ &= f(x_3)\cdot\mu(X)+(f(x_1)-f(x_3))\cdot\mu(\{x_1,x_2\})+(f(x_2)-f(x_1))\cdot\mu(\{x_2\}) \\ &= 5\times 1+(8-5)\times 0.6+(10-8)\times 0.2 \\ &= 5\times 1+3\times 0.6+2\times 0.2 \\ &= 7.2 \quad . \end{aligned}$$

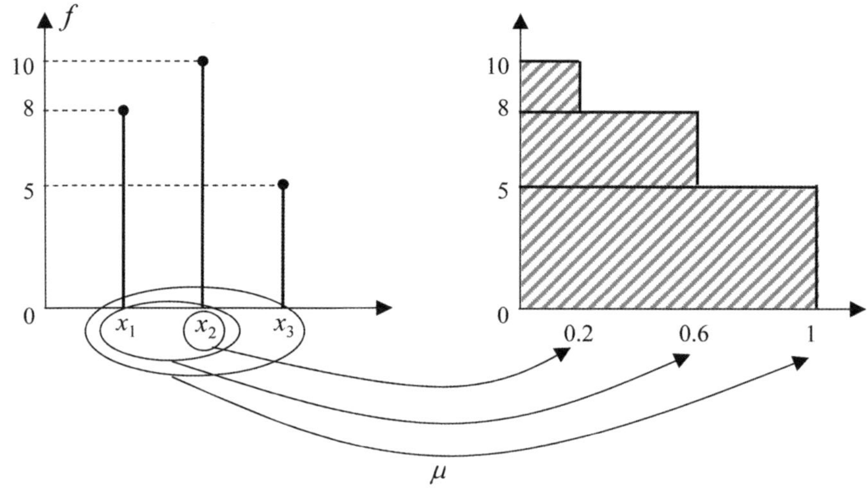

Fig. 5.2 The calculation of the Choquet integral defined on a finite set $\{x_1, x_2, x_3\}$.

Alternatively, we may use (5.4*) to calculate the same result as follows.

$$
\begin{aligned}
(C)\int f\, d\mu &= [\mu(\{x_1^*, x_2^*, x_3^*\}) - \mu(\{x_2^*, x_3^*\})]\cdot f(x_1^*) \\
&\quad + [\mu(\{x_2^*, x_3^*\}) - \mu(\{x_3^*\})]\cdot f(x_2^*) + \mu(\{x_3^*\})\cdot f(x_3^*) \\
&= [\mu(X) - \mu(\{x_1, x_2\})]\cdot f(x_3) + [\mu(\{x_1, x_2\}) - \mu(\{x_2\})]\cdot f(x_1) \\
&\quad + \mu(\{x_2\})\cdot f(x_2) \\
&= (1-0.6)\times 5 + (0.6-0.2)\times 8 + 0.2\times 10 \\
&= 0.4\times 5 + 0.4\times 8 + 0.2\times 10 \\
&= 7.2\ .
\end{aligned}
$$

We should know that, once the integrand f is given, the calculation of its Choquet integral only involves the value of μ at the sets in a chain from the universal set to the empty set, but not all sets in the power set. In Example 5.7, the chain in lattice $(\mathscr{P}(X), \subseteq)$ is

$$(\{x_1, x_2, x_3\},\ \{x_1, x_2\},\ \{x_2\},\ \varnothing) \text{ (see Figure 5.3).}$$

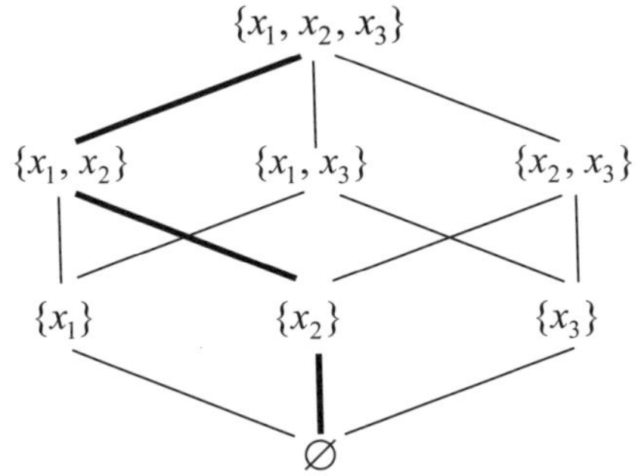

Fig. 5.3 The chain used in the calculation of the Choquet integral in Example 5.7.

Formula (5.4) is effective when the number of attributes is not large and the Choquet integral is calculated by hand. However, when an inverse problem of information fusion is considered, it is not convenient since the expression of the Choquet integral is not in an explicit linear form of unknown parameters that are the values of μ. In fact, rearranging the order of attributes is not a linear operation. Thus, the linear algebraic method cannot be used to estimate the values of μ based on the observed data set (see Chapters 9-11). Hence, it is necessary to introduce an alternate calculation formula for the Choquet integral as follows. For given nonnegative function f on a monotone measure space $(X, \mathscr{P}(X), \mu)$, where X is a finite universal set, the Choquet integral of f with respect to μ can be calculated by

$$(\mathrm{C})\int f d\mu = \sum_{j=1}^{2^n-1} z_j \mu_j \,, \tag{5.5}$$

where $\mu_j = \mu(\bigcup_{j_i=1}\{x_i\})$ if j is expressed in terms of binary digits $j_n j_{n-1} \cdots j_1$ for every $j = 1, 2, \cdots, 2^n - 1$ and

$$z_j = \begin{cases} \min\limits_{i|\mathrm{frc}(j/2^i)\in[1/2,1)} f(x_i) - \max\limits_{i:\mathrm{frc}(j/2^i)\in[0,1/2)} f(x_i) & \text{if it is} > 0 \text{ or } j = 2^n - 1 \\ 0, & \text{otherwise} \end{cases}$$

$$\text{for } j = 1, 2, \cdots, 2^n - 1. \tag{5.6}$$

In expression (5.6), $\text{frc}(j/2^i)$ denotes the fractional part of $j/2^i$, and we need the convention that the maximum taken on the empty set is zero. The expression can also be written in a simpler form via the replacement

$$\{i \mid \text{frc}(j/2^i) \in [1/2, 1)\} = \{i \mid j_i = 1\}$$

and

$$\{i \mid \text{frc}(j/2^i) \in [0, 1/2)\} = \{i \mid j_i = 0\}.$$

The significance of this alternate formula is that the value of the Choquet integral is now expressed as a linear function of the values of μ. Hence, when the data set of the values of the integrand f and the corresponding integration value are available, an algebraic method can be used to estimate the optimal values of μ. So, in data mining, such as in nonlinear multiregressions, this new calculation formula is more convenient than formula (5.4).

As for the validation of this new formula, rewriting the old formula (5.4) as

$$(\text{C})\int f d\mu = \sum_{i=1}^{n}[f(x_i^*) - f(x_{i-1}^*)] \cdot \mu(\{x_i^*, x_{i+1}^*, \cdots, x_n^*\}) = \sum_{j=1}^{2^n-1} a_{E_j} \cdot \mu(E_j),$$

where

$$E_j = \bigcup_{j_i=1}\{x_i\}$$

and

$$a_{E_j} = \begin{cases} f(x_i^*) - f(x_{i-1}^*) & \text{if } E_j = \{x_i^*, x_{i+1}^*, \cdots, x_n^*\} \text{ for some } i = 1, 2, \cdots, n \\ 0 & \text{otherwise} \end{cases}$$

$$= \begin{cases} \min\limits_{x \in E_j} f(x) - \max\limits_{x \notin E_j} f(x) & \text{if } E_j = \{x_i^*, x_{i+1}^*, \cdots, x_n^*\} \text{ for some } i = 1, 2, \cdots, n \\ 0 & \text{otherwise} \end{cases}$$

$$=\begin{cases}\min_{x\in E_j} f(x)-\max_{x\notin E_j} f(x) & \text{if } E_j=\{x_i^*, x_{i+1}^*, \cdots, x_n^*\} \text{ for some } i=1,2,\cdots,n\\ 0 & \min_{x\in E_j} f(x)-\max_{x\notin E_j} f(x)\le 0\end{cases}$$

for $j=1,2,\cdots,2^n-1$, and noticing that $j_i=1$ if and only if $x_i \in E_j$, we can see that the new formula is equivalent to the old one. In the above expression for a_{Ej}, we also need the convention that

$$\max_{x\notin X} f(x)=\max_{\varnothing} f(x)=0 .$$

In addition, we should note that in the above expression the function is defined in two parts. They overlap when $x_{i-1}^* = x_i^*$ for some $i=1,2,\cdots,n$. Fortunately, they are both zero at the overlapped j and, therefore, these two parts are consistent.

The Choquet integral of a nonnegative measurable function f with respect to monotone measure μ has the following basic properties, where we assume that all involved functions and sets are measurable:

(CIP1) $(C)\int_E f\, d\mu \ge 0$;

(CIP2) $(C)\int_E 1\, d\mu = (C)\int \chi_E\, d\mu = \mu(E)$;

(CIP3) $(C)\int_E f\, d\mu = (C)\int \chi_E \cdot f\, d\mu$;

(CIP4) $(C)\int_E cf\, d\mu = c\cdot(C)\int_E f\, d\mu$ for any nonnegative constant c.

These properties, which are similar to those of the Lebesgue integral, can be obtained from the definition of the Choquet integral directly. However, the Choquet integral is not linear, though it has property (CIP4). In fact, $(C)\int_E (f+g)\, d\mu \ne (C)\int_E f\, d\mu + (C)\int_E g\, d\mu$ generally. This can be verified by the following Example.

Example 5.8 Let $X = \{a, b\}$, $\mathscr{F} = \mathscr{P}(X)$, and

$$\mu(E) = \begin{cases} 0 & \text{if } E = \emptyset \\ 1 & \text{otherwise} \end{cases}.$$

In this case, any function on X is measurable. Considering two functions,

$$f(x) = \begin{cases} 0 & \text{if } x = a \\ 1 & \text{if } x = b \end{cases}$$

and

$$g(x) = \begin{cases} 0 & \text{if } x = b \\ 1 & \text{if } x = a, \end{cases}$$

we have

$$\text{(C)} \int f \, d\mu = \int_0^\infty \mu(\{x \mid f(x) \geq \alpha\}) d\alpha = \int_0^1 \mu(\{b\}) d\alpha = 1 \times 1 = 1$$

and

$$\text{(C)} \int g \, d\mu = \int_0^\infty \mu(\{x \mid g(x) \geq \alpha\}) d\alpha = \int_0^1 \mu(\{a\}) d\alpha = 1 \times 1 = 1 .$$

Since $f + g = 1$, a constant function on X, we obtain

$$\text{(C)} \int (f + g) \, d\mu = \text{(C)} \int 1 \, d\mu = 1 \cdot \mu(X) = 1 .$$

Thus,

$$\text{(C)} \int (f + g) \, d\mu \neq \text{(C)} \int f \, d\mu + \text{(C)} \int g \, d\mu .$$

This shows that the Choquet integral is not linear with respect to its integrand in general.

The nonlinearity of the Choquet integral comes from the nonadditivity of the involved monotone measure. Though the Choquet integral loses the linearity in general, it still has the monotonicity and the translatability, which the Lebesgue integral also holds and are implied by its linearity, shown in the next theorem.

Theorem 5.5 Let f and g be nonnegative measurable functions on $(X, \mathscr{F})$. The Choquet integral with respect to monotone measure μ holds the monotonicity (CIP5) and the translatability (CIP6):

(CIP5) $\quad (\mathrm{C})\int_E f\, d\mu \le (\mathrm{C})\int_E g\, d\mu \quad \text{if } f \le g \quad \text{on } E;$

(CIP6) $\quad (\mathrm{C})\int_E (f + c)\, d\mu = (\mathrm{C})\int_E f\, d\mu + c \cdot \mu(E)$ for any constant c

satisfying $f + c \ge 0$.

Proof. There is no loss of generality in assuming $E = X$. To (CIP5), from $f \le g$, we know that $\{x \mid f(x) \ge \alpha\} \subseteq \{x \mid g(x) \ge \alpha\}$ and, therefore, $\mu(\{x \mid f(x) \ge \alpha\}) \le \mu(\{x \mid g(x) \ge \alpha\})$. Hence,

$$(\mathrm{C})\int f\, d\mu = \int_0^\infty \mu(\{x \mid f(x) \ge \alpha\})\, d\alpha \le \int_0^\infty \mu(\{x \mid g(x) \ge \alpha\})\, d\alpha = (\mathrm{C})\int g\, d\mu .$$

As for (CIP6), noticing that $f(x) + c \ge \alpha$ for every $x \in X$ when α is between 0 and c, we have

$$\begin{aligned}(\mathrm{C})\int (f + c)\, d\mu &= \int_0^\infty \mu(\{x \mid f(x) + c \ge \alpha\})\, d\alpha \\ &= \int_c^\infty \mu(\{x \mid f(x) + c \ge \alpha\})\, d\alpha + \int_0^c \mu(\{x \mid f(x) + c \ge \alpha\})\, d\alpha \\ &= \int_c^\infty \mu(\{x \mid f(x) \ge \alpha - c\})\, d(\alpha - c) + \int_0^c \mu(X)\, d\alpha\end{aligned}$$

$$= \int_0^\infty \mu(\{x \mid f(x) \ge \alpha\})\, d\alpha + \int_0^c \mu(X)\, d\alpha$$
$$= (\mathrm{C})\int f\, d\mu + c \cdot \mu(X) \quad .$$

The proof is now complete. □

The above discussion on the Choquet integral is restricted to nonnegative measurable functions. Now we consider a more general case, where the integrand is not necessarily nonnegative. A natural idea is, similar to the Lebesgue integral shown in Section 5.3, to decompose a function to its positive part and negative part, that is, express measurable function $f: X \to (-\infty, \infty)$ as $f = f^+ - f^-$, where

$$f^+(x) = \begin{cases} f(x) & \text{if } f(x) \ge 0 \\ 0 & \text{otherwise} \end{cases}$$

and

$$f^-(x) = \begin{cases} -f(x) & \text{if } f(x) < 0 \\ 0 & \text{otherwise} \end{cases} \quad .$$

Both f^+ and f^- are nonnegative measurable functions. Their Choquet integrals are well defined. Hence, we may define the Choquet integral of *f*, without any loss of generality, on *X* as follows.

Definition 5.19 Let $f: X \to (-\infty, \infty)$ be a measurable function. The *symmetric Choquet integral* of *f* with respect to monotone measure μ on *X*, denoted by $(\mathrm{C_s})\int f\, d\mu$, is defined as

$$(\mathrm{C_s})\int f\, d\mu = (\mathrm{C})\int f^+\, d\mu - (\mathrm{C})\int f^-\, d\mu ,$$

provided not both terms on the right-hand side are infinite.

From Definition 5.19, we may see that $(-f)^+ = f^-$ and $(-f)^- = f^+$ for any given function *f*. So,

$$(\mathrm{C_s})\int(-f)\,d\mu = (\mathrm{C})\int(-f)^+\,d\mu - (\mathrm{C})\int(-f)^-\,d\mu$$
$$= (\mathrm{C})\int f^-\,d\mu - (\mathrm{C})\int f^+\,d\mu$$
$$= -(\mathrm{C_s})\int f\,d\mu .$$

This is just the reason why people use word "symmetric" to such type of Choquet integrals. Unfortunately, the symmetric Choquet integral loses the translatability in general. We can see it from the following example.

Example 5.9 Let $X = \{a, b\}$, $\mathscr{F} = \mathscr{P}(X)$, and

$$\mu(E) = \begin{cases} 1 & \text{if } E = X \\ 0 & \text{otherwise} \end{cases} .$$

Considering function

$$f(x) = \begin{cases} 0 & \text{if } x = a \\ -1 & \text{if } x = b \end{cases}$$

with $f^+ = 0$ and $f^- = -f$. Noting that

$$0 \le f(x) + 1 = \begin{cases} 1 & \text{if } x = a \\ 0 & \text{if } x = b \end{cases},$$

we have

$$(\mathrm{C_s})\int f\,d\mu = (\mathrm{C})\int f^+\,d\mu - (\mathrm{C})\int f^-\,d\mu = 0 - (\mathrm{C})\int f^-\,d\mu = -\mu(\{b\}) = 0 ,$$

$$(\mathrm{C_s})\int (f+1)\, d\mu = \mu(\{a\}) = 0\,.$$

So,

$$(\mathrm{C_s})\int (f+1)\, d\mu \neq (\mathrm{C_s})\int f\, d\mu + 1\cdot \mu(X)\,,$$

that is, the symmetric Choquet is not translatable.

Anyway, the translatability is one of major requirements to an aggregation tool in information fusion. Though the symmetric Choquet integral still holds property (CIP4), some decisions based on information fusion using such an integral will depends on the selection of the origin and the unit that measures the received information. For instance, to measure the temperature, there are two common systems: Celsius degree and Fahrenheit degree. The different selection of the temperature system may lead to a different decision if the symmetric Choquet integral is used as an aggregation tool in information fusion. Hence, it is necessary to find a way for defining the Choquet integral with signed integrand such that the translatability can be reserved. The following definition is an ideal approach, where we simply consider the integral taken on X. There is no difficulty for generalizing it to be taken on any measurable subset E of X.

Definition 5.20 Let $f: X \to (-\infty, \infty)$ be a measurable function on monotone measure space $(X, \mathscr{F}, \mu)$. The *translatable Choquet integral* of f with respect to monotone measure μ on X, denoted by $(\mathrm{C_t})\int f\, d\mu$, is defined as

$$(\mathrm{C_t})\int f\, d\mu = \int_{-\infty}^{0} [\mu(F_\alpha) - \mu(X)]\, d\alpha + \int_{0}^{\infty} \mu(F_\alpha)\, d\alpha\,,$$

where $F_\alpha = \{x \mid f(x) \geq \alpha\}$ for $\alpha \in (-\infty, \infty)$, provided not both terms in the right-hand side of the formula are infinite.

The next theorem shows the reason why such type of integral is said to be translatable.

Theorem 5.6 Let $f : X \to (-\infty, \infty)$ be a measurable function on monotone measure space $(X, \mathscr{F}, \mu)$. Then

$$(\mathrm{C_t})\int (f + c)\, d\mu = (\mathrm{C_t})\int f\, d\mu + c \cdot \mu(X)$$

for any real number c.

Proof. By using some well known properties of the Riemann integral, we have

$$\begin{aligned}
&(\mathrm{C_t})\int (f + c)\, d\mu \\
&= \int_{-\infty}^{0} [\mu(\{x \mid f(x) + c \ge \alpha\}) - \mu(X)]\, d\alpha + \int_{0}^{\infty} \mu(\{x \mid f(x) + c \ge \alpha\})\, d\alpha \\
&= \int_{-\infty}^{0} [\mu(\{x \mid f(x) \ge \alpha - c\}) - \mu(X)]\, d\alpha + \int_{0}^{\infty} \mu(\{x \mid f(x) \ge \alpha - c\})\, d\alpha \\
&= \int_{-\infty}^{0} [\mu(\{x \mid f(x) \ge \alpha - c\}) - \mu(X)]\, d(\alpha - c) \\
&\quad + \int_{0}^{\infty} \mu(\{x \mid f(x) \ge \alpha - c\})\, d(\alpha - c) \\
&= \int_{-\infty}^{-c} [\mu(\{x \mid f(x) \ge \beta\}) - \mu(X)]\, d\beta + \int_{-c}^{\infty} \mu(\{x \mid f(x) \ge \beta\})\, d\beta \\
&= \int_{-\infty}^{0} [\mu(\{x \mid f(x) \ge \beta\}) - \mu(X)]\, d\beta + \int_{0}^{\infty} \mu(\{x \mid f(x) \ge \beta\})\, d\beta \\
&\quad - \int_{-c}^{0} [\mu(\{x \mid f(x) \ge \beta\}) - \mu(X)]\, d\beta + \int_{-c}^{0} \mu(\{x \mid f(x) \ge \beta\})\, d\beta \\
&= \int_{-\infty}^{0} [\mu(\{x \mid f(x) \ge \beta\}) - \mu(X)]\, d\beta + \int_{0}^{\infty} \mu(\{x \mid f(x) \ge \beta\})\, d\beta + \int_{-c}^{0} \mu(X)\, d\beta \\
&= (\mathrm{C_t})\int f\, d\mu + c \cdot \mu(X),
\end{aligned}$$

where $\beta = \alpha - c$. The equality of translatability is now proved. □

The translatable Choquet integral also keeps properties (CIP3), (CIP4), and (CIP5). Since the symmetric Choquet integral is never used in applications of nonlinear integrals discussed in this book, from now on, we omit the subscript "t" from the symbol of the translatable Choquet integral of f with respect to μ as well as omit word "translatable" from its full name, that is, write $(C)\int f\, d\mu$ and still called it the Choquet integral, if there is no confusion.

As for the calculation formula of the translatable Choquet integral when the universal set X is finite, it is totally the same as (5.4).

Example 5.10 We still use the monotone measure space (X, $\mathscr{P}(X)$, μ) given in Example 5.7, where $X = \{x_1, x_2, x_3\}$, $\mathscr{F} = \mathscr{P}(X)$, and

$$\mu(E) = \begin{cases} 0 & \text{if } E = \varnothing \\ 0.5 & \text{if } E = \{x_1\} \\ 0.2 & \text{if } E = \{x_2\} \\ 0.6 & \text{if } E = \{x_1, x_2\} \\ 0.4 & \text{if } E = \{x_3\} \\ 0.7 & \text{if } E = \{x_1, x_3\} \\ 0.9 & \text{if } E = \{x_2, x_3\} \\ 1 & \text{if } E = X \quad . \end{cases}$$

Now let function $g : X \to (-\infty, \infty)$ be

$$g(x) = \begin{cases} 1 & \text{if } x = x_1 \\ 3 & \text{if } x = x_2 \\ -2 & \text{if } x = x_3 \,. \end{cases}$$

For function g we still have $x_1^* = x_3$, $x_2^* = x_1$, and $x_3^* = x_2$, but it is not nonnegative. By using formula (5.4), the Choquet integral of g with respect to μ is

$$(C)\int g\,d\mu = (g(x_1^*)-0)\cdot\mu(\{x_1^*,x_2^*,x_3^*\})+(g(x_2^*)-g(x_1^*))\cdot\mu(\{x_2^*,x_3^*\})$$
$$+(g(x_3^*)-g(x_2^*))\cdot\mu(\{x_3^*\})$$
$$= g(x_3)\cdot\mu(X)+(g(x_1)-g(x_3))\cdot\mu(\{x_1,x_2\})+(g(x_2)-g(x_1))\cdot\mu(\{x_2\})$$
$$=(-2)\times 1+3\times 0.6+2\times 0.2$$
$$=0.2\ .$$

Since $g = f-7$, we may use the translatability of the Choquet integral and the result in Example 5.7 to verify the result. By using (CIP6), it should hold that

$$(C)\int g\,d\mu = (C)\int f\,d\mu - 7\cdot\mu(X) = 7.2-7\times 1 = 0.2\,.$$

This coincides with the obtained result.

When the universal set X is finite, the Choquet integral can be generalized for efficiency measure and signed efficiency measure without any essential difficulty. In fact, if f is a real-valued function on $(X, \mathscr{P}(X), \mu)$ where μ is an efficiency measure, then $\mu(F_\alpha)$ is a function of bounded variation with respect to α. Hence, formula

$$(C)\int f\,d\mu = \int_{-\infty}^{0}[\mu(F_\alpha)-\mu(X)]\,d\alpha + \int_{0}^{\infty}\mu(F_\alpha)\,d\alpha$$

can still well define the Choquet integral provided not both terms in the right-hand side of the formula are infinite. When μ is a signed efficiency measure, since μ can be decomposed as a difference of two efficiency measures μ^+ and μ^-: $\mu=\mu^+-\mu^-$, we have

$$(C)\int f\,d\mu = (C)\int f\,d\mu^+ - (C)\int f\,d\mu^-\,.$$

Hence, the Choquet integral of real-valued function f with respect to signed efficiency measure μ is well defined provided not both terms in the right-hand side of the formula are infinite. The calculation formulas (5.4)-(5.6) are still available and properties (CIP2), (CIP3), (CIP4), and (CIP6) still hold in this case.

Example 5.11 Let $X=\{x_1, x_2, x_3\}$ and $\mathscr{F}=\mathscr{P}(X)$. We still use function g given in Example 5.10, that is,

$$g(x)=\begin{cases} 1 & \text{if } x=x_1 \\ 3 & \text{if } x=x_2 \\ -2 & \text{if } x=x_3 \,. \end{cases}$$

But the monotone measure μ is replaced by a signed efficiency measure ν given as

$$\nu(E)=\begin{cases} 0 & \text{if } E=\emptyset \\ -0.5 & \text{if } E=\{x_1\} \\ 0.2 & \text{if } E=\{x_2\} \\ -0.6 & \text{if } E=\{x_1, x_2\} \\ 0.4 & \text{if } E=\{x_3\} \\ -0.3 & \text{if } E=\{x_1, x_3\} \\ 0.9 & \text{if } E=\{x_2, x_3\} \\ 0.3 & \text{if } E=X \quad . \end{cases}$$

Thus,

$$\begin{aligned}(\mathrm{C})\int g\, d\nu &= (g(x_1^*)-0)\cdot\nu(\{x_1^*, x_2^*, x_3^*\})+(g(x_2^*)-g(x_1^*))\cdot\nu(\{x_2^*, x_3^*\}) \\ &\quad +(g(x_3^*)-g(x_2^*))\cdot\nu(\{x_3^*\}) \\ &= g(x_3)\cdot\nu(X)+(g(x_1)-g(x_3))\cdot\nu(\{x_1, x_2\})+(g(x_2)-g(x_1))\cdot\nu(\{x_2\})\end{aligned}$$

$$= (-2)\times 0.3 + 3\times(-0.6) + 2\times 0.2$$
$$= -2 \ .$$

The Choquet integral has more useful properties, which the Lebesgue integral has, such as the continuity and the monotonicity with respect to the integrand. They are shown in the next two theorems.

Theorem 5.7 (Continuity) Let $f_1 : X \to (-\infty, \infty)$ and $f_2 : X \to (-\infty, \infty)$ be bounded measurable functions on measurable space $(X, \mathscr{F})$ and $\mu : \mathscr{F} \to [0, \infty)$ be a monotone measure on $\mathscr{F}$. Assume that Choquet integrals $(C)\int f_1 d\mu$ and $(C)\int f_2 d\mu$ exist. Then, for any given $\varepsilon > 0$, there exists $\delta > 0$, such that $|(C)\int f_1 d\mu - (C)\int f_2 d\mu| < \varepsilon$ whenever $|f_1 - f_2| < \delta$.

Proof. Let M be the bound of f_1 and f_2, that is, $f_1 \le M$ and $f_2 \le M$. Denote $\{x \mid f_1(x) \ge \alpha\}$ and $\{x \mid f_2(x) \ge \alpha\}$ by $F_\alpha^{(1)}$ and $F_\alpha^{(2)}$ respectively. For any given $\varepsilon > 0$, taking $\delta = \varepsilon / 2\mu(X)$, we have $F_\alpha^{(1)} \subseteq F_{\alpha-\delta}^{(2)}$ and, therefore, $\mu(F_\alpha^{(1)}) \le \mu(F_{\alpha-\delta}^{(2)})$ if $|f_1 - f_2| < \delta$. Similarly, $\mu(F_{\alpha+\delta}^{(2)}) \le \mu(F_\alpha^{(1)})$ if $|f_1 - f_2| < \delta$. Thus,

$$\begin{aligned}
&(C)\int f_1 d\mu - (C)\int f_2 d\mu \\
&= \int_{-\infty}^{0} [\mu(F_\alpha^{(1)}) - \mu(X)] d\alpha + \int_0^\infty \mu(F_\alpha^{(1)}) d\alpha - \int_{-\infty}^{0} [\mu(F_\alpha^{(2)}) - \mu(X)] d\alpha \\
&\quad - \int_0^\infty \mu(F_\alpha^{(2)}) d\alpha \\
&= \int_{-M}^{0} [\mu(F_\alpha^{(1)}) - \mu(X)] d\alpha + \int_0^M \mu(F_\alpha^{(1)}) d\alpha - \int_{-M}^{0} [\mu(F_\alpha^{(2)}) - \mu(X)] d\alpha \\
&\quad - \int_0^M \mu(F_\alpha^{(2)}) d\alpha \\
&= \int_{-M}^{0} [\mu(F_\alpha^{(1)}) - \mu(F_\alpha^{(2)})] d\alpha + \int_0^M [\mu(F_\alpha^{(2)}) - \mu(F_\alpha^{(1)})] d\alpha \\
&\le \int_{-M}^{0} [\mu(F_{\alpha-\delta}^{(2)}) - \mu(F_\alpha^{(2)})] d\alpha + \int_0^M [\mu(F_\alpha^{(2)}) - \mu(F_{\alpha+\delta}^{(2)})] d\alpha \\
&= \int_{-M-\delta}^{-\delta} \mu(F_\alpha^{(2)}) d\alpha - \int_{-M}^{0} \mu(F_\alpha^{(2)}) d\alpha + \int_0^M \mu(F_\alpha^{(2)}) d\alpha - \int_\delta^{M+\delta} \mu(F_\alpha^{(2)}) d\alpha
\end{aligned}$$

$$\le \int_{-M-\delta}^{-M} \mu(F_\alpha^{(2)})d\alpha + \int_0^\delta \mu(F_\alpha^{(2)})\,d\alpha$$
$$\le 2\delta \cdot \mu(X)$$
$$= \varepsilon\,.$$

In the same way, we can show that

$$(\mathrm{C})\int f_2 d\mu - (\mathrm{C})\int f_1 d\mu \le \varepsilon\,.$$

Consequently,

$$|\,(\mathrm{C})\int f_1 d\mu - (\mathrm{C})\int f_2 d\mu\,| \le \varepsilon\,.$$

The proof is now complete. □

The monotonicity of the Choquet integral of nonnegative measurable functions shown in Theorem 5.5 can be generalized to the case where the integrand functions may not be nonnegative.

Theorem 5.8 (Monotonicity) Let $f_1 : X \to (-\infty, \infty)$ and $f_2 : X \to (-\infty, \infty)$ be measurable functions on measurable space $(X, \mathscr{F})$ and $\mu : \mathscr{F} \to [0, \infty)$ be a monotone measure on $\mathscr{F}$. Assume that Choquet integrals $(\mathrm{C})\int f_1 d\mu$ and $(\mathrm{C})\int f_2 d\mu$ exist. Then $(\mathrm{C})\int f_1 d\mu \le (\mathrm{C})\int f_2 d\mu$ if $f_1 \le f_2$.

Proof. From $f_1 \le f_2$, we know that $F_\alpha^{(1)} \le F_\alpha^{(2)}$ and, therefore, $\mu(F_\alpha^{(1)}) \le \mu(F_\alpha^{(2)})$ for every $\alpha \in (-\infty, \infty)$ by the monotonicity of μ. Thus,

$$\begin{aligned}(\mathrm{C})\int f_1 d\mu &= \int_{-\infty}^0 [\mu(F_\alpha^{(1)}) - \mu(X)]d\alpha + \int_0^\infty \mu(F_\alpha^{(1)})\,d\alpha \\ &\le \int_{-\infty}^0 [\mu(F_\alpha^{(2)}) - \mu(X)]d\alpha + \int_0^\infty \mu(F_\alpha^{(2)})\,d\alpha \\ &= (\mathrm{C})\int f_2 d\mu\,.\end{aligned}$$

The proof is now complete. □

When X is finite, even if μ is only a signed efficiency measure, the Choquet integral $(\mathrm{C})\int f\, d\mu$ exists for any real-valued function f defined on X, and it is also continuous with respect to the integrand. However, the monotonicity of μ is essential to the monotonicity of the Choquet integral with respect to the integrand.

5.5 Upper and Lower Integrals

We have seen that the Choquet integral is a generalization of the Lebesgue integral. Its definition is just one of the equivalent definition of the Lebesgue integral. That is to say, in case we consider using nonadditive measures to replace classical additive measure in some systems, though the original definition of the Lebesgue integral fails, we still can use some of its equivalent definitions to define nonlinear integrals as aggregation tools in systems. By such an idea, this chapter presents other tow types of nonlinear integrals, the upper integral and the lower integral.

Throughout this section, we assume that $(X, \mathscr{F}, \mu)$ is an efficiency measure space, that is, μ is an efficiency measure on measurable space $(X, \mathscr{F})$, $f: X \to [0, \infty)$ and $g: X \to [0, \infty)$ are nonnegative measurable functions.

Definition 5.21 Given a nonnegative measurable function $f: X \to [0, \infty)$ and a set $E \in \mathscr{F}$, the *upper integral* of f with respect to μ on E, in symbol $(\mathrm{U})\int_E f\, d\mu$, is defined as

$$(\mathrm{U})\int_E f\, d\mu = \lim_{\varepsilon \to 0+} U_\varepsilon ,$$

where

$$U_\varepsilon = \sup\Big\{ \sum_{j=1}^{\infty} \lambda_j \cdot \mu(E_j) \Big| \; f \ge \sum_{j=1}^{\infty} \lambda_j \cdot \chi_{E_j} \ge f - \varepsilon, E_j \in \mathscr{F} \cap E, \lambda_j \ge 0, j = 1, 2, \cdots \Big\}$$

for $\varepsilon > 0$, in which $\mathscr{F} \cap E = \{F \cap E \mid F \in \mathscr{F}\}$. Similarly, the lower integral of f with respect to μ on E, $(\mathrm{L})\int_E f\, d\mu$, is defined as

$$(\mathrm{L})\int_E f\, d\mu = \lim_{\varepsilon \to 0+} L_\varepsilon ,$$

where

$$L_\varepsilon = \inf\{ \sum_{j=1}^{\infty} \lambda_j \cdot \mu(E_j) \mid f \le \sum_{j=1}^{\infty} \lambda_j \cdot \chi_{E_j} \le f + \varepsilon, E_j \in \mathscr{F} \cap E, \lambda_j \ge 0, j = 1, 2, \cdots \}$$

for $\varepsilon > 0$.

Similar to the Lebesgue integral and the Choquet integral, we omit the subscript E in the symbol of the integral when $E = X$.

If the universal set is finite, i.e., $X = \{x_1, x_2, \cdots, x_n\}$, the supremum and the infimum in Definition 5.21 are accessible. Hence, the upper integral of f with respect to μ, $(\mathrm{U})\int f\, d\mu$, can be reduced as

$$(\mathrm{U})\int f\, d\mu = \sup\{ \sum_{j=1}^{2^n - 1} \lambda_j \cdot \mu(E_j) \mid \sum_{j=1}^{2^n - 1} \lambda_j \chi_{E_j} = f \}, \qquad (5.7)$$

where $\lambda_j \ge 0$ and $E_j = \bigcup_{i \mid j_i = 1} \{x_i\}$ if j is expressed in binary digits as $j_n j_{n-1} \cdots j_1$ for every $j = 1, 2, \cdots, 2^n - 1$. The value of $(\mathrm{U})\int f\, d\mu$ then is just the solution of the following linear programming problem:

maximize $\quad z = \sum_{j=1}^{2^n - 1} \lambda_j \cdot \mu_j$

subject to $\quad \sum_{j=1}^{2^n - 1} \lambda_j \chi_{E_j}(x_i) = f(x_i), \quad i = 1, 2, \cdots, n$

$$\lambda_j \geq 0, \ \ j = 1, 2, \ \cdots, 2^n - 1$$

where $\lambda_1, \lambda_2, \cdots, \lambda_{2^n-1}$ are unknown parameters, $\mu_j = \mu(E_j)$ for $j = 1, 2, \cdots, 2^n - 1$. The above n constraints can be also rewritten as

$$\sum_{j \mid x \in E_j \subseteq X} \lambda_j = f(x) \quad \forall x \in X \, .$$

By knowledge on the linear programming, the above maximum can be accessed by at most n nonzero-valued λ_j, that is, the solution can be expressed as

$$\sum_{i=1}^{n} \lambda_{j_i} \mu_{j_i} \, ,$$

where $\{j_1, j_2, \cdots, j_n\}$ is a subset of $\{1, 2, \ \cdots, 2^n - 1\}$.

Example 5.12 We use monotone measure μ and nonnegative function f given in Example 5.7. The upper integral of f with respect to μ, $(\mathrm{U})\int f \, d\mu$, is the solution of the following linear programming problem:

maximize $\quad z = 0.5\lambda_1 + 0.2\lambda_2 + 0.6\lambda_3 + 0.4\lambda_4 + 0.7\lambda_5 + 0.9\lambda_6 + \lambda_7$

subject to $\quad \lambda_1 + \lambda_3 + \lambda_5 + \lambda_7 = 8$

$$\lambda_2 + \lambda_3 + \lambda_6 + \lambda_7 = 10$$

$$\lambda_4 + \lambda_5 + \lambda_6 + \lambda_7 = 5$$

$$\lambda_j \geq 0, \ \ j = 1, 2, \cdots, 7$$

By using the simplex method, a solution of this linear programming problem can be obtained as $\lambda_1 = 8$, $\lambda_2 = 5$, and $\lambda_6 = 5$ with $z = 9.5$.

Similar to the upper integral, the lower integral of f with respect to μ, $(\mathrm{L})\int f\, d\mu$, can be reduced as

$$(\mathrm{L})\int f\, d\mu = \inf\{\sum_{j=1}^{2^n-1} \lambda_j \cdot \mu(E_j) \mid \sum_{j=1}^{2^n-1} \lambda_j \chi_{E_j} = f\} \tag{5.8}$$

Its value is just the solution of the following linear programming problem:

minimize $\quad z = \sum_{j=1}^{2^n-1} \lambda_j \cdot \mu_j$

subject to $\quad \sum_{j=1}^{2^n-1} \lambda_j \chi_{E_j}(x_i) = f(x_i), \quad i = 1, 2, \cdots, n$

$\quad \lambda_j \geq 0, \quad j = 1, 2, \cdots, 2^n - 1$

where $\lambda_1, \lambda_2, \cdots, \lambda_{2^n-1}$ are unknown parameters, $\mu_j = \mu(E_j)$ for $j = 1, 2, \cdots, 2^n - 1$. The above minimum can be accessed by at most n nonzero-valued λ_j, that is, the solution can be expressed as

$$\sum_{i=1}^{n} \lambda_{j_i'} \mu_{j_i'},$$

where $\{j_1', j_2', \cdots, j_n'\}$ is a subset of $\{1, 2, \cdots, 2^n - 1\}$.

Example 5.13 We still use monotone measure μ and nonnegative function f given in Example 5.7. The lower integral of f with respect to

μ, $(\mathrm{L})\int f\,d\mu$, is the solution of the following linear programming problem:

$$\text{minimize} \qquad z = 0.5\lambda_1 + 0.2\lambda_2 + 0.6\lambda_3 + 0.4\lambda_4 + 0.7\lambda_5 + 0.9\lambda_6 + \lambda_7$$

$$\text{subject to} \qquad \lambda_1 + \lambda_3 + \lambda_5 + \lambda_7 = 8$$

$$\lambda_2 + \lambda_3 + \lambda_6 + \lambda_7 = 10$$

$$\lambda_4 + \lambda_5 + \lambda_6 + \lambda_7 = 5$$

$$\lambda_j \geq 0, \quad j = 1, 2, \cdots, 7$$

Similar to Example 5.12, by using the simplex method, a solution of this linear programming problem can be obtained as $\lambda_2 = 7$, $\lambda_3 = 3$, and $\lambda_5 = 5$ with $z = 6.7$.

The upper and the lower integrals have some common properties that the Lebesgue integral with a nonnegative integrand has:

(ULIP1) $(\mathrm{U})\int_E f\,d\mu = (\mathrm{U})\int f \cdot \chi_E\,d\mu$ and $(\mathrm{L})\int_E f\,d\mu = (\mathrm{L})\int f \cdot \chi_E\,d\mu$;

(ULIP2) $(\mathrm{U})\int f\,d\mu \geq 0$ and $(\mathrm{L})\int f\,d\mu \geq 0$;

(ULIP3) if $f \leq g$, then $(\mathrm{U})\int f\,d\mu \leq (\mathrm{U})\int g\,d\mu$ and, moreover,

$(\mathrm{L})\int f\,d\mu \leq (\mathrm{L})\int g\,d\mu$ provided μ is a monotone measure;

(ULIP4) $(\mathrm{U})\int c \cdot f\,d\mu = c \cdot (\mathrm{U})\int f\,d\mu$ and $(\mathrm{L})\int c \cdot f\,d\mu = c \cdot (\mathrm{L})\int f\,d\mu$

for any constant $c \geq 0$.

Moreover, we have

(ULIP5) $(\mathrm{U})\int f\, d\mu \geq (\mathrm{L})\int f\, d\mu$.

However, neither the upper integral nor the lower integral is linear, that is, we may have

$$(\mathrm{U})\int (f+g)\, d\mu \neq (\mathrm{U})\int f\, d\mu + (\mathrm{U})\int g\, d\mu$$

and

$$(\mathrm{L})\int (f+g)\, d\mu \neq (\mathrm{L})\int f\, d\mu + (\mathrm{L})\int g\, d\mu$$

for some monotone measure μ and nonnegative measurable functions f and g.

Example 5.14 Let $X = \{x_1, x_2, x_3\}$ and $\mathscr{F} = \mathscr{P}(X)$. Monotone measure μ is defined as

$$\mu(E) = \begin{cases} 0 & \text{if } E = \varnothing \\ 3 & \text{if } E = \{x_1\} \\ 3 & \text{if } E = \{x_2\} \\ 1 & \text{if } E = \{x_3\} \\ 5 & \text{otherwise} \end{cases}.$$

Considering functions

$$f(x) = \begin{cases} 1 & \text{if } x = x_1 \\ 1 & \text{if } x = x_2 \\ 0 & \text{if } x = x_3 \end{cases}$$

and

$$g(x)=\begin{cases}0 & \text{if } x=x_1\\ 0 & \text{if } x=x_2\\ 1 & \text{if } x=x_3\,,\end{cases}$$

we obtain

$$(\mathrm{U})\int f\,d\mu=1\cdot\mu(x_1)+1\cdot\mu(x_2)=1\times 3+1\times 3=6\,,$$

$$(\mathrm{U})\int g\,d\mu=1\cdot\mu(x_3)=1\times 1=1\,,$$

and

$$(\mathrm{U})\int(f+g)\,d\mu=1\cdot\mu(x_1)+1\cdot\mu(\{x_2,x_3\})=1\times 3+1\times 5=8\,.$$

That is, we have

$$(\mathrm{U})\int(f+g)\,d\mu>(\mathrm{U})\int f\,d\mu+(\mathrm{U})\int g\,d\mu\,.$$

Similarly,

$$(\mathrm{L})\int f\,d\mu=1\cdot\mu(\{x_1,x_2\})=1\times 5=5\,,$$

$$(\mathrm{L})\int g\,d\mu=1\cdot\mu(x_3)=1\times 1=1\,,$$

and

$$(\mathrm{L})\int(f+g)\,d\mu=1\cdot\mu(\{x_1,x_2,x_3\})=1\times 5=5\,.$$

That is,

$$(\mathrm{L})\int(f+g)\,d\mu<(\mathrm{L})\int f\,d\mu+(\mathrm{L})\int g\,d\mu\,.$$

The results in Example 5.14 suggest the following general inequalities as a property of the upper and the lower integrals.

(ULIP6) $(\mathrm{U})\int(f+g)\,d\mu\ge(\mathrm{U})\int f\,d\mu+(\mathrm{U})\int g\,d\mu\,;$

$$\text{(L)}\int (f+g)\,d\mu \le \text{(L)}\int f\,d\mu + \text{(L)}\int g\,d\mu .$$

Another important point is that the upper and the lower integrals do not have a property like (LIP2) or (CIP2) the Lebesgue integral and the Choquet integral hold.

Example 5.15 Let $X = \{x_1, x_2\}$ and $\mathscr{F} = \mathscr{P}(X)$. Set function μ is defined as

$$\mu(E) = \begin{cases} 0 & \text{if } E = \varnothing \\ 1 & \text{otherwise}. \end{cases}$$

Clearly, μ is a monotone measure. Taking constant 1 as the integrand, we have

$$\text{(U)}\int 1\,d\mu = 1\cdot\mu(\{x_1\}) + 1\cdot\mu(\{x_2\}) = 1\times 1 + 1\times 1 = 2 \ne \mu(X).$$

It is easy to cite a similar counterexample for the lower integral. This is left to the reader as an exercise. Though the equalities do not hold, we still have the inequalities expressed as one more property of the upper and the lower integrals:

(ULIP7) $\text{(L)}\int 1\,d\mu \le \mu(X) \le \text{(U)}\int 1\,d\mu$.

Finally, we show another inequality for the upper integral as one of its properties in the following theorem.

Theorem 5.9 Let $X = \{x_1, x_2, \cdots, x_n\}$ and μ be monotone measures on $\mathscr{P}(X)$. Then,

(ULIP8) $\text{(U)}\int 1\,d\mu \le n\cdot\mu(X)$.

Proof. Consider each $\sum_{j=1}^{2^n-1}\lambda_j \cdot \mu(E_j)$ satisfying $\sum_{j=1}^{2^n-1}\lambda_j\chi_{E_j}(x)=1$ for every $x \in X$. Since $\sum_{j=1}^{2^n-1}\lambda_j\chi_{E_j}(x)=1$ means $\sum_{j|x_i\in A_j}\lambda_j=1$ for every x_i, $i=1,2,\cdots,n$, we have

$$\begin{aligned}\sum_{j=1}^{2^n-1}\lambda_j \cdot \mu(E_j) &\le \sum_{j=1}^{2^n-1}\lambda_j \cdot \mu(X)\\ &= \mu(X)\cdot\sum_{j=1}^{2^n-1}\lambda_j\\ &\le \mu(X)\cdot\sum_{i=1}^{n}(\sum_{x_i\in E_j}\lambda_j)\\ &= \mu(X)\cdot\sum_{i=1}^{n}1\\ &= n\cdot\mu(X) \quad .\end{aligned}$$

Hence,

$$(\mathrm{U})\int 1\, d\mu = \sup\{\sum_{j=1}^{2^n-1}\lambda_j \cdot \mu(E_j) \mid \sum_{j=1}^{2^n-1}\lambda_j\chi_{E_j}=1\} \le n\cdot\mu(X)\,.$$

The proof is now complete. □

Unlike the Choquet integral, the upper integral is not translatable, even

$$(\mathrm{U})\int(f+c)\, d\mu \ne (\mathrm{U})\int(f+c)\, d\mu + c\cdot(\mathrm{U})\int 1\, d\mu\,,$$

in general. This can be seen from the following example.

Example 5.16 Let $X=\{x_1,x_2,x_3\}$ and $\mathscr{F}=\mathscr{P}(X)$. Set function μ is defined as

$$\mu(E)=\begin{cases}0 & \text{if } |E|\le 1 \text{ or } E=\{x_1, x_3\}\\ 1 & \text{otherwise}\end{cases}.$$

Obviously, μ is a monotone measure on $(X, \mathscr{F})$. Taking

$$f(x)=\begin{cases}1 & \text{if } x=x_2\\ 0 & \text{otherwise}\end{cases},$$

we have

$$(\mathrm{U})\int(f+1)\,d\mu=1\cdot\mu(\{x_1, x_2\})+1\cdot\mu(\{x_2, x_3\})=1+1=2.$$

However, $(\mathrm{U})\int f\,d\mu=0$ and $(\mathrm{U})\int 1\,d\mu=1$. Consequently,

$$(\mathrm{U})\int(f+1)\,d\mu>(\mathrm{U})\int f\,d\mu+(\mathrm{U})\int 1\,d\mu.$$

A similar conclusion is also valid for the lower integral.

In the definitions of the upper and the lower integrals, the efficiency measure can be replaced by a signed efficiency measure. In this case, properties (ULIP2), (ULIP4), and (ULIP8) may not hold.

5.6 *r*-Integrals on Finite Spaces

In the previews sections, four different types of integrals defined on signed efficiency measure spaces or on classical measure spaces have been presented. They are the Lebesgue integral, the Choquet integral, the upper integral, and the lower integral. In this section, we use a unified point of view to inspect them when the universal set is finite and the integrand is nonnegative.

Let $X=\{x_1, x_2, \cdots, x_n\}$, f be a nonnegative function on X, and μ be a signed efficiency measure on $\mathscr{P}(X)$.

Definition 5.22 A set function $\pi : \mathscr{P}(X) - \{\varnothing\} \to [0, \infty)$ is called a *partition* of f if

$$f(x) = \sum_{E | x \in E \subseteq X} \pi(E)$$

for every $x \in X$.

Taking the characteristic function of a crisp set or the membership function of a fuzzy set as f, it is easy to see that the concept of partition in Definition 5.22 is a generalization of the classical partition for crisp sets and the fuzzy partition for fuzzy sets.

Definition 5.23 Each type of integrals with respect to μ is characterized by a rule r, by which, for any given nonnegative function f, a partition π of f can be obtained. In this case, we say that rule r *partitions* function f. Regarding both π and μ as $(2^n - 1)$-dimensional vectors, the value of the *integral* of f under rule r, denoted by $(r)\int f \, d\mu$, is the inner product of vectors π and μ, that is, $(r)\int f \, d\mu = \pi \cdot \mu$, where (r) is used to indicate the type of integral.

The above definition provides a flexible aggregation tool in information fusion and data mining. It is generally called an *r-integral*, and simply, an *integral* when the partitioning rule r has been uniquely chosen and there is no confusion.

The Choquet integral is a special r-integral. The partitioning rule corresponding to the Choquet integral can be described as follows: for any given nonnegative function $f : X \to [0, \infty)$, partition $\pi : \mathscr{P}(X) - \{\varnothing\} \to [0, \infty)$ is obtained by

$$\pi(E) = \begin{cases} f(x_i^*) - f(x_{i-1}^*) & \text{if } E = \{x_i^*, x_{i+1}^*, \cdots, x_n^*\} \text{ for some } i = 1, 2, \cdots, n \\ 0 & \text{otherwise} \end{cases}$$

for every $E \in \mathscr{P}(X) - \{\emptyset\}$, where $(x_1^*, x_2^*, \cdots, x_n^*)$ is a permutation of $\{x_1, x_2, \cdots, x_n\}$ such that $f(x_1^*) \le f(x_2^*) \le \cdots \le f(x_n^*)$ and $f(x_0^*) = 0$ as the convention made in Section 5.4. It is easy to verify that

$$\sum_{E|x \in E \subseteq X} \pi(E) = f(x) \qquad \forall x \in X .$$

This partitioning rule takes the coordination of the attributes into account maximally, that is, the manner of the partition is to make the coordination among the attributes in X as much as possible. It is evident that there are only at most n sets E with $\pi(E) > 0$ in such a partition.

Example 5.17 The data in Example 5.7 are used here again. The partition of f corresponding to the Choquet integral is illustrated in Figure 5.4 where the black part, the dark grey part, and the white part show $\pi(X) = 5$, $\pi(\{x_1, x_2\}) = 3$, and $\pi(\{x_2\}) = 2$ respectively. The values of π at other sets are zeros. Geometrically, this partitioning rule divides function f horizontally.

We have seen that the Choquet integral locates at the one extreme in terms of the coordination among attributes. To show another extreme, we need to generalize the classical Lebesgue integral such that it can be taken with respect to any signed efficiency measure.

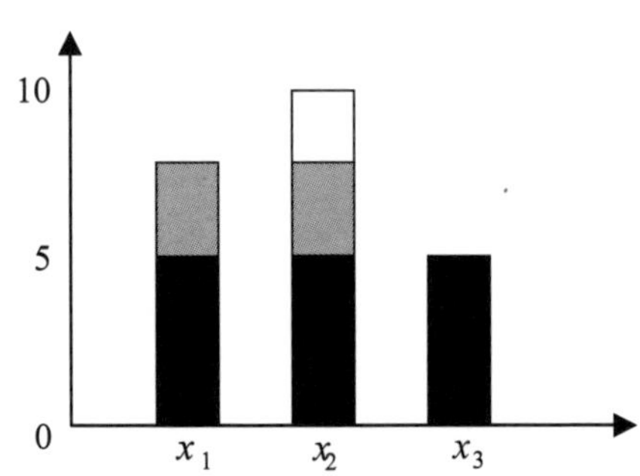

Fig. 5.4 The partition of f corresponding to the Choquet integral in Example 5.17.

Definition 5.24 The Lebesgue integral of nonnegative function *f* with respect to signed efficiency measure μ on set $E \subseteq X$, denoted by $\int_E f\, d\mu$, is defined as

$$\int_E f\, d\mu = \int_E f\, d\mu',$$

where μ' is the additive measure on $\mathscr{P}(X)$ determined by $\mu'(\{x_i\}) = \mu(\{x_i\})$, $i = 1, 2, \cdots, n$.

Example 5.18 We use the data given in Example 5.7 again. By the additivity, the corresponding additive measure μ' is obtained as

$$\mu'(E) = \begin{cases} 0 & \text{if } E = \varnothing \\ 0.5 & \text{if } E = \{x_1\} \\ 0.2 & \text{if } E = \{x_2\} \\ 0.7 & \text{if } E = \{x_1, x_2\} \\ 0.4 & \text{if } E = \{x_3\} \\ 0.9 & \text{if } E = \{x_1, x_3\} \\ 0.6 & \text{if } E = \{x_2, x_3\} \\ 1.1 & \text{if } E = X \end{cases}.$$

Thus, the Lebesgue integral of *f* with respect to monotone measure μ is

$$\int f\, d\mu = \int f\, d\mu' = f(x_1) \cdot \mu'(x_1) + f(x_2) \cdot \mu'(x_2) + f(x_3) \cdot \mu'(x_3)$$
$$= 8 \times 0.5 + 10 \times 0.2 + 5 \times 0.4 = 8.$$

Figure 5.5 illustrates the Lebesgue integral of function *f* with respect to μ, from which we can see that the corresponding partition of *f* is made vertically.

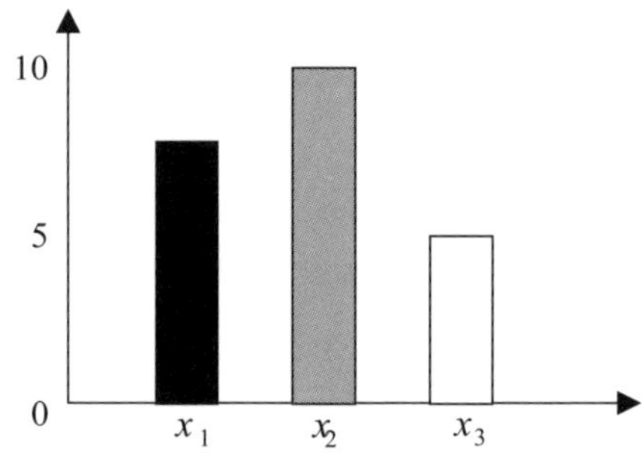

Fig. 5.5 The partition of *f* corresponding to the Lebesgue integral in Example 5.18.

According to Definition 5.24, the Lebesgue integral of a function with respect to signed efficiency measure μ only depends on the values of μ at singletons in $\mathscr{P}(X)$, ignoring the values at other sets. It is also a special type of *r*-integral and is another extreme in terms of the coordination among attributes. The Lebesgue integral takes no coordination into account at all, that is, the manner of the partition is to avoid any coordination. Based on such a point of view, we shall well understand why the Choquet integral coincides with the Lebesgue integral when there is no interaction among the contribution rates of attributes, that is, if no interaction exists objectively, the integration value should be always the same no matter how much coordination is considered. This conclusion is formalized as the following theorem.

Theorem 5.10 If the signed efficiency measure is additive, then any *r*-integral is the Lebesgue integral.

Proof. Let μ be an additive signed efficiency measure (i.e., a signed measure) on the power set of $X = \{x_1, x_2, \cdots, x_n\}$. For any given nonnegative function *f* defined on *X*, let π be the partition of *f* obtained by the corresponding rule *r*. Then, according to Definition 5.23, we have

$$
\begin{aligned}
(r)\int f\,d\mu &= \sum_{j=1}^{2^n-1} \pi(E_j)\cdot\mu(E_j) \\
&= \sum_{j=1}^{2^n-1} [\pi(E_j)\cdot \sum_{x_i\in E_j}\mu(\{x_i\})] \\
&= \sum_{i=1}^{n} [\mu(\{x_i\})\cdot \sum_{E|x_i\in E}\pi(E)] \\
&= \sum_{i=1}^{n} [\mu(\{x_i\})\cdot f(x_i)] \\
&= \int f\,d\mu .
\end{aligned}
$$

So, the r-integral coincides with the Lebesgue integral when μ is additive.

□

The above theorem also shows that the concept of r-integral is a generalization of the classical Lebesgue integral. From this theorem, we can say that any partitioning rule, by which a special r-integral can be obtained, corresponding to an equivalent definition of the classical Lebesgue integral. Indeed, the upper integral and the lower integral discussed in Section 5.5 are also two types of r-integral and, therefore, are generalizations of the classical Lebesgue integral.

Recall expression (5.7) for the upper integral, the value of $(\mathrm{U})\int f\,d\mu$ is just the optimal value of z in the linear programming problem

maximize $\quad z = \sum_{j=1}^{2^n-1} \lambda_j \cdot \mu_j$

subject to $\quad \sum_{j=1}^{2^n-1} \lambda_j \chi_{E_j}(x_i) = f(x_i), \quad i = 1, 2, \cdots, n$

$\lambda_j \geq 0, \quad j = 1, 2, \cdots, 2^n - 1$

where $\lambda_1, \lambda_2, \cdots, \lambda_{2^n-1}$ are unknown parameters and $\mu_j = \mu(E_j)$ for $j = 1, 2, \cdots, 2^n - 1$.

The above n constraints can be also rewritten as

$$\sum_{j|x \in E_j \subseteq X} \lambda_j = f(x) \quad \forall x \in X . \tag{5.9}$$

If we define set function $\pi : \mathscr{P}(X) \to [0, \infty)$ by $\pi(E_j) = \pi_j = \lambda_j$ for $j = 1, 2, \cdots, 2^n - 1$, expression (5.9) shows that π is a partition of f and, therefore, $z = \sum_{j=1}^{2^n-1} \pi_j \cdot \mu_j$ is an r-integral. Since the maximum in the linear programming problem is accessible, the upper integral is also a special type of r-integral. Its corresponding partitioning rule is "divide the integrand in such a way so that the integration value is maximized".

By a knowledge on the linear programming, the above maximum can be accessed by at most n nonzero-valued λ_j, that is, the value of the upper integral $(\mathrm{U})\int f\, d\mu$ can be expressed as

$$\sum_{i=1}^{n} \lambda_{j_i} \mu_{j_i} ,$$

where $\{j_1, j_2, \cdots, j_n\}$ is a subset of $\{1, 2, \cdots, 2^n - 1\}$.

Example 5.19 Use the data in Examples 5.7 again. The value of the upper integral of f with respect to monotone measure μ, $(\mathrm{U})\int f\, d\mu$, is z = 9.5 that is shown in the solution of the linear programming problem in Example 5.12, where $\lambda_1 = 8$, $\lambda_2 = 5$, and $\lambda_6 = 5$. These λ's form a partition π of function f, illustrated in Figure 5.6.

Similarly, from expression (5.8), the value of the lower integral $(\mathrm{L})\int f\, d\mu$ is just the optimal value of z in the linear programming problem

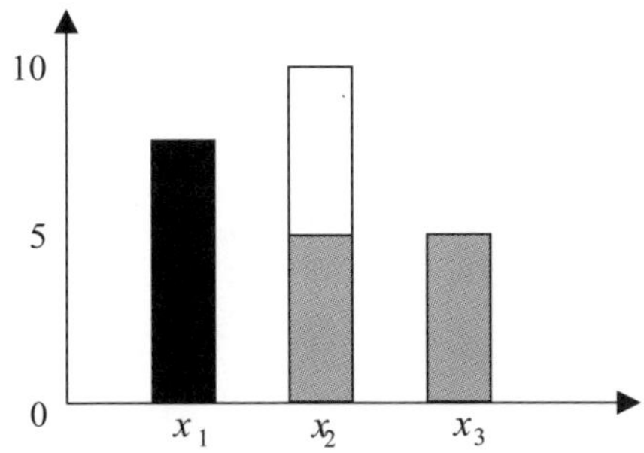

Fig. 5.6 The partition of *f* corresponding to the upper integral in Example 5.19.

minimize $$z = \sum_{j=1}^{2^n-1} \lambda_j \cdot \mu_j$$

subject to $$\sum_{j=1}^{2^n-1} \lambda_j \chi_{E_j}(x_i) = f(x_i), \quad i = 1, 2, \cdots, n$$

$$\lambda_j \geq 0, \quad j = 1, 2, \cdots, 2^n - 1$$

The minimum can be accessed by at most n nonzero-valued λ_j. Hence, the value of the lower integral $(\text{L})\int f\, d\mu$ can be expressed as

$$\sum_{i=1}^{n} \lambda_{j_i} \mu_{j_i},$$

where $\{j_1, j_2, \cdots, j_n\}$ is a subset of $\{1, 2, \cdots, 2^n - 1\}$. It is just a special r-integral. The corresponding partitioning rule is "divide the integrand in such a way so that the integration value is minimized".

Example 5.20 Using the data given in Example 5.7, we know that the value of the lower integral of f with respect to monotone measure μ, $(\text{L})\int f\, d\mu$, is z = 6.7 that is shown in the solution of the linear programming problem in Example 5.13, where $\lambda_2 = 7$, $\lambda_3 = 3$, and $\lambda_5 = 5$. These λ's form a partition π of function f, illustrated in Figure 5.7.

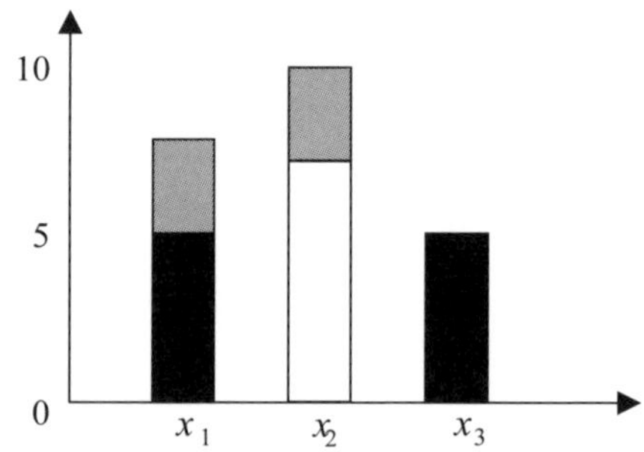

Fig. 5.7 The partition of *f* corresponding to the lower integral in Example 5.20.

Generally, for any nonnegative function *f*, any signed efficiency measure μ on finite measurable space $(X, \mathscr{P}(X))$, and any partitioning rule *r*, we have

$$(\mathrm{L})\int f\, d\mu \le (r)\int f\, d\mu \le (\mathrm{U})\int f\, d\mu\,.$$

As a summary of this section, let see the following example.

Example 5.21 Three workers, x_1, x_2, and x_3, are hired for manufacturing a certain kind of wooden toys. The universal set is taken as $X = \{x_1, x_2, x_3\}$. The individual and joint efficiencies (the number of produced toys per hour) of these three workers are shown in Example 4.12 as a monotone measure μ defined on the power set of *X* by

$$\mu(E) = \begin{cases} 0 & \text{if } E = \varnothing \\ 5 & \text{if } E = \{x_1\} \\ 6 & \text{if } E = \{x_2\} \\ 14 & \text{if } E = \{x_1, x_2\} \\ 7 & \text{if } E = \{x_3\} \\ 13 & \text{if } E = \{x_1, x_3\} \\ 9 & \text{if } E = \{x_2, x_3\} \\ 17 & \text{if } E = X \end{cases}.$$

Someday, they are hired to work for 6, 3, and 4 hours respectively. The working hours can be regarded as a function $f: X \to [0, \infty)$. If these three workers work separately, then the total number of toys they produce during this day can be expressed as the Lebesgue integral (see Figure 5.8(a))

$$\int f \, d\mu = 6 \times 5 + 3 \times 6 + 4 \times 7 = 76 .$$

If, working together, they start their work at the same time, say 9:00, and x_2 leaves at 12:00 while x_3 leaves at 13:00, then the total number of toys they produce during this day can be expressed as the Choquet integral (see Figure 5.8(b))

$$(\mathrm{C})\int f \, d\mu = 3 \times 17 + 1 \times 13 + 2 \times 5 = 74 .$$

The third case is that there is an excellent manager who knows the individual and joint efficiencies of these three workers well. The manager arranges their work in a certain coordination manner such that the toys produced by them during this day are as many as possible. This is just a linear programming problem:

$$
\begin{aligned}
\text{maximize} \quad & z = 5a_1 + 6a_2 + 14a_3 + 7a_4 + 13a_5 + 9a_6 + 17a_7 \\
\text{subject to} \quad & a_1 + a_3 + a_5 + a_7 = 6 \\
& a_2 + a_3 + a_6 + a_7 = 3 \\
& a_4 + a_5 + a_6 + a_7 = 4 \\
& a_j \geq 0, \ \ j = 1, 2, \cdots, 7
\end{aligned}
$$

Using the simplex method, a solution of this linear programming problem can be obtained as $a_3 = 3$, $a_4 = 1$, and $a_5 = 3$ with $z = 88$. That is, the manager arranges x_1 and x_2 to work together for 3 hours, x_1 and x_3 to work together for 3 hours, and x_3 works alone for one hour. Then, the total amount of the produced toys will be the maximal 88. It is just the upper integral (see Figure 5.8(c))

$$(\mathrm{U})\int f\,d\mu = 3\times 14 + 3\times 13 + 1\times 7 = 88.$$

This number represents the potential of the team of these three workers during this day. Finally, let's consider the most conservative estimation for the total number of the toys that can be produced by these workers during this day. This is another linear programming problem:

minimize $\quad z = 5a_1 + 6a_2 + 14a_3 + 7a_4 + 13a_5 + 9a_6 + 17a_7$

subject to $\quad a_1 + a_3 + a_5 + a_7 = 6$

$$a_2 + a_3 + a_6 + a_7 = 3$$

$$a_4 + a_5 + a_6 + a_7 = 4$$

$$a_j \ge 0,\ j = 1, 2, \cdots, 7$$

Its solution is $a_1 = 6$, $a_4 = 1$, and $a_6 = 3$ with $z = 64$. The corresponding arrangement is that x_2 and x_3 work together for 3 hours, x_1 works alone for 6 hours, and x_3 works alone for one hour. The total amount of produced toys will be the least as 64. It is just the lower integral (see Figure 5.8(d))

$$(\mathrm{L})\int f\,d\mu = 3\times 9 + 6\times 5 + 1\times 7 = 64.$$

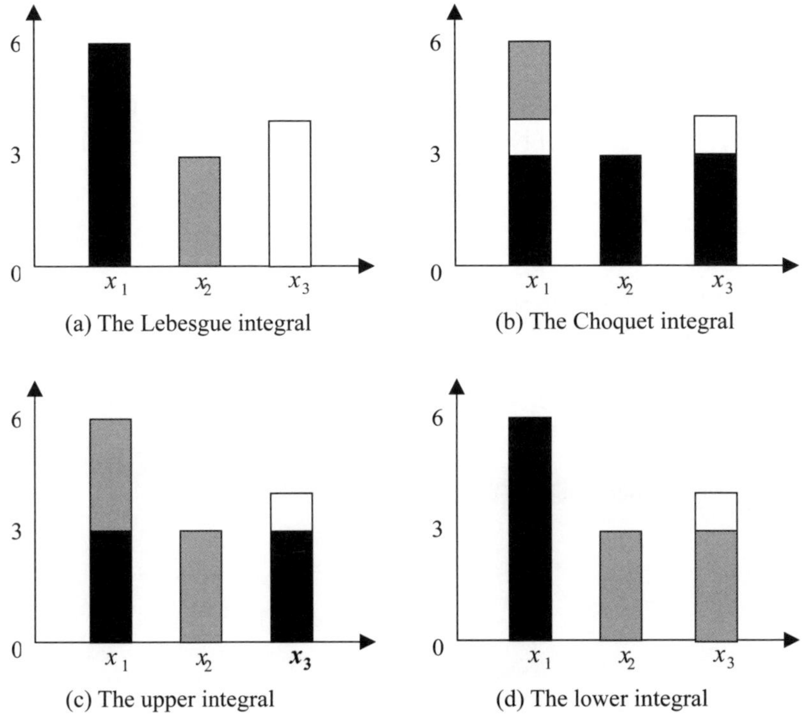

Fig. 5.8 The partitions corresponding to various types of nonlinear integrals in Example 5.21.

Up to now, we have seen that, among the integrals of a given nonnegative function with respect to a monotone measure (even a signed efficiency measure) on a finite set, the Lebesgue integral and the Choquet integral form an extreme pair in terms of the coordination among the attributes, while the upper integral and the lower integral form another extreme pair that is in terms of the integration amount. Generally, we have

$$(\mathrm{L})\int f\, d\mu \le \int f\, d\mu \le (\mathrm{U})\int f\, d\mu \tag{5.10}$$

and

$$(\mathrm{L})\int f\,d\mu \le (\mathrm{C})\int f\,d\mu \le (\mathrm{U})\int f\,d\mu \tag{5.11}$$

for any function f and signed efficiency measure μ on $(X, \mathscr{P}(X))$. Inequality (5.10) and (5.11) are also confirmed by Example 5.21.

Exercises

Exercise 5.1 Let f and g be measurable functions on measurable space $(X, \mathscr{F})$. Show that $\max(f, g)$ and $\min(f, g)$ are also measurable.

Exercise 5.2 Prove that any elementary function on measurable space $(X, \mathscr{F})$ is measurable.

Exercise 5.3 Is function f on [0, 1] defined by

$$f(x) = \begin{cases} 0 & \text{if } x = 0 \\ \frac{1}{x}\sin\frac{1}{x} & \text{if } x \in (0, 1] \end{cases}$$

Riemann integrable? Why?

Exercise 5.4 Let f be a nonnegative measurable function on measure space $(X, \mathscr{F}, \mu)$. Prove that, if the Lebaesgue integral $\int f\,d\mu = 0$, then there exists set $E \in \mathscr{F}$ with $\mu(E) = 0$ such that $f(x) = 0$ for all $x \notin E$.

Exercise 5.5 Let $X = \{x_1, x_2, x_3\}$, $\mathscr{F} = \mathscr{P}(X)$, and monotone measure μ on $\mathscr{P}(X)$ be given as

$$\mu(E) = \begin{cases} 0 & \text{if } E = \varnothing \\ 0.5 & \text{if } E = \{x_1\} \\ 0.2 & \text{if } E = \{x_2\} \\ 0.6 & \text{if } E = \{x_1, x_2\} \\ 0.4 & \text{if } E = \{x_3\} \\ 0.7 & \text{if } E = \{x_1, x_3\} \\ 0.9 & \text{if } E = \{x_2, x_3\} \\ 1 & \text{if } E = X \end{cases}.$$

Given function $f : X \to [0, \infty)$ as

$$f(x) = \begin{cases} 10 & \text{if } x = x_1 \\ 2 & \text{if } x = x_2 \\ 5 & \text{if } x = x_3 \,, \end{cases}$$

Calculate $(\mathrm{C})\int f\, d\mu$.

Exercise 5.6 Let f and g be measurable function on monotone measure space $(X, \mathscr{F}, \mu)$. Prove that, if $f \le g$, then $(\mathrm{C})\int f\, d\mu \le (\mathrm{C})\int g\, d\mu$, where the Choquet integrals are translatable.

Exercise 5.7 Let $X = \{x_1, x_2\}$, $\mathscr{F} = \mathscr{P}(X)$,

$$f(x) = \begin{cases} t_1 & \text{if } x = x_1 \\ t_2 & \text{if } x = x_2 \end{cases},$$

and monotone measure μ_1, μ_2, and μ_3 on $\mathscr{P}(X)$ be given as

$$\mu_1(E) = \begin{cases} 0 & \text{if } E = \varnothing \\ 0.5 & \text{if } E = \{x_1\} \\ 0.3 & \text{if } E = \{x_2\} \\ 1 & \text{if } E = X \end{cases},$$

$$\mu_2(E) = \begin{cases} 0 & \text{if } E = \varnothing \\ 0.5 & \text{if } E = \{x_1\} \\ 0.5 & \text{if } E = \{x_2\} \\ 1 & \text{if } E = X \end{cases},$$

and

$$\mu_3(E) = \begin{cases} 0 & \text{if } E = \varnothing \\ 0.5 & \text{if } E = \{x_1\} \\ 0.8 & \text{if } E = \{x_2\} \\ 1 & \text{if } E = X \end{cases},$$

respectively. Regarding $(\mathrm{C})\int f\, d\mu_k$ as a function of parameters t_1 and t_2, $k = 1, 2, 3$, find the contours $(\mathrm{C})\int f\, d\mu_1 = 2$, $(\mathrm{C})\int f\, d\mu_2 = 2$, and $(\mathrm{C})\int f\, d\mu_3 = 2$.

Draw their figures.

Exercise 5.8 For monotone measure μ_j, $j = 1, 2, 3$, and function *f* given in Exercise 5.7, find the track of the vertices of the contours when the value of the Choquet integral $(\mathrm{C})\int f\, d\mu$ varies in $(-\infty, \infty)$. If the integrand *f* is replaced by $w \cdot f$, where

$$w(x) = \begin{cases} 0.2 & \text{if } x = x_1 \\ 0.8 & \text{if } x = x_2 \end{cases},$$

what is the track?

Exercise 5.9 Construct an example showing $(\mathrm{L})\int 1\, d\mu \neq \mu(X)$.

Exercise 5.10 Prove property (ULIP6).

Chapter 6

Information Fusion

In decision making, when some high dimensional information (an observation for a set of several attributes) is available, usually, according to a specified decision target, people need to aggregate it into lower dimensional space (even to be a one-dimensional datum, i.e., a number) so that a reasonable decision can be easily made. Such a procedure is called *information fusion*. For different decision targets, people may choose different aggregation tools. The previous chapter provides various integrals that can serve as the aggregation tool in information fusion. This chapter presents some basic knowledge on information fusion.

6.1 Information Sources and Observations

Let $X = \{x_1, x_2, \cdots, x_n\}$ be the set of all considered information sources. Set X is taken as the universal set in our discussion. Each information source, x_i, is called an *attribute*. The numerical (may be categorical sometimes) information received from all attributes once can be regarded as a function f defined on X, that is, $f : X \to (-\infty, \infty)$ and is called an *observation* or a *record* of the attributes. They are written as $f(x_1)$, $f(x_2)$, $\cdots$, $f(x_n)$. If we observe these attributes l times, then l functions, f_1, f_2, $\cdots$, f_l, are obtained. They form a data set as follows.

x_1	x_2	$\cdots$	x_n
$f_1(x_1)$	$f_1(x_2)$	$\cdots$	$f_1(x_n)$
$f_2(x_1)$	$f_2(x_2)$	$\cdots$	$f_2(x_n)$
$\vdots$	$\vdots$		$\vdots$
$f_l(x_1)$	$f_l(x_2)$	$\cdots$	$f_l(x_n)$

In the data set, the j-th row is the j-th record of attributes $x_1, x_2, \cdots$, and x_n, $j = 1, 2, \cdots, l$. Through out the book, we assume that any data set we discussed is complete, that is, all $f_j(x_i)$, $i = 1, 2, \cdots, n$; $j = 1, 2, \cdots, l$ are available, where integer l is called the *size* of the data.

Example 6.1 A student takes three courses, Calculus, Linear Algebra, and Elementary Physics, in his first semester after enrolling a university. They are 5 credits, 3 credits, and 4 credits courses respectively. At the end of the semester, the student obtain grade *B*, *A*, and *C* correspondingly. Here, the three courses can be regarded as three information sources (attributes), denoted by x_1, x_2, and x_3 respectively. The grades *B*, *A*, and *C* are received categorical information from these three information sources. The received categorical values *B*, *A*, and *C* correspond to numerical values 3, 4, and 2 respectively. Thus, categorical (*B*, *A*, *C*) or numerical (3, 4, 2) form an observation of attributes x_1, x_2, and x_3.

Example 6.2 To investigate different types of iris, people collect its 150 flowers and measure their sepal length, sepal width, petal length, and petal width. Thus, a data set consisting of 4 attributes and 150 records is obtained (see columns 2-5 and 8-11 of Table 6.1). For instance, $f_{35}(x_2) = 3.1$, while $f_{127}(x_4) = 1.8$. It is a complete data set. As for the integers in the sixth and twelfth column in Table 6.1, they indicate the types of iris. This data set has been used to test the classifiers in a number of works including [Xu et al 2003].

Table 6.1 Iris data (from ftp://ftp.ics.uci.edu/pub/machine-learning-databases).

Sample no	Sepal length (x_1)	Sepal width (x_2)	Petal length (x_3)	Petal width (x_4)	Class	Sample no	Sepal length (x_1)	Sepal width (x_2)	Petal length (x_3)	Petal width (x_4)	Class
1	5.1	3.5	1.4	0.2	1	76	6.6	3.0	4.4	1.4	2
2	4.9	3.0	1.4	0.2	1	77	6.8	2.8	4.8	1.4	2
3	4.7	3.2	1.3	0.2	1	78	6.7	3.0	5.0	1.7	2
4	4.6.	3.1	1.5	0.2	1	79	6.0	2.9	4.5	1.5	2
5	5	3.6	1.4	0.2	1	80	5.7	2.6	3.5	1.0	2
6	5.4	3.9	1.7	0.4	1	81	5.5	2.4	3.8	1.1	2
7	4.6	3.4	1.4	0.3	1	82	5.5	2.4	3.7	1.0	2
8	5.0	3.4	1.5	0.2	1	83	5.8	2.7	3.9	1.2	2
9	4.4	2.9	1.4	0.2	1	84	6.0	2.7	5.1	1.6	2
10	4.9	3.1	1.5	0.1	1	85	5.4	3.0	4.5	1.5	2
11	5.4	3.7	1.5	0.2	1	86	6.0	3.4	4.5	1.6	2
12	4.8	3.4	1.6	0.2	1	87	6.7	3.1	4.7	1.5	2
13	4.8	3.0	1.4	0.1	1	88	6.3	2.3	4.4	1.3	2
14	4.3	3.0	1.1	0.1	1	89	5.6	3.0	4.1	1.3	2
15	5.8	4.0	1.2	0.2	1	90	5.5	2.5	4.0	1.3	2
16	5.7	4.4	1.5	0.4	1	91	5.5	2.6	4.4	1.2	2
17	5.4	3.9	1.3	0.4	1	92	6.1	3.0	4.6	1.4	2
18	5.1	3.5	1.4	0.3	1	93	5.8	2.6	4.0	1.2	2
19	5.7	3.8	1.7	0.3	1	94	5.0	2.3	3.3	1.0	2
20	5.1	3.8	1.5	0.3	1	95	5.6	2.7	4.2	1.3	2
21	5.4	3.4	1.7	0.2	1	96	5.7	3.0	4.2	1.2	2
22	5.1	3.7	1.5	0.4	1	97	5.7	2.9	4.2	1.3	2
23	4.6	3.6	1.0	0.2	1	98	6.2	2.9	4.3	1.3	2
24	5.1	3.3	1.7	0.5	1	99	5.1	2.5	3.0	1.1	2
25	4.8	3.4	1.9	0.2	1	100	5.7	2.8	4.1	1.3	2
26	5.0	3.0	1.6	0.2	1	101	6.3	3.3	6.0	2.5	3
27	5.0	3.4	1.6	0.4	1	102	5.8	2.7	5.1	1.9	3
28	5.2	3.5	1.5	0.2	1	103	7.1	3.0	5.9	2.1	3
29	5.2	3.4	1.4	0.2	1	104	6.3	2.9	5.6	1.8	3
30	4.7	3.2	1.6	0.2	1	105	6.5	3.0	5.8	2.2	3
31	4.8	3.1	1.6	0.2	1	106	7.6	3.0	6.6	2.1	3
32	5.4	3.4	1.5	0.4	1	107	4.9	2.5	4.5	1.7	3
33	5.2	4.1	1.5	0.1	1	108	7.3	2.9	6.3	1.8	3
34	5.5	4.2	1.4	0.2	1	109	6.7	2.5	5.8	1.8	3
35	4.9	3.1	1.5	0.1	1	110	7.2	3.6	6.1	2.5	3
36	5.0	3.2	1.2	0.2	1	111	6.5	3.2	5.1	2.0	3
37	5.5	3.5	1.3	0.2	1	112	6.4	2.7	5.3	1.9	3

38	4.9	3.1	1.5	0.1	1	113	6.8	3.0	5.5	2.1	3
39	4.4	3.0	1.3	0.2	1	114	5.7	2.5	5.0	2.0	3
40	5.1	3.4	1.5	0.2	1	115	5.8	2.8	5.1	2.4	3
41	5.0	3.5	1.3	0.3	1	116	6.4	3.2	5.3	2.3	3
42	4.5	2.3	1.3	0.3	1	117	6.5	3.0	5.5	1.8	3
43	4.4	3.2	1.3	0.2	1	118	7.7	3.8	6.7	2.2	3
44	5.0	3.5	1.6	0.6	1	119	7.7	2.6	6.9	2.3	3
45	5.1	3.8	1.9	0.4	1	120	6.0	2.2	5.0	1.5	3
46	4.8	3.0	1.4	0.3	1	121	6.9	3.2	5.7	2.3	3
47	5.1	3.8	1.6	0.2	1	122	5.6	2.8	4.9	2.0	3
48	4.6	3.2	1.4	0.2	1	123	7.7	2.8	6.7	2.0	3
49	5.3	3.7	1.5	0.2	1	124	6.3	2.7	4.9	1.8	3
50	5.0	3.3	1.4	0.2	1	125	6.7	3.3	5.7	2.1	3
51	7.0	3.2	4.7	1.4	2	126	7.2	3.2	6.0	1.8	3
52	6.4	3.2	4.5	1.5	2	127	6.2	2.8	4.8	1.8	3
53	6.9	3.1	4.9	1.5	2	128	6.1	3.0	4.9	1.8	3
54	5.5	2.3	4.0	1.3	2	129	6.4	2.8	5.6	2.1	3
55	6.5	2.8	4.6	1.5	2	130	7.2	3.0	5.8	1.6	3
56	5.7	2.8	4.5	1.3	2	131	7.4	2.8	6.1	1.9	3
57	6.3	3.3	4.7	1.6	2	132	7.9	3.8	6.4	2.0	3
58	4.9	2.4	3.3	1.0	2	133	6.4	2.8	5.6	2.2	3
59	6.6	2.9	4.6	1.3	2	134	6.3	2.8	5.1	1.5	3
60	5.2	2.7	3.9	1.4	2	135	6.1	2.6	5.6	1.4	3
61	5.0	2.0	3.5	1.0	2	136	7.7	3.0	6.1	2.3	3
62	5.9	3.0	4.2	1.5	2	137	6.3	3.4	5.6	2.4	3
63	6.0	2.2	4.0	1.0	2	138	6.4	3.1	5.5	1.8	3
64	6.1	2.9	4.7	1.4	2	139	6.0	3.0	4.8	1.8	3
65	5.6	2.9	3.6	1.3	2	140	6.9	3.1	5.4	2.1	3
66	6.7	3.1	4.4	1.4	2	141	6.7	3.1	5.6	2.4	3
67	5.6	3.0	4.5	1.5	2	142	6.9	3.1	5.1	2.3	3
68	5.8	2.7	4.1	1.0	2	143	5.8	2.7	5.1	1.9	3
69	6.2	2.2	4.5	1.5	2	144	6.8	3.2	5.9	2.3	3
70	5.6	2.5	3.9	1.1	2	145	6.7	3.3	5.7	2.5	3
71	5.9	3.2	4.8	1.8	2	146	6.7	3.0	5.2	2.3	3
72	6.1	2.8	4.0	1.3	2	147	6.3	2.5	5.0	1.9	3
73	6.3	2.5	4.9	1.5	2	148	6.5	3.0	5.2	2.0	3
74	6.1	2.8	4.7	1.2	2	149	6.2	3.4	5.4	2.3	3
75	6.4	2.9	4.3	1.3	2	150	5.9	3.0	5.1	1.8	3

6.2 Integrals Used as Aggregation Tools

The weighted average, or more general, the weighted sum is the most common aggregation tool used in information fusion. From Example 5.5 we have seen that the weight sum is just the Lebesgue integral of the received information with respect to the additive measure determined by the weights.

Example 6.3 Recalling Example 6.1 and using its data that contains only one record, the question is what the current GPA of the student is. Since the total credits of the courses the student takes is 12, we have

$$\text{GPA} = \sum_{i=1}^{3} f(x_i) \cdot w_i = 3 \times \frac{5}{12} + 4 \times \frac{3}{12} + 2 \times \frac{4}{12} = \frac{35}{12} \approx 2.92 ,$$

where

$$f(x) = \begin{cases} 3 & \text{if } x = x_1 \\ 4 & \text{if } x = x_2 \\ 2 & \text{if } x = x_3 \end{cases}$$

is the grade record that the student obtained, and $w_1 = 5, w_2 = 3, w_3 = 4$ denote the credits of three courses. If μ is the classical additive measure determined by $\mu(\{x_i\}) = w_i$, $i = 1, 2, 3$, or, directly,

$$\mu(E) = \sum_{i | x_i \in E} w_i ,$$

then the GPA is just the Lebesgue integral of function f with respect to μ, i.e., $\text{GPA} = \int f \, d\mu = 35/12$.

Generally, given n attributes $x_1, x_2, \cdots, x_n$ and an observation f for fusing, the aggregation value, as another attribute (the *target attribute*) denoted by y, is a functional of f. The relation between them can be regarded as an input-output system, in which the values of n attributes $x_1, x_2, \cdots$, and x_n are the input and the value of the target attribute is the output. Example 6.3 presents a linear system, where y depends on $x_1, x_2, \cdots$, and x_n linearly. In such a system, each attribute makes contributions towards the target linearly and independently. Here the independency means that there is no interaction among the contribution rates from attributes towards the target. Due to such an independency, the joint contribution from $x_1, x_2, \cdots$, and x_n towards the target is a simple sum of their individual contributions and, as discussed in Section 5.3, can be expressed as the Lebesgue integral on a discrete set (the universal set).

In real information fusion problems, many input-output systems do not have such linearity due to the interaction among the contributions of all given attributes towards the target. Such an interaction has been seen in the discussion of monotone measures in Section 4.3. It is totally different from the concept of correlationship in statistics. Let us recall Example 4.12.

Example 6.4 In Example 4.12, the universal set $X = \{x_1, x_2, x_3\}$ consists of three workers x_1, x_2, and x_3, and monotone measure

$$\mu(E) = \begin{cases} 0 & \text{if } E = \emptyset \\ 5 & \text{if } E = \{x_1\} \\ 6 & \text{if } E = \{x_2\} \\ 14 & \text{if } E = \{x_1, x_2\} \\ 7 & \text{if } E = \{x_3\} \\ 13 & \text{if } E = \{x_1, x_3\} \\ 9 & \text{if } E = \{x_2, x_3\} \\ 17 & \text{if } E = X \end{cases}$$

represents the individual and joint contribution rates from these workers towards the target "total amount of produced toys". As shown in Example 4.12, the nonadditivity of μ describes the interaction among

Table 6.2 Data of working times in Example 6.4.

x_1	x_2	x_3
8	0	4
4	4	4
0	8	4
2	6	4
6	2	4

the contribution rates from these three workers towards the total amount of their produced toys. The working time (the number of hours for working) of a worker in a specified day is a record of that worker. If 5 records have been made for these three workers during some week as shown in Table 6.2, then, in statistics, the correlation coefficient of x_1 and x_2 is $r_{12} = -1$, while the correlation coefficient of x_1 and x_3 is $r_{13} = 0$. They describe the relation between the appearing record values of two attributes involved. So, the interaction among the contribution rates from all given attributes towards the target is totally different from the concept of correlationship in statistics.

The following example of synthetic evaluation and decision making shows that the method of classical weighted sum (or say, the Lebesgue integral with respect to an additive measure), as an aggregation tool in information fusion, fails when the above-mentioned interaction cannot be ignored.

Example 6.5 There are three used TV sets on sale at the same price. We want to evaluate the global quality of TV sets based on an estimation on two factors "picture" and "sound", denoted by x_1 and x_2, separately to each TV set, and then choose the best to buy. Now, factors "picture" and "sound" are attributes, while the global quality is the target.

First, we assume that the weights of two factors are $w_1 = 0.7$ and $w_2 = 0.3$ respectively. Now, for each factor and each TV set, an adjudicator gives the scores in Table 6.3.

Using the method of weighted average, we get synthetic evaluations of the three TV sets:

Table 6.3 The scores of TV sets in Example 6.5.

TV Set No.	x_1 (picture)	x_2 (sound)
1	1	0
2	0	1
3	0.45	0.45

$$E_{w1} = w_1 \times 1 + w_2 \times 0 = 0.7, \quad E_{w2} = w_1 \times 0 + w_2 \times 1 = 0.3,$$
$$E_{w3} = w_1 \times 0.45 + w_2 \times 0.45 = 0.45.$$

According to these results, the first TV set is the best. Such a result is hardly acceptable since it does not agree with our intuition: A TV set without any sound is not practical at all, even though it has an excellent picture. It is significant to realize that the cause of this counterintuitive result is not an improper choice of the weights. For example, if we chose $w_1 = 0.4$ and $w_2 = 0.6$, we would have obtained $E_{w1} = 0.4$, $E_{w2} = 0.6$, and $E_{w3} = 0.45$. Now, the second TV set is identified as the best one, which is also counterintuitive: A TV set with good sound but no picture is not a real TV set, but just a radio. We may conclude that, according to our intuition, the third TV set should be identified as the best one: among the three TV sets, only the third one is really practical, even though neither picture nor sound are perfect. Unfortunately, when using the method of weighted average, no choice of the weights would lead to this expected result under the given scores.

The crux of this problem is that the method of weighted mean is based on an implicit assumption that the factors $x_1, x_2, \cdots$, and x_n are "independently contribute to the global quality". That is, their effects are viewed as additive. This, however, is not justifiable in some real problems. In this example, the joint importance of picture and sound is much higher then the sum of importance associated with picture and sound alone. If we adopt a monotone measure to characterize the importance of the two factors and, relevantly, use the Choquet integral as a synthetic evaluator of the quality of the three TV sets, a satisfactory result may be obtained. For instance, given the importance $\mu(\{x_1\}) = 0.3$,

$\mu(\{x_2\}) = 0.1$, $\mu(X) = 1$, and $\mu(\varnothing) = 0$ as a monotone measure, and using the Choquet integral, we obtain the following synthetic evaluations:

$$E_{c1} = (C)\int f_1\, d\mu = 0 \times 1 + 1 \times 0.3 = 0.3\,,$$

$$E_{c2} = (C)\int f_2\, d\mu = 0 \times 1 + 1 \times 0.1 = 0.1\,,$$

$$E_{c3} = (C)\int f_3\, d\mu = 0.45 \times 1 + 0 \times 0.45 = 0.45\,,$$

where

$$f_1(x) = \begin{cases} 1 & \text{if } x = x_1 \\ 0 & \text{if } x = x_2 \end{cases},$$

$$f_2(x) = \begin{cases} 0 & \text{if } x = x_1 \\ 1 & \text{if } x = x_2 \end{cases},$$

and

$$f_3(x) = \begin{cases} 0.45 & \text{if } x = x_1 \\ 0.45 & \text{if } x = x_2 \end{cases}.$$

Hence, we get a reasonable conclusion: the third TV set is the best, which agrees with our intuition. When some other type of nonlinear integrals is chosen, a similar result can be obtained.

In fact, not only the Choquet integral with respect to a monotone measure (even to a signed efficiency measure) can be used, but also the other types, such as the upper integral and the lower integral, can be adopted as aggregation tools in information fusion. Example 5.21 shows that four different types of integrals can be used in information fusion, where the input information is the working time (hours) of workers and the output (the target) is the total amount of produced toys during the working time. Example 5.21 also explains the intuitive meaning of these types of integrals and the relevant results. The most important points in information fusion are:

(1) using a nonadditive set function to describe the interaction among the contribution rates from attributes towards the target, and
(2) choosing a suitable nonlinear integral as an aggregation tool.

6.3 Uncertainty Associated with Set Functions

Let $X = \{x_1, x_2, \cdots, x_n\}$. In this section, we assume that set function μ is a nontrivial monotone measure on $(X, \mathscr{P}(X))$. Here, the word "nontrivial" means that there exists at least one set $E \subseteq X$ such that $\mu(E) > 0$. Since μ is monotone, this requirement is simply equivalent to $\mu(X) > 0$. We have seen from Section 6.2 that, due to the nonadditivity of μ, for a given nonnegative function f, different types of integrals may result in different aggregation values. This may be viewed as the uncertainty associated with monotone measure μ. Since the upper integral and the lower integral are two extremes in regard to the aggregation value, we may numerically estimate the uncertainty associated with the monotone measure by the difference of the upper integral and the lower integral.

Definition 6.1 Given a monotone measure μ on $(X, \mathscr{P}(X))$, the *degree of the relative uncertainty* associated with μ is defined by

$$\gamma_\mu = \frac{(\mathrm{U})\int 1\, d\mu - (\mathrm{L})\int 1\, d\mu}{\mu(X)}.$$

It is evident that, when μ is a classical measure, the upper integral coincides with the lower integral and, therefore, $\gamma_\mu = 0$.

Theorem 6.1 For any monotone measure μ on $(X, \mathscr{P}(X))$, $0 \le \gamma_\mu \le n$.

Proof. On the one hand, from

$$(\mathrm{U})\int 1\,d\mu \geq (\mathrm{L})\int 1\,d\mu\,,$$

we obtain

$$\gamma_\mu = \frac{(\mathrm{U})\int 1\,d\mu - (\mathrm{L})\int 1\,d\mu}{\mu(X)} \geq 0\,.$$

On the other hand, from Theorem 5.9 and Definition 6.1, since $(\mathrm{L})\int 1\,d\mu \geq 0$, we have

$$\begin{aligned}
\gamma_\mu &= \frac{(\mathrm{U})\int 1\,d\mu - (\mathrm{L})\int 1\,d\mu}{\mu(X)} \\
&\leq \frac{(\mathrm{U})\int 1\,d\mu}{\mu(X)} \\
&\leq \frac{n \cdot \mu(X)}{\mu(X)} \\
&= n\,.
\end{aligned}$$

□

To present an estimate formula for the difference between the upper integral and the lower integral of a given nonnegative function, we need the following lemma.

Lemma 6.1 For any given monotone measure μ and a bounded nonnegative function *f*,

$$(\mathrm{U})\int f\,d\mu - (\mathrm{L})\int f\,d\mu \leq (\mathrm{U})\int c\,d\mu - (\mathrm{L})\int c\,d\mu\,,$$

where *c* may be any upper bound of *f*.

Proof. From the expressions of the upper integral and the lower integral on a finite set given in Section 5.5, we know that there are $\lambda_j \ge 0$ and $\nu_j \ge 0$, $j = 1, 2, \cdots, 2^n - 1$, satisfying

$$\sum_{j|x \in E_j \subseteq X} \lambda_j = f(x) \text{ and } \sum_{j|x \in E_j \subseteq X} \nu_j = f(x)$$

for every $x \in X$, such that

$$(\mathrm{U})\int f\, d\mu = \sum_{j=1}^{2^n - 1} \lambda_j \cdot \mu(E_j)$$

and

$$(\mathrm{L})\int f\, d\mu = \sum_{j=1}^{2^n - 1} \nu_j \cdot \mu(E_j).$$

For nonnegative function $c - f$, we can find $\lambda'_j \ge 0$ and $\nu'_j \ge 0$, $j = 1, 2, \cdots, 2^n - 1$, satisfying

$$\sum_{j|x \in E_j \subseteq X} \lambda'_j = c - f(x) \text{ and } \sum_{j|x \in E_j \subseteq X} \nu'_j = c - f(x)$$

for every $x \in X$, such that

$$(\mathrm{U})\int (c - f)\, d\mu = \sum_{j=1}^{2^n - 1} \lambda'_j \cdot \mu(E_j)$$

and

$$(\mathrm{L})\int (c - f)\, d\mu = \sum_{j=1}^{2^n - 1} \nu'_j \cdot \mu(E_j).$$

Since

$$\sum_{j|x \in E_j \subseteq X} \lambda_j + \sum_{j|x \in E_j \subseteq X} \lambda'_j = \sum_{j|x \in E_j \subseteq X} (\lambda_j + \lambda'_j) = c$$

and

$$\sum_{j|x\in E_j\subseteq X} \nu_j + \sum_{j|x\in E_j\subseteq X} \nu'_j = \sum_{j|x\in E_j\subseteq X} (\nu_j + \nu'_j) = c\,,$$

we have

$$\sum_{j=1}^{2^n-1}(\lambda_j + \lambda'_j)\cdot \mu(E_j) \le (\mathrm{U})\int c\, d\mu$$

and

$$\sum_{j=1}^{2^n-1}(\nu_j + \nu'_j)\cdot \mu(E_j) \ge (\mathrm{L})\int c\, d\mu\,.$$

Thus, from

$$(\mathrm{L})\int (c-f)\, d\mu \le (\mathrm{U})\int (c-f)\, d\mu\,,$$

we obtain

$$\begin{aligned}
(\mathrm{U})\int f\, d\mu - (\mathrm{L})\int f\, d\mu &= \sum_{j=1}^{2^n-1}\lambda_j\cdot\mu(E_j) - \sum_{j=1}^{2^n-1}\nu_j\cdot\mu(E_j)\\
&\le \sum_{j=1}^{2^n-1}\lambda_j\cdot\mu(E_j) + \sum_{j=1}^{2^n-1}\lambda'_j\cdot\mu(E_j) - \sum_{j=1}^{2^n-1}\nu_j\cdot\mu(E_j) - \sum_{j=1}^{2^n-1}\nu'_j\cdot\mu(E_j)\\
&= \sum_{j=1}^{2^n-1}(\lambda_j+\lambda'_j)\cdot\mu(E_j) - \sum_{j=1}^{2^n-1}(\nu_j+\nu'_j)\cdot\mu(E_j)\\
&\le (\mathrm{U})\int c\, d\mu - (\mathrm{L})\int c\, d\mu.
\end{aligned}$$

□

Theorem 6.2 Given a monotone measure μ on $(X, \mathscr{P}(X))$ and any nonnegative function f on X, we have

$$0 \le (\mathrm{U})\int f\, d\mu - (\mathrm{L})\int f\, d\mu \le \gamma_\mu \cdot \mu(X)\cdot \max_{x\in X} f(x)\,.$$

Proof. Let $c = \max_{x\in X} f(x)$. From the definition of γ_μ, Lemma 6.1, and properties (ULIP3) and (ULIP5) given in Section 5.5, we have

$$0 \le (\mathrm{U})\int f\, d\mu - (\mathrm{L})\int f\, d\mu \le (\mathrm{U})\int c\, d\mu - (\mathrm{L})\int c\, d\mu$$
$$\le c \cdot [(\mathrm{U})\int 1\, d\mu - (\mathrm{L})\int 1\, d\mu] = \gamma_\mu \cdot \mu(X) \cdot \max_{x\in X} f(x).$$

□

Example 6.6 The data and some results in Examples 5.21 are used here. It is easy to obtain $(\mathrm{U})\int 1\, d\mu = 21$ and $(\mathrm{L})\int 1\, d\mu = 14$. Then we have

$$\gamma_\mu = \frac{21-14}{17} = \frac{7}{17}.$$

From $(\mathrm{U})\int f\, d\mu - (\mathrm{L})\int f\, d\mu = 88 - 64 = 22$, $\max_{x\in X} f(x) = 6$, and $\mu(X) = 17$, Theorem 6.2 is verified: $22 \le (7/17)\times 6 \times 17 = 42$.

Theorem 6.2 can be used to estimate the uncertainty associated with the monotone measure in an aggregation process if the coordination manner is unknown.

6.4 The Inverse Problem of Information Fusion

From Section 6.2, we have seen that nonlinear integrals can be used as aggregation tools in information fusion. Given the set of information sources $X = \{x_1, x_2, \cdots, x_n\}$, any discussed nonlinear integral with respect to a known signed efficiency measure μ defined on $\mathscr{P}(X)$ can be regarded as a nonlinear n-input one-output system. Once the input, an observation of attributes $x_1, x_2, \cdots, x_n$ is available, denoted by f as a function defined on X, an output can be obtained by calculating the value of the integral of f with respect to μ. The output is the value of the fusion target and is denoted by y, that is, $y = (r)\int f\, d\mu$, where (r) is the indicator of the adopted type of nonlinear integral. Signed efficiency measure μ consists of 2^n values with individual corresponding set in $\mathscr{P}(X)$, among them $\mu(\emptyset) = 0$ is fixed. Thus, the input-output system has $2^n - 1$ parameters. The system is identified with the type of the nonlinear integral and these $2^n - 1$ structural parameters that are represented by μ. So, the information fusion is a procedure of finding the value of the target y when the values of attributes $x_1, x_2, \cdots, x_n$, the type of the nonlinear integral, and the values of $2^n - 1$ parameters are all known. Here, the nonlinear integral is used as an aggregation tool.

The most interesting one of the inverse problems to information fusion is to find the values of signed efficiency measure μ when the type of the nonlinear integral is fixed and the values of observations from attributes $x_1, x_2, \cdots, x_n$ as well as the corresponding values of the target are known. That is, knowing some pairs of the input and the output of the above-mentioned system with a given type of aggregation tool, we want to estimate the structural parameters of the system. Obviously, only knowing one pair of the input and the output is not sufficient to obtain a reasonable estimation of the parameters. A data set with large enough size is necessary for the purpose of estimating the values of signed efficiency measure μ. Thus, the data set, generally, has a form as follows.

x_1	x_2	$\cdots$	x_n	y
$f_1(x_1)$	$f_1(x_2)$	$\cdots$	$f_1(x_n)$	y_1
$f_2(x_1)$	$f_2(x_2)$	$\cdots$	$f_2(x_n)$	y_2
$\vdots$	$\vdots$		$\vdots$	$\vdots$
$f_l(x_1)$	$f_l(x_2)$	$\cdots$	$f_l(x_n)$	y_l

Table 6.1 shows an example for such a form of data sets, which we use in Chapter 10.

If an n-input one-output system is linear, the system can be expressed as

$$y = a_0 + a_1 x_1 + a_2 x_2 + \cdots + a_n x_n ,$$

where $x_1, x_2, \cdots, x_n$ are the input and y is the output, while coefficients $a_0, a_1, \cdots, a_n$, which identify the linear system, are parameters of the system. Once the above-mentioned data set is available, the value of $a_0, a_1, \cdots, a_n$ can be estimated by the least square method: find the values of $a_0, a_1, \cdots, a_n$ such that

$$\hat{\sigma}^2 = \frac{1}{l}\sum_{j=1}^{l}[y_j - (a_0 + a_1 f_j(x_1) + a_2 f_j(x_2) + \cdots + a_n f_j(x_n))]^2 \qquad (6.1)$$

is minimized. This optimization is just the inverse problem of the information fusion when the aggregation tool is a linear weighted sum. A linear algebraic method can be used for solving this quadratic optimization problem (see Section 9.1).

However, when the aggregation tool in the information fusion is a nonlinear integral $(r)\int f\,d\mu$, the input-output system can be expressed as

$$y = (r)\int f\,d\mu\,,$$

where the values of μ are unknown parameters. Now the most interesting inverse problem of the information fusion is: once the above-mentioned data set is available, how to estimate the values of μ. Similar to the linear case, these unknown parameters should be determined by minimizing

$$\hat{\sigma}^2 = \frac{1}{l}\sum_{j=1}^{l}[y_j - (r)\int f_j\,d\mu]^2\,. \qquad (6.2)$$

Unfortunately, the linear algebraic method fails for solving such a nonlinear optimization problem generally due to the nonlinearity of the r-integral. In this case, we have to use some numerical methods to obtain an approximately optional solution. These numerical computation methods, called soft computing techniques, are discussed in the next chapter.

Chapter 7

Optimization and Soft Computing

There are a number of traditional methods for solving optimization problems. In this chapter, two new numerical methods, genetic algorithm and pseudo gradient search, are discussed. Both of them are called soft computing techniques. Through either of them, usually, an approximately optimal solution may be obtained. Unlike the most traditional and common optimization methods, these methods avoid requiring some rather strong conditions, such as the differentiability to the objective function in the problem. A hybrid method of genetic algorithm and pseudo gradient search is more effective for solving optimization problems.

7.1 Basic Concepts of Optimization

Consider m numerical variables $t_1, t_2, \cdots, t_m$ and a target $z = z(t_1, t_2, \cdots, t_m)$ defined on a subset D of the m-dimensional Euclidian space R^m. Given $F \subseteq D$, we want to find a point $(t_1^*, t_2^*, \cdots, t_m^*)$ subject to the restriction $(t_1^*, t_2^*, \cdots, t_m^*) \in F$ such that $z(t_1^*, t_2^*, \cdots, t_m^*)$ is the smallest value of z in F, that is, $z(t_1^*, t_2^*, \cdots, t_m^*) \le z(t_1, t_2, \cdots, t_m)$ for any $(t_1, t_2, \cdots, t_m) \in F$. This is a *minimization* problem, where z is called the *objective function* (or *target*), set F is called the *feasible region*, any point $(t_1, t_2, \cdots, t_m) \in F$ is called a *feasible point*, point $(t_1^*, t_2^*, \cdots, t_m^*)$ is called the *global minimizer* (simply called *minimizer* if there is no confusion), denoted by $\arg\min z(t_1, t_2, \cdots, t_m)$, of the minimization problem, and value $z(t_1^*, t_2^*, \cdots, t_m^*)$ is called the *minimum* of z in F. A point $(t_1', t_2', \cdots, t_m') \in F$ is called a *local minimizer*, if there exists an open set O containing $(t'_1, t'_2, \cdots, t'_m)$ such that $z(t'_1, t'_2, \cdots, t'_m) \le z(t_1, t_2, \cdots, t_m)$ for any point $(t_1, t_2, \cdots, t_m) \in O \cap F$. Similarly, we have concepts of *maximization*, *global maximizer*, *maximum*, and *local maximizer* correspondingly. Both of minimization and maximization are called *optimization*.

Formally, a minimization problem in m-dimensional Euclidian space can, usually, be expressed as follows.

$$\begin{array}{ll} \text{min} & z = z(t_1, t_2, \cdots, t_m) \\ \text{subject to} & g_k(t_1, t_2, \cdots, t_m) \ge \alpha_k, \quad k = 1, 2, \cdots, r \end{array} \tag{7.1}$$

where α_k, $k = 1, 2, \cdots, r$, are constants and the inequalities describe the restriction, that is, the feasible region of the minimization problem is

$$F = \{(t_1, t_2, \cdots, t_m) \mid g_k(t_1, t_2, \cdots, t_m) \ge \alpha_k, k = 1, 2, \cdots, r\},$$

in which, r is a nonnegative integer. When $r = 0$, the minimization problem is said to be *unconstrained*. For any maximization problem, $\max z = z(t_1, t_2, \cdots, t_m)$ can be rewritten as $\min z' = -z(t_1, t_2, \cdots, t_m)$, that is, it can be converted into a form of (7.1). Furthermore, if there is some inequality with "less than or equal to", say, $g(t_1, t_2, \cdots, t_m) \le \alpha$ in the restriction, we can rewrite it as $-g(t_1, t_2, \cdots, t_m) \ge -\alpha$. Finally, if some restriction condition is an equality, e.g., $g(t_1, t_2, \cdots, t_m) = \alpha$, then we can separate it into two inequalities $g(t_1, t_2, \cdots, t_m) \ge \alpha$ and $-g(t_1, t_2, \cdots, t_m) \ge -\alpha$. Thus, we call the form shown in (7.1) the *standard form* of an optimization problem. We should note that inequalities "greater than" and "less than" are never used in restrictions for optimization problem because they will lead to no solution usually.

In (7.1), if all functions z and g's are linear, then the optimization problem is called a *linear programming* problem; otherwise, i.e., if at least one of functions z and g's is nonlinear, it is called a *nonlinear programming* problem.

7.2 Genetic Algorithms

Genetic algorithm is one of the soft computing techniques used in optimization. It is a global parallel random search method. Genetic algorithm mimics the natural evolution process of a species in a given environment to obtain an optimal (or approximate optimal) solution after a large number of generations.

Suppose that a given optimization problem (say, a minimization problem) involves m variables (unknown parameters) with feasible region F. A genetic algorithm for solving the optimization problem may contain the following components.

(1) **Encoding genes**. Using suitable transformations, the m unknown parameters are respectively represented by m binary bit strings, which are randomly and independently generated obeying the uniform distribution, such that the feasible region F is well covered with a reasonable distribution. Each binary bit string is called a *gene*. The length of each binary bit string is determined according to the required precision of the corresponding parameter in the solution of the optimization problem. For example, if the required precision of a parameter is 10^{-3}, then 10 bits are needed in the corresponding gene since $2^{-9} > 10^{-3} > 2^{-10}$. Adopting a real coding to replace the binary coding is also workable.

(2) **Chromosomes**. According to a fixed order, link the m genes to form a *chromosome*. It is also a string of bits. Its length is the sum of the lengths of all genes. A chromosome represents a candidate of the set consisting of values of all unknown parameters in the optimization problem. So, each chromosome should be in F.

(3) **Population**. The *population* is a set of chromosomes. The number of chromosomes in the population is called the *size*, denoted by s, of the population. The size is usually a large positive even integer, such as 100 or 500. The population is initialized by generating chromosomes randomly. Keeping the size, the population will be renewed in the evolution process. Relative to the population, each chromosome is called an *individual*.

(4) **Fitness function**. According to the goal of the optimization, choose a criterion and, by which, a suitable *fitness function* is constructed to measure the goodness of each chromosome. Usually, the fitness is a straight monotone function of the value of the objective function. Decode each individual if it is necessary and then calculate its fitness.

(5) **Genetic operators**. Design several *genetic operators* for producing new chromosomes using some existing chromosomes that are selected as the parents. The following common operators are suggested to be used, though the user may design new genetic operators according to the need of the given optimization problem.

  (a) Two-point crossover. The *crossover* is a binary operator. For each two selected chromosomes, randomly choose two points (each point here is a location between two successive bits) to separate each of them into three pieces, and then interchange their middle piece to form two new chromosomes (see Figure 7.1(a)).

  (b) Three-point mutation. The three-point *mutation* is a unary operator. For each selected chromosome, randomly select three bits and toggle their 0 and 1 to obtain a new chromosome (see Figure 7.1(b)).

  (c) Two-point realignment. The two-point *realignment* is also a unary operator. For each selected chromosome, randomly choose two points (the same as in (a), each point here is also a location between two successive bits) to separate it into three pieces, and then realign them in a randomly selected order and direction (see Figure 7.1(c)). There are totally 48 different ways to the realignment. They have the same chance to be selected, i.e., each way has a chance of 1/48 to be selected.

(6) **Selecting parents**. According to the individuals' fitness, randomly select individuals as *parents*. The higher the fitness is, the more chance to be selected, which is known as fitness proportionate selection.

(7) **Produce offspring**. According to a selected probability distribution, randomly choose a genetic operator to produce individuals from the selected parents.

(8) **Renewing the population**. After producing a certain number, e. g., the same as the initial population size *s*, of new individuals, add them into the population. Then, according to the fitness, select the best *s* individuals to form a new population, called a new *generation*.

(9) **Recurrence and stopping**. Based on the new population, recycle procedure (4)-(8) until the stopping condition is satisfied. The *stopping condition* may be chosen from anyone of the following or their combinations.

   (a) The number of generations (or the number of produced individuals) reaches a given positive integer.

   (b) After producing a given number of generations (or individuals), the fitness of the best individual in the population has not been changed (or not been significantly improved).

   (c) The population is identical.

   (d) The value of the objective function for the best individual in the current population falls in a given small region, such as the error being smaller than a given small positive number ε.

(10) **Outputting the result**. Once stopping condition satisfied, output the best individual (after decoding) in the current population.

A general flow chat of the genetic algorithm is shown in Figure 7.2. Since the genetic algorithm is a global search method, we need not worry about falling in a valley of local minimizer. This is the advantage of the genetic algorithm. However, the search process of a genetic algorithm is usually rather slow, and there even exists a phenomenon so-called *prematurity*, that is, after several generation, the fitness of the best individual has not been significantly improved, though it is still a little far from the global minimizer. To speed up the search procedure and reduce the prematurity, we may adopt two additional approaches as follows. One is to keep the diversity of the population, that is, to select some individuals not very good but with much different feature from the good individuals in the new generation to avoid the population being identical or similar too early. Another is to set up some adaptive mechanisms in the search procedure, i.e., some involved probability distributions are dynamically adjusted. For example, at the beginning of the search procedure, the probability of taking mutation may be smaller than later, and the probability of choosing higher bits may be larger than lower bits for mutation and vice versa towards the end.

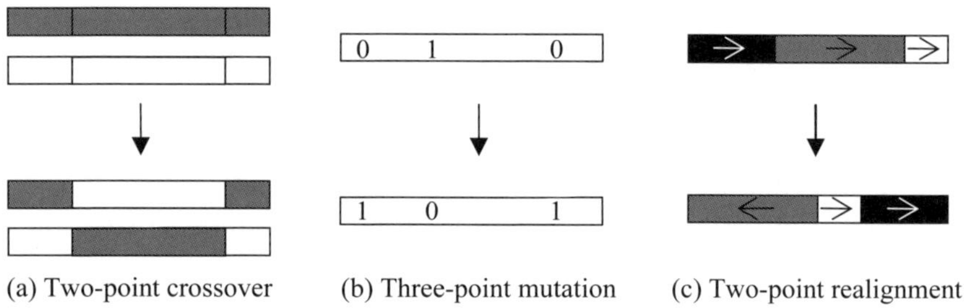

Fig. 7.1 Illustration of genetic operators.

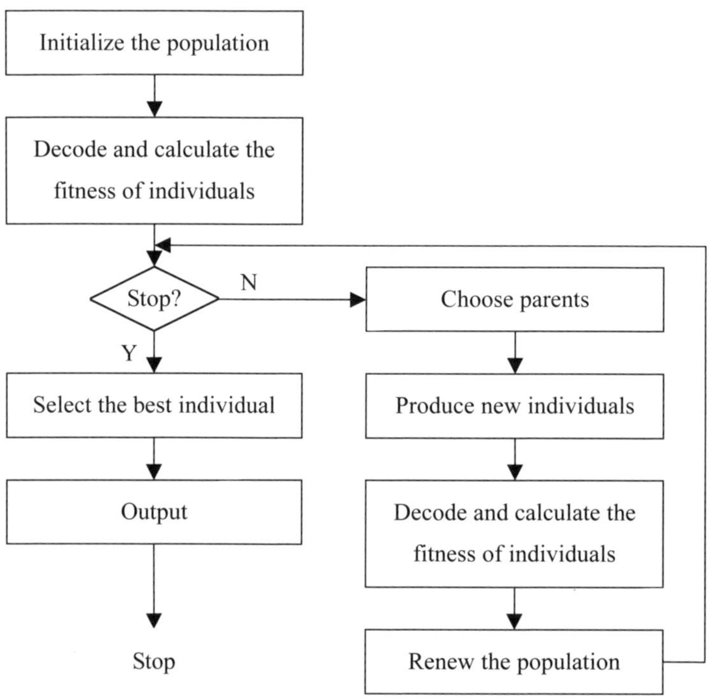

Fig. 7.2 The flowchart of genetic algorithms.

7.3 Pseudo Gradient Search

For solving optimization problems, the well known gradient search method requires that the objective function is differentiable with respect to the unknown parameters. However, many optimization problems do not satisfy such a requirement. To get a more general search method, a natural idea is to replace the differential with a difference, which can be obtained through a learning procedure, to determine a prefer search direction, called *pseudo gradient*. Then, along this direction, a much better point can be found by another learning procedure. This point is used as the starting point of the next iteration. Repeating this procedure leads to obtaining an approximate local extremer (minimizer or maximizer). In the procedures, the feasibility should be always kept. Such a search method is called the *pseudo gradient search*. Like the gradient search method, it is a local search method.

Let $z = z(t_1, t_2, \cdots, t_m)$ be the objective function in the given minimization problem with feasible region *F*. The search space is *m*-dimensional. Starting from a given *initial point* $t^{(0)} = (t_1^{(0)}, t_2^{(0)}, \cdots, t_m^{(0)}) \in F$, we want to search for a minimizer $t^* = (t_1^*, t_2^*, \cdots, t_m^*) \in F$. The procedure of the pseudo gradient search can be described by the following steps.

(1) Choose an initial point $t^{(0)} = (t_1^{(0)}, t_2^{(0)}, \cdots, t_m^{(0)}) \in F$.

(2) Initialize $\delta = \delta_0 / 2\sqrt{m}$ and $\alpha = 2$, where $\delta_0 > 0$ is the required solution precision.

(3) For each $j = 1, 2, \cdots, m$, calculate

$$\Delta_{j+} = z(t_1^{(0)}, t_2^{(0)}, \cdots, t_j^{(0)}, \cdots, t_m^{(0)}) - z(t_1^{(0)}, t_2^{(0)}, \cdots, t_j^{(0)} + \delta, \cdots, t_m^{(0)})$$

if $(t_1^{(0)}, t_2^{(0)}, \cdots, t_j^{(0)} + \delta, \cdots, t_m^{(0)}) \in F$; otherwise, let $\Delta_{j+} = 0$. Similarly, calculate

$$\Delta_{j-} = z(t_1^{(0)}, t_2^{(0)}, \cdots, t_j^{(0)}, \cdots, t_m^{(0)}) - z(t_1^{(0)}, t_2^{(0)}, \cdots, t_j^{(0)} - \delta, \cdots, t_m^{(0)})$$

if $(t_1^{(0)}, t_2^{(0)}, \cdots, t_j^{(0)} - \delta, \cdots, t_m^{(0)}) \in F$; otherwise, let $\Delta_{j-} = 0$.

(4) Let

$$\Delta_j = \begin{cases} \max[\Delta_{j+}, \Delta_{j-}, 0] & \text{if } \Delta_{j+} \geq \Delta_{j-} \\ -\max[\Delta_{j+}, \Delta_{j-}, 0] & \text{otherwise} \end{cases}, \quad j = 1, 2, \cdots, m.$$

If $\Delta_j = 0$ for all $j = 1, 2, \cdots, m$, then go to step (5); otherwise, go to step (6).

(5) Calculate

$$\Delta_{jk++} = \begin{cases} 0 & \text{if } (t_1^{(0)}, t_2^{(0)}, \cdots, t_j^{(0)} + \delta, \cdots, t_k^{(0)} + \delta, \cdots, t_m^{(0)}) \notin F \\ z(t_1^{(0)}, t_2^{(0)}, \cdots, t_j^{(0)}, \cdots, t_k^{(0)}, \cdots, t_m^{(0)}) & \\ \quad - z(t_1^{(0)}, t_2^{(0)}, \cdots, t_j^{(0)} + \delta, \cdots, t_k^{(0)} + \delta, \cdots, t_m^{(0)}) & \text{otherwise} \end{cases},$$

$$\Delta_{jk-+} = \begin{cases} 0 & \text{if } (t_1^{(0)}, t_2^{(0)}, \cdots, t_j^{(0)} - \delta, \cdots, t_k^{(0)} + \delta, \cdots, t_m^{(0)}) \notin F \\ z(t_1^{(0)}, t_2^{(0)}, \cdots, t_j^{(0)}, \cdots, t_k^{(0)}, \cdots, t_m^{(0)}) & \\ \quad - z(t_1^{(0)}, t_2^{(0)}, \cdots, t_j^{(0)} - \delta, \cdots, t_k^{(0)} + \delta, \cdots, t_m^{(0)}) & \text{otherwise} \end{cases},$$

$$\Delta_{jk+-} = \begin{cases} 0 & \text{if } (t_1^{(0)}, t_2^{(0)}, \cdots, t_j^{(0)} + \delta, \cdots, t_k^{(0)} - \delta, \cdots, t_m^{(0)}) \notin F \\ z(t_1^{(0)}, t_2^{(0)}, \cdots, t_j^{(0)}, \cdots, t_k^{(0)}, \cdots, t_m^{(0)}) & \\ \quad - z(t_1^{(0)}, t_2^{(0)}, \cdots, t_j^{(0)} + \delta, \cdots, t_k^{(0)} - \delta, \cdots, t_m^{(0)}) & \text{otherwise} \end{cases},$$

$$\Delta_{jk--} = \begin{cases} 0 & \text{if } (t_1^{(0)}, t_2^{(0)}, \cdots, t_j^{(0)} - \delta, \cdots, t_k^{(0)} - \delta, \cdots, t_m^{(0)}) \notin F \\ z(t_1^{(0)}, t_2^{(0)}, \cdots, t_j^{(0)}, \cdots, t_k^{(0)}, \cdots, t_m^{(0)}) & \\ \quad - z(t_1^{(0)}, t_2^{(0)}, \cdots, t_j^{(0)} - \delta, \cdots, t_k^{(0)} - \delta, \cdots, t_m^{(0)}) & \text{otherwise} \end{cases},$$

and $\Delta_{jk\pm\mp} = \max[\Delta_{jk++}, \Delta_{jk-+}, \Delta_{jk+-}, \Delta_{jk--}]$ for $j, k = 1, 2, \cdots, m$ with $j < k$, where $\pm\mp$ is one of $++$, $-+$, $+-$, and $--$ such that the maximum is reached. Find

$$\max_{j,k=1,2,\cdots,m;\ j<k} \Delta_{jk\pm\mp}.$$

If it is not positive, then go to (10); otherwise, find

$$(j_0, k_0) = \arg\max_{j,k=1,2,\cdots,m;\ j<k} \Delta_{jk\pm\mp},$$

where argmax denotes the maximizer, and use

$$(t_1^{(0)}, t_2^{(0)}, \cdots, t_{j_0}^{(0)} \pm \delta, \cdots, t_{k_0}^{(0)} \mp \delta, \cdots, t_m^{(0)})$$

to replace

$$t^{(0)} = (t_1^{(0)}, t_2^{(0)}, \cdots, t_{j_0}^{(0)}, \cdots, t_{k_0}^{(0)}, \cdots, t_m^{(0)}),$$

where $\mp$ and $\pm$ are either + or – that are recorded in the subscript of $\Delta_{jk\pm\mp}$. Then go back to (2).

(6) Form the pseudo gradient direction $\Delta = (\Delta_1, \Delta_2, \cdots, \Delta_m)$ and calculate $|\Delta| = [\sum_{j=1}^{m} (\Delta_j)^2]^{1/2}$.
(7) Replace δ by $\alpha\delta$. If the new $\delta \le \delta_0 / 2\sqrt{m}$, then go back to (2); otherwise, from point $t^{(0)}$ and along direction Δ, take a step with length δ to reach point t^*, that is,

$$t^* = (t_1^{(0)} + \frac{\delta\Delta_1}{|\Delta|}, t_2^{(0)} + \frac{\delta\Delta_2}{|\Delta|}, \cdots, t_m^{(0)} + \frac{\delta\Delta_m}{|\Delta|}).$$

(8) If $t^* \notin F$, then go to (9). Otherwise, calculate $z(t^*)$ and check whether $z(t^*) \le z(t^{(0)})$ or not. If yes, replace $t^{(0)}$ by t^*, then go back to (7); if no, go to (9).
(9) Let $\alpha = 1/2$ and go back to (7).
(10) Stop. Output $t^{(0)}$ and some required information on the search procedure.

In the algorithm above, each iteration formed by steps (2)-(9) has a search direction described by the pseudo gradient. Along with this direction, the initial point (or the point obtained at the end of the previous iteration) is updated by a much good point, though not being an approximate best point in this direction. We may add some search steps such that the updated point is an approximate best point in this direction. However, it is not necessary since its benefit will be totally covered by the next iteration in the current algorithm.

In comparison with genetic algorithms, the pseudo gradient search is much faster. The quick search speed is the advantage of the pseudo gradient search method. However, like other local search methods, it is hard to avoid falling into a local extremer or obtaining a saddle. Even the user does not know whether the found point is a global extremer approximately or not. This is the weakness of the pseudo gradient search method.

7.4 A Hybrid Search Method

As we have seen that the genetic algorithm is a global random search method while the pseudo gradient search is a local search method. The advantage of the former is no risk of falling in a local extremer, while its weakness is the slow search speed and the risk of prematurity. Unlike the former, the latter has a fast search speed but cannot avoid the risk of falling in a local extremer or staying at a point close to a saddle of the objective function.

A natural idea to improve these two search methods is to combine them together. Once an optimization problem is given, we may first use a genetic algorithm to find a relatively good point in the feasible region, then, taking this point as the initial point, turn to a pseudo gradient search to continue the search procedure. Usually, we may obtain a satisfactory approximate optimal solution.

In such a hybrid search procedure, we may appropriately slacken the stopping conditions in the part of the genetic algorithm to reduce the total running time of the program. Some successful experiments using the combination of a genetic algorithm and a special iterative search, which is a simplified version of the pseudo gradient search, are presented in [Spilde and Wang 2005].

Chapter 8

Identification of Set Functions

Identification of set functions is a technique, based on given data, to determine a set function satisfying some given requirements. There are two different kinds of identification. One is to construct a specified type of set function, such as a λ-measure or a belief measure, based on a given set function with a revision as slight as possible. Another is, regarding a given type of nonlinear integral as an input-output system, to determine the involved set function, such as a monotone measure or a signed efficiency measure, based on some observations of the input and the corresponding output of the system. The former can be called the *revising* for set functions that are discussed in Sections 8.1-8.3, while the latter is called the *fitting* and discussed in Sections 8.4-8.7.

From this chapter through the book, the universal set X is always finite. Let $X = \{x_1, x_2, \cdots, x_n\}$.

8.1 Identification of λ-Measures

Given set function $\mu : \mathscr{P}(X) \to [0, \infty)$ with $\mu(X) > 0$, we want to find a λ-measure g: $\mathscr{P}(X) \to [0, \infty)$, such that $g(X) = \mu(X)$ and the difference between g and μ is as small as possible. To be consistent with the other methods, we may choose the squared error as the difference, that is, determine λ-measure g such that

$$\sum_{E \subset X} [g(E) - \mu(E)]^2$$

is minimized. Adopting the sum of absolute difference is also feasible since a genetic algorithm discussed in section 7.2 may be used to search the numerical optimal solution and, therefore, we need not worry about the differentiability of the objective function in the optimization problem. However, we still prefer to use the total squared error as the objective function in the optimization such that a comparison with traditional methods in some special cases can be easily made if any.

A soft computing technique, for instance, a genetic algorithm, may be adopted to solve this identification problem. From Section 4.4, we have known that, when $g(X) > 0$ is given, a λ-measure g is identified by its values at all singletons, $g(\{x_i\})$, $i = 1, 2, \cdots, n$. Noting that each binary bit string represents a real number in [0, 1) and any λ-measure is monotone, we may directly use a gene to represent $g(\{x_i\})/g(X)$ for each $i = 1, 2, \cdots, n$. Let $g(X) = b$ and $g(\{x_i\})/g(X) = g_i$, $i = 1, 2, \cdots, n$. The objective function may be taken as

$$z(g_1, g_2, \cdots, g_n) = \sum_{E \subset X} [g(E) - \mu(E)]^2 , \tag{8.1}$$

where

$$g(E) = \frac{1}{\lambda}\Big(\prod_{i | x_i \in E} [1 + \lambda b g_i] - 1\Big), \tag{8.2}$$

in which λ is the unique feasible solution of equation

$$1 + b\lambda = \prod_{i=1}^{n} (1 + b g_i \lambda) . \tag{8.3}$$

Once a chromosome $(g_1, g_2, \cdots, g_n)$ is available, we should check whether only one $i \in \{1, 2, \cdots, n\}$ such that $g_i \neq 0$. If yes, this chromosome should be abandoned. Otherwise, according to the genes g_i, $i = 1, 2, \cdots, n$, we can calculate the value of λ by solving equation (8.3) and, then, calculate $g(E)$ for each set $E \subset X$ through (8.2).

Finally, from (8.1), we may obtain the value of the objective function z. As for the fitness of the chromosome, we may choose, for example,

$$f(z)=\frac{1}{z+0.1}.$$

After obtaining the approximate optimal value of each g_i, $i=1,2,\cdots,n$, we need still use (8.2) and (8.3) to calculate the values of the λ-measure g and the corresponding value of λ.

This identification problem can be generalized as follows. Assume that we have l observations, which are not accurate, for the values of a λ-measure g: $\mathscr{P}(X)\to[0,\infty)$, where l is a positive integer. These observation are denoted as l set functions μ_j: $\mathscr{P}(X)\to[0,\infty)$, $j=1,2,\cdots,l$. Now we want to optimally determine λ-measure g in the following meaning: the total squared error

$$z(g_1,g_2,\cdots,g_n)=\sum_{j=1}^{l}\sum_{E\subset X}[g(E)-\mu_j(E)]^2 \tag{8.4}$$

is minimized, where $g(E)$ has the same meaning as in (8.2) with (8.3). The algorithm is totally the same as the original one except the objective function shown in (8.1) being replaced by (8.4).

8.2 Identification of Belief Measures

Similar to the generalized model of the identification of λ-measures discussed in the previous section, given l set function μ_j: $\mathscr{P}(X)\to[0,1]$ with $\mu_j(\varnothing)=0$ and $\mu_j(X)=1$, $j=1,2,\cdots,l$, now we want to determine a belief measure bel: $\mathscr{P}(X)\to[0,1]$, such that the difference between bel and all μ_j is as small as possible, for example, such that

$$z=\sum_{j=1}^{l}\sum_{E\subset X}[bel(E)-\mu_j(E)]^2$$

is minimized.

This minimization problem can also be implemented by the genetic algorithm discussed in Section 7.2. From Section 4.7, we know that a belief measure is uniquely determined from its corresponding basic probability assignment m: $\mathscr{P}(X) \to [0, 1]$ with $m(\emptyset) = 0$ and

$$\sum_{E \subseteq X} m(E) = 1 \tag{8.5}$$

by

$$bel(E) = \sum_{F \subseteq E} m(F). \tag{8.6}$$

So, we just need to arrange the values of m as the genes in chromosomes. Considering the constraint (8.5) and each gene is a real number in [0, 1), after decoding, we may let chromosome $(g_1, g_2, \cdots, g_{2^n-1})$ represent a basic probability assignment m by $m(\{x_1\}) = g_1/c$, $m(\{x_2\}) = g_2/c$, $m(\{x_1, x_2\}) = g_3/c$, $m(\{x_3\}) = g_4/c$, $m(\{x_1, x_3\}) = g_5/c$, $\cdots$, and $m(X) = g_{2^n-1}/c$, where $c = \sum_{j=1}^{2^n-1} g_j$. For each chromosome created in the genetic algorithm, using (8.6), we may convert it to a belief measure. Then the value of z can be calculated. The fitness of a chromosome can be chosen in the same way as shown in Section 8.1.

8.3 Identification of Monotone Measures

Given set function $\mu: \mathscr{P}(X) \to [0, \infty)$ with $\max\{\mu(E) \mid E \subseteq X\} > 0$, we want to find a monotone measure ν: $\mathscr{P}(X) \to [0, \infty)$, such that the total squared error

$$z = \sum_{E \subseteq X} [\nu(E) - \mu(E)]^2 \tag{8.7}$$

is minimized. Under this criterion, the optimal solution ν cannot be obtained by only reordering the values of μ generally, even if $\mu(X) \geq \mu(E)$ for all $E \in \mathscr{P}(X)$. This can be seen from the following counterexample.

Example 8.1 Let $X = \{x_1, x_2, x_3\}$ and

$$\mu(E) = \begin{cases} 0 & \text{if } E = \varnothing \\ 0.7 & \text{if } E = \{x_1\} \\ 0.2 & \text{if } E = \{x_2\} \\ 0.5 & \text{if } E = \{x_1, x_2\} \\ 0.4 & \text{if } E = \{x_3\} \\ 0.8 & \text{if } E = \{x_1, x_3\} \\ 0.9 & \text{if } E = \{x_2, x_3\} \\ 1 & \text{if } E = X \end{cases} .$$

Set function μ is nonnegative. It is not monotone since $\mu\{x_1, x_2\} < \mu\{x_1\}$. If we construct set function ν from μ by exchanging the values of $\mu\{x_1, x_2\}$ and $\mu\{x_1\}$, i.e., let

$$\nu(E) = \begin{cases} 0 & \text{if } E = \varnothing \\ 0.5 & \text{if } E = \{x_1\} \\ 0.2 & \text{if } E = \{x_2\} \\ 0.7 & \text{if } E = \{x_1, x_2\} \\ 0.4 & \text{if } E = \{x_3\} \\ 0.8 & \text{if } E = \{x_1, x_3\} \\ 0.9 & \text{if } E = \{x_2, x_3\} \\ 1 & \text{if } E = X \end{cases} ,$$

then ν is a monotone revision of μ. However, it is not the optimal monotone revision of μ under the criterion of minimizing (8.7). In fact, according (8.7), the total squared error of ν is $z(\nu) = 0.2^2 + 0.2^2 = 0.08$. If we take

$$\nu^*(E)=\begin{cases}0 & \text{if } E=\varnothing\\ 0.6 & \text{if } E=\{x_1\}\\ 0.2 & \text{if } E=\{x_2\}\\ 0.6 & \text{if } E=\{x_1,x_2\}\\ 0.4 & \text{if } E=\{x_3\}\\ 0.8 & \text{if } E=\{x_1,x_3\}\\ 0.9 & \text{if } E=\{x_2,x_3\}\\ 1 & \text{if } E=X\end{cases}$$

as a monotone revision of μ, then its total squared error is $z(\nu^*)=0.1^2+0.1^2=0.02$, which is much smaller than $z(\nu)$.

In Example 8.1, ν^* is obtained only via taking the average of a pair of two μ's values, which violates the required monotonicity, to replace the pair. To obtain a monotone revision of a given nonnegative set function vanishing at the empty set, in general case, is not so simple. It is convenient to use a soft computing technique, such as a genetic algorithm, with an embedded reordering algorithm to obtain an approximate numerical optimal solution for this identification, especially, when the generalized model that is similar to the ones discussed in Sections 8.1 and 8.2 is considered. The genetic algorithm with an embedded reordering algorithm can be also used for identification of monotone measures based on an input-output nonlinear integral system that is discussed in Section 8.6. One of the advantages of using soft computing techniques is that the algorithm, except the formula of the fitness function, does not depend on the choice of the optimization criterion.

The generalized identification model for monotone measures is expressed as follows.

Given l rough observations (records) for a monotone measure $\nu:\mathscr{P}(X)\to[0,\infty)$ with $\nu(X)>0$, denoted by μ_j, $j=1,2,\cdots,l$, where each μ_j is a nonnegative set function defined on $\mathscr{P}(X)$ and l is a positive integer, we want to find a monotone measure ν on $\mathscr{P}(X)$ such that

$$z = \sum_{j=1}^{l} \sum_{E \subseteq X} [\nu(E) - \mu_j(E)]^2 \tag{8.8}$$

is minimized. Equation (8.7) can be regarded as a special case of (8.8) with $l = 1$.

We assume that not all observations are trivial, i.e., not all values of every observed set function μ_j are zeros. A genetic algorithm now is used to solve this identification problem. It consists of two parts: the main algorithm and the embedded reordering algorithm.

8.3.1 *Main algorithm*

The genetic algorithm shown in Section 7.2 can be adopted with a few adjustments as follows. Each chromosome now consists of $2^n - 1$ genes, denoted as $g_1, g_2, \cdots, g_{2^n-1}$, representing the values of a set function at all nonempty subsets of *X* respectively. Each gene consists λ bits and is decoded as a real number in [0,1]. The target of the optimization is the total squared error *z* shown in (8.8). The decoding formulas are

$$\mu(\{x_1\}) = \frac{g_1}{1 - g_1},$$

$$\mu(\{x_2\}) = \frac{g_2}{1 - g_2},$$

$$\mu(\{x_1, x_2\}) = \frac{g_3}{1 - g_3},$$

$$\mu(\{x_3\}) = \frac{g_4}{1 - g_4},$$

$$\mu(\{x_1, x_3\}) = \frac{g_5}{1 - g_5},$$

$$\vdots$$

$$\mu(X) = \frac{g_{2^n - 1}}{1 - g_{2^n - 1}}.$$

After decoding each chromosome, the following reordering algorithm should be used to convert the nonnegative set function μ to a monotone measure v. Then, according (8.8), calculate the total squared error for v.

8.3.2 *Reordering algorithm*

Assume that set function $\mu : \mathscr{P}(X) \to [0, \infty)$ with $\mu(\varnothing) = 0$ is given.

(1) According to the lattice structure of the power set $\mathscr{P}(X)$ (see Figure 8.1 when $n = 4$), divide all 2^n subsets of X into $n+1$ layers: the empty set is at layer 0 (the bottom layer); all singletons are at layer 1; all sets consisting of two points are at layer 2; $\cdots$; the universal set is at layer n (the top layer). That is, any set consisting of k points is arranged in layer k, $k = 0, 1, 2, \cdots, n$. The class of all sets at layer k is denoted as $\mathscr{L}_k$.

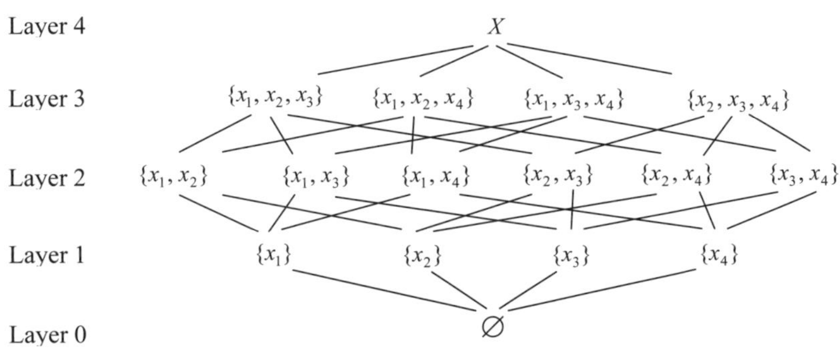

Fig. 8.1 The lattice structure for the power set of a universal set with 4 attributes.

(2) Initialize $k = 1$.

(3) For layer k, rearrange these $C(n, k)$ sets according to their values of μ in a nonincreasing order, denoted as $\{ E_{kj} | j = 1, 2, \cdots, C(n, k)\}$.

(4) Starting from $j = 1$, find the set $E^*_{kj} = \arg\min\{\mu(E) \mid E \supset E_{kj}\}$ from layer $k+1$, i.e., E^*_{kj} is the set with the smallest value of μ among all sets that include E_{kj} in $\mathscr{L}_{k+1}$. Exchange the values of $\mu(E_{kj})$ and $\mu(E^*_{kj})$ if $\mu(E_{kj}) > \mu(E^*_{kj})$; otherwise, they are unchanged. Then do this for the next j and continue this procedure until $j = C(n, k) - 1$.

(5) Add 1 to k and check whether $k = n$. If yes, go to (6); if no, go to (3).

(6) Check whether for at least one value of k the exchange in step (4) has been done. If yes, go to (2); if no, go back to the main algorithm.

Once the embedded reordering algorithm finishes and diverts back to the main algorithm, the actual set function μ has been reordered to be a monotone measure v on $\mathscr{P}(X)$. The complexity of the reordering algorithm above can still be reduced a little. In an iteration beginning from (2), if the first $r+1$ layers (form layer 0 to layer r) are not involved for any exchange, then k can be initialized with r in the next iteration.

If all considered set functions, including the monotone measure v and its observations μ_j, $j = 1, 2, \cdots, l$, are regular, i.e.,

$$v(X) = \mu_1(X) = \mu_2(X) = \cdots = \mu_l(X) = 1,$$

then each chromosome only needs $2^n - 2$ genes. In this case, the total squared error (8.8) is reduced to be

$$z = \sum_{j=1}^{l} \sum_{E \subset X} [v(E) - \mu_j(E)]^2. \tag{8.9}$$

8.4 Identification of Signed Efficiency Measures by a Genetic Algorithm

Now, we want to obtain a set function $\mu : \mathscr{P}(X) \to (-\infty, \infty)$ with no special requirement except vanishing at the empty set, i.e., $\mu(\emptyset) = 0$. This means that μ is a signed efficiency measure. Of course, in this case, since there is not enough restriction on μ that can be used to form optimization criterion, the identification model must be essentially different from those discussed in Sections 8.1-8.3 that depend on the characters (e.g., the λ-rule) of the target set function. Thus, what type of the data set we should have and what corresponding optimization criterion we should adopt are new problems being faced.

In this case, an input-output system described by a nonlinear integral can be adopted to determine a signed efficiency measure. We only need a data set consisting of sufficiently many input-output records for the system. This is just an inverse problem of the information fusion, where a nonlinear integral aggregates the received information, discussed in Section 6.4.

Suppose that an r-integral is taken as the aggregation tool, where r indicates the type of nonlinear integrals shown in the previous chapters and is known. The input-output system can be expressed as

$$y = (r)\int f \, d\mu ,$$

where $f : X \to [0, \infty)$ (or $f : X \to (-\infty, \infty)$ such that the nonlinear integral $(r)\int f \, d\mu$ is well defined) is the input, y is the output and μ is a signed efficiency measure. The system is fully described by the type of the nonlinear integral and the involved signed efficiency measure. Now, only μ is unknown. Based on a given data set

$$\begin{array}{ccccc}
x_1 & x_2 & \cdots & x_n & y \\
\hline
f_1(x_1) & f_1(x_2) & \cdots & f_1(x_n) & y_1 \\
f_2(x_1) & f_2(x_2) & \cdots & f_2(x_n) & y_2 \\
\vdots & \vdots & & \vdots & \vdots \\
f_l(x_1) & f_l(x_2) & \cdots & f_l(x_n) & y_l
\end{array} \tag{8.10}$$

where $f_j(x_i)$ is the j-th record (observation) of attribute x_i and l is the data size that should be larger than $2^n - 1$, we may determine the unknown values of μ according to the criterion that

$$z = \sum_{j=1}^{l} [y_j - (r)\int f_j \, d\mu]^2$$

is minimized. To solve this optimization problem, a genetic algorithm can be adopted. In the genetic algorithm, each chromosome consists of $2^n - 1$ binarily coded genes $g_1, g_2, \cdots, g_{2^n - 1}$, and each gene represents a real number in [0, 1). Except those chromosomes involving at least one zero gene, the values of μ corresponding to the chromosome are calculated from

$$\mu(\{x_1\}) = \frac{(g_1 - 0.5)}{g_1(1 - g_1)},$$

$$\mu(\{x_2\}) = \frac{(g_2 - 0.5)}{g_2(1 - g_2)},$$

$$\mu(\{x_1, x_2\}) = \frac{(g_3 - 0.5)}{g_3(1 - g_3)},$$

$$\mu(\{x_3\}) = \frac{(g_4 - 0.5)}{g_4(1 - g_4)},$$

$$\mu(\{x_1, x_3\}) = \frac{(g_5 - 0.5)}{g_5(1 - g_5)},$$

$$\vdots$$

$$\mu(X) = \frac{(g_{2^n-1} - 0.5)}{g_{2^n-1}(1 - g_{2^n-1})}.$$

Any chromosome involving zero genes should be abandoned. The other components of the genetic algorithm are similar to those in Sections 8.1-8.3. Running such a genetic algorithm, an approximate numerical optimal solution of the signed efficiency measure can be obtained.

From this identification problem, we can see the advantage of using genetic algorithms. To solve an inverse problem of information fusion, no matter how complex the aggregation tool is, we can use genetic algorithms to search the optimal solution whenever the aggregation is computable. Thus, solving an inverse problem is converted to repeatedly solving the original problems, the aggregations. As shown in this section, to determine a signed efficiency measure, we just need to implement the aggregation of the input and compare the output with the corresponding given values repeatedly for various input-output pairs, by which, the set function is then optimized.

8.5 Identification of Signed Efficiency Measures by the Pseudo Gradient Search

The identification problem of signed efficiency measures shown in Section 8.4 can also be solved by the pseudo gradient search method discussed in Section 7.3.

Let the same data set adopted in Section 8.3 be available. To determine a signed efficiency measure $\mu: \mathscr{P}(X) \to (-\infty, \infty)$ with $\mu(\emptyset) = 0$, we need to find the values of $2^n - 1$ unknown parameters $\mu_1, \mu_2, \cdots, \mu_{2^n-1}$. So, the search space is $(2^n - 1)$-dimensional, that is, the integer m used in Section 7.3 is $2^n - 1$. The objective function is still

$$z = \sum_{j=1}^{l} [y_j - (r)\int f_j \, d\mu]^2$$

that should be minimized, where "r" at the front of the integral indicates the chosen type of the nonlinear integral. Usually, the size of the data set should not be less than the number of the unknown parameters, i.e., $l \geq 2^n - 1$. Otherwise, we may face the case that there are infinitely many optimal signed efficiency measures with $z = 0$.

In this optimization problem, since the feasible region F is the whole $(2^n - 1)$-dimensional Euclidean space $R^{2^n - 1}$ and the shape of the objective function is not very complex, using the pseudo gradient search shown in Section 7.3 is convenient. The algorithm of the pseudo gradient search can now be simplified as follows.

(1) Choose $t^{(0)} = (t_1^{(0)}, t_2^{(0)}, \cdots, t_{2^n-1}^{(0)})$ as the initial point.

(2) Initialize $\delta = \delta_0 / 2\sqrt{2^n - 1}$ and $\alpha = 2$, where $\delta_0 > 0$ is the required solution precision.

(3) For each $j = 1, 2, \cdots, 2^n - 1$, calculate

$$\Delta_{j+} = z(t_1^{(0)}, t_2^{(0)}, \cdots, t_j^{(0)}, \cdots, t_{2^n-1}^{(0)}) - z(t_1^{(0)}, t_2^{(0)}, \cdots, t_j^{(0)} + \delta, \cdots, t_{2^n-1}^{(0)}) .$$

Similarly, calculate

$$\Delta_{j-} = z(t_1^{(0)}, t_2^{(0)}, \cdots, t_j^{(0)}, \cdots, t_{2^n-1}^{(0)}) - z(t_1^{(0)}, t_2^{(0)}, \cdots, t_j^{(0)} - \delta, \cdots, t_{2^n-1}^{(0)}) .$$

(4) Let

$$\Delta_j = \begin{cases} \max[\Delta_{j+}, \Delta_{j-}, 0] & \text{if } \Delta_{j+} \geq \Delta_{j-} \\ -\max[\Delta_{j+}, \Delta_{j-}, 0] & \text{otherwise} \end{cases}, \quad j = 1, 2, \cdots, 2^n - 1 .$$

If $\Delta_j = 0$ for all $j = 1, 2, \cdots, 2^n - 1$, then go to step (9); otherwise, go to step (5).

(5) Form the pseudo gradient direction $\Delta = (\Delta_1, \Delta_2, \cdots, \Delta_{2^n-1})$ and calculate $|\Delta| = [\sum_{j=1}^{2^n-1} (\Delta_j)^2]^{1/2}$.

(6) Replace δ by $\alpha\delta$. If the new $\delta \le \delta_0 / 2\sqrt{2^n - 1}$, then go back to (2); otherwise, from point $t^{(0)}$ and along direction Δ, take a step with length δ to reach point t^*, that is,

$$t^* = (t_1^{(0)} + \frac{\delta\Delta_1}{|\Delta|}, t_2^{(0)} + \frac{\delta\Delta_2}{|\Delta|}, \cdots, t_m^{(0)} + \frac{\delta\Delta_{2^n-1}}{|\Delta|}) .$$

(7) Calculate $z(t^*)$ and check whether $z(t^*) \le z(t^{(0)})$ or not. If yes, replace $t^{(0)}$ by t^*, then go back to (6); if no, go to (8).

(8) Let $\alpha = 1/2$ and go back to (6).

(9) Stop. Output $t^{(0)}$ and some required information on the search procedure.

Similar to the situation in Section 7.3, in the algorithm above, each iteration formed by steps (2)-(8) has a search direction described by the pseudo gradient. Along this direction, the initial point (or the point obtained at the end of the previous iteration) is updated by a much better point, though not being an approximate best point in this direction. Adding some search steps such that the updated point is an approximate best point in this direction is possible. However, it is not necessary since its benefit will be totally covered by the next iteration in the current relatively simple algorithm.

8.6 Identification of Signed Efficiency Measures Based on the Choquet Integral by an Algebraic Method

When the Choquet integral is chosen as the aggregation tool in the input-output system, the identification of signed efficiency measure becomes easy to be solved. Due to the advantage of linear expression (5.5) with (5.6), an algebraic method can be used to solve this identification problem with a precise solution. In fact, if data set (8.10) is available, the identification of signed efficiency measure μ can be expressed as an optimization problem of determining set function $\mu : \mathscr{P}(X) \to (-\infty, \infty)$ with $\mu(\emptyset) = 0$ such that

$$z = \sum_{j=1}^{l} [y_j - (\mathrm{C})\int f_j \, d\mu]^2$$

is minimized. By using (5.5) and (5.6), we can see that its optimal solution is just the least square solution of linear system

$$\sum_{j=1}^{2^n - 1} z_{kj} \mu_j = y_k, \ k = 1, 2, \cdots, l,$$

where

$$z_{kj} = \begin{cases} \min\limits_{i | \mathrm{frc}(j/2^i) \in [1/2, 1)} f_k(x_i) - \max\limits_{i : \mathrm{frc}(j/2^i) \in [0, 1/2)} f_k(x_i), & \text{if it is} > 0 \text{ or } j = 2^n - 1 \\ 0, & \text{otherwise} \end{cases}$$

for $k = 1, 2, \cdots, l$; $j = 1, 2, \cdots, 2^n - 1$. Each z_{kj} can be calculated from the given data set (8.10) in advance. Hence, the precise solution of the optimization problem is

$$\begin{bmatrix} \mu_1 \\ \mu_2 \\ \vdots \\ \mu_m \end{bmatrix} = (\mathbf{Z}^T \mathbf{Z})^{-1} \mathbf{Z}^T \mathbf{Y},$$

where $m = 2^n - 1$,

$$\mathbf{Z}=\begin{bmatrix} 1 & z_{11} & \cdots & z_{1m} \\ 1 & z_{21} & \cdots & z_{2m} \\ \vdots & & & \\ 1 & z_{l1} & \cdots & z_{lm} \end{bmatrix},$$

$$\mathbf{Z}^T=\begin{bmatrix} 1 & 1 & \cdots & 1 \\ z_{11} & z_{21} & \cdots & z_{l1} \\ \vdots & & & \\ z_{1m} & z_{2m} & \cdots & z_{lm} \end{bmatrix},$$

i.e., the transposed matrix of **Z**, and

$$\mathbf{Y}=\begin{bmatrix} y_1 \\ y_2 \\ \vdots \\ y_l \end{bmatrix}.$$

The details on the least square solution of a linear system are shown in Section 9.1.

8.7 Identification of Monotone Measures Based on *r*-Integrals by a Genetic Algorithm

To determine a monotone measure based on an input-output system consisting of a nonlinear integral, the difference from determining a signed efficiency measure discussed in Sections 8.4-8.6 is the restriction on the set function. Now, the determined set function should be nonnegative and monotone. Based on an existing data set with the same form as (8.10) shown in Section 8.4, to determine a monotone measure defined on the power set of a given finite universal set optimally, the genetic algorithm discussed in Section 8.4 with the same objective function

$$z=\sum_{j=1}^{l}[y_j-(r)\int f_j\,d\mu]^2$$

may still be used. However, the decode formulas should be taken as those in Section 8.3 to keep the nonnegativity. The reordering algorithm, where a max-min strategy is adopted to reduce the computation complexity, shown in Section 8.3 should also be embedded to guarantee the monotonicity for the obtained set functions represented by chromosomes.

We should notice that, even the Choquet integral is chosen as the nonlinear integral in this identification problem, the algebraic method shown in Section 8.6 cannot be used here. The least square solution obtained in Section 8.6 violates the nonnegativity and the monotonicity generally. After adjusting the solution to be nonnegative and reordering it to be monotone, the result will usually no longer have the least squared error. So, we have to take a relatively complex soft computing technique to obtain an approximately optimal solution for this identification problem.

Chapter 9

Multiregression Based on Nonlinear Integrals

Regression is one of the major techniques in statistics and data mining. Based on a set of observations involving a number of variables (attributes), regression provides an approach to determine the unknown parameters in an input-output system and, therefore, find a linear or nonlinear relation between one variable (output) and the other variables (input). It can be regarded as a generalization of identification of signed efficiency measure (or classical signed measure). Once the relation on how one variable (*target attribute*) depends on other variables (*predictive attributes*) is known, we can use it to predict the value of the target variable if a new observation of predictive variables has been obtained in some way.

9.1 Linear Multiregression

Suppose that there are $n+1$ attributes: $x_1, x_2, \cdots, x_n$ and y in a database. We want to know how y depends on these x's. Using the same setup as before, let $X = \{x_1, x_2, \cdots, x_n\}$. Regarding y as a random variable, the simplest relation between y and $x_1, x_2, \cdots, x_n$ is a *linear regression* expressed as

$$y = a_0 + a_1x_1 + a_2x_2 + \cdots + a_nx_n + N(0, \sigma^2), \qquad (9.1)$$

where $a_0, a_1, a_2, \cdots,$ and a_n are unknown constants, called *regression coefficients*, and $N(0, \sigma^2)$ is a normally distributed random variable with mean 0 and variance σ^2. The variance σ^2 is also unknown. Each observation (record) of attributes $x_1, x_2, \cdots, x_n$ is a function defined on *X*. Using such a model needs a basic assumption that there is no interaction among the contributions from $x_1, x_2, \cdots, x_n$ towards target *y*. Under this assumption, to estimate these regression coefficients as well as σ^2, we need a data set consisting of sufficiently many observations of $x_1, x_2, \cdots, x_n$ and corresponding values of *y*. It has the same form as shown in (8.10):

x_1	x_2	$\cdots$	x_n	y
$f_1(x_1)$	$f_1(x_2)$	$\cdots$	$f_1(x_n)$	y_1
$f_2(x_1)$	$f_2(x_2)$	$\cdots$	$f_2(x_n)$	y_2
$\vdots$	$\vdots$		$\vdots$	$\vdots$
$f_l(x_1)$	$f_l(x_2)$	$\cdots$	$f_l(x_n)$	y_l

where the size of the data set, *l*, should be much larger than *n* (say, $l \geq 5n$) to avoid possible over fitting. Once a proper data set is available, we may use the least square method to determine regression coefficients $a_0, a_1, a_2, \cdots,$ and a_n, that is, by minimizing the total squared error:

$$z = \sum_{j=1}^{l} [y_j - (a_0 + a_1 x_{j1} + a_2 x_{j2} + \cdots + a_n x_{jn})]^2 , \tag{9.2}$$

where $x_{ji} = f_j(x_i)$, $j = 1, 2, \cdots, l$, $i = 1, 2, \cdots, n$, we may obtain an estimation of unknown parameters $a_0, a_1, a_2, \cdots,$ and a_n. Usually, we denote

$$E(y_j) = a_0 + a_1 x_{j1} + a_2 x_{j2} + \cdots + a_n x_{jn}$$

and call it the *expected value* of y_j.

To solve the optimization problem with objective function expressed by (9.2), some knowledge in calculus and linear algebra is needed. Regarding z as a function of variables $a_0, a_1, a_2, \cdots,$ and a_n, it is quadratic and concave up. So, its minimum exists uniquely. A necessary (in fact, also sufficient) condition for $a_0, a_1, a_2, \cdots,$ and a_n being the minimizer is

$$\frac{\partial z}{\partial a_i} = 0, \quad i = 0, 1, 2, \cdots, n,$$

that is,

$$-2\sum_{j=1}^{l}[y_j - (a_0 + a_1 x_{j1} + a_2 x_{j2} + \cdots + a_n x_{jn})] = 0,$$

$$-2\sum_{j=1}^{l} x_{ji}[y_j - (a_0 + a_1 x_{j1} + a_2 x_{j2} + \cdots + a_n x_{jn})] = 0, \quad i = 1, 2, \cdots, n.$$

Rearranging them, we obtain a system of linear equations

$$\begin{aligned}
a_0 + \sum_{j=1}^{l} x_{j1} a_1 + \sum_{j=1}^{l} x_{j2} a_2 + \cdots + \sum_{j=1}^{l} x_{jn} a_n &= \sum_{j=1}^{l} y_j \\
\sum_{j=1}^{l} x_{j1} a_0 + \sum_{j=1}^{l} x_{j1} x_{j1} a_1 + \sum_{j=1}^{l} x_{j1} x_{j2} a_2 + \cdots + \sum_{j=1}^{l} x_{j1} x_{jn} a_n &= \sum_{j=1}^{l} x_{j1} y_j \\
\sum_{j=1}^{l} x_{j2} a_0 + \sum_{j=1}^{l} x_{j2} x_{j1} a_1 + \sum_{j=1}^{l} x_{j2} x_{j2} a_2 + \cdots + \sum_{j=1}^{l} x_{j2} x_{jn} a_n &= \sum_{j=1}^{l} x_{j2} y_j \qquad (9.3) \\
&\vdots \\
\sum_{j=1}^{l} x_{jn} a_0 + \sum_{j=1}^{l} x_{jn} x_{j1} a_1 + \sum_{j=1}^{l} x_{jn} x_{j2} a_2 + \cdots + \sum_{j=1}^{l} x_{jn} x_{jn} a_n &= \sum_{j=1}^{l} x_{jn} y_j,
\end{aligned}$$

where $a_0, a_1, a_2, \cdots,$ and a_n are unknown variables of the system of linear equations. Denoting

$$\mathbf{X}=\begin{bmatrix} 1 & x_{11} & \cdots & x_{1n} \\ 1 & x_{21} & \cdots & x_{2n} \\ \vdots & & & \\ 1 & x_{l1} & \cdots & x_{ln} \end{bmatrix}, \quad \mathbf{Y}=\begin{bmatrix} y_1 \\ y_2 \\ \vdots \\ y_l \end{bmatrix}, \quad \mathbf{A}=\begin{bmatrix} a_0 \\ a_1 \\ \vdots \\ a_n \end{bmatrix},$$

and

$$\mathbf{X}^T=\begin{bmatrix} 1 & 1 & \cdots & 1 \\ x_{11} & x_{21} & \cdots & x_{l1} \\ \vdots & & & \\ x_{1n} & x_{2n} & \cdots & x_{ln} \end{bmatrix},$$

we may rewrite the system of linear equations (9.3) in a matrix form

$$\mathbf{X}^T\mathbf{X}\mathbf{A}= \mathbf{X}^T\mathbf{Y}.$$

Since $\mathbf{X}^T\mathbf{X}$ is always nonsingular, its inverse matrix $(\mathbf{X}^T\mathbf{X})^{-1}$ exists. Thus, the solution of system (9.3) is

$$\begin{bmatrix} \hat{a}_0 \\ \hat{a}_1 \\ \vdots \\ \hat{a}_n \end{bmatrix} = (\mathbf{X}^T\mathbf{X})^{-1}\,\mathbf{X}^T\mathbf{Y}.$$

Values $\hat{a}_0, \hat{a}_1, \hat{a}_2, \cdots,$ and $\hat{a}_n$ are called the *least square estimation* of $a_0, a_1, a_2, \cdots,$ and a_n respectively. In addition,

$$\hat{\sigma}^2 = \frac{1}{l-n-1}\sum_{j=1}^{l}[y_j - (\hat{a}_0 + \hat{a}_1 x_{j1} + \hat{a}_2 x_{j2} + \cdots + \hat{a}_n x_{jn})]^2$$

can be used as an estimation of the variance σ^2.

In the above discussion, symbol x_i is used to denote an attribute as well as a general observation of this attribute. However, when we want to emphasize that it is an observation, symbol $f(x_i)$ is adopted, especially, when we use an integral as the aggregation tool in information fusion, the observation (received information) of attributes is the integrand and, therefore, must be presented as a function defined on the set of all predictive attributes. Thus, in the linear multiregression model (9.1), $a_1x_1 + a_2x_2 + \cdots + a_nx_n$ should be rewritten as

$$a_1f(x_1) + a_2f(x_2) + \cdots + a_nf(x_n),$$

a weighted sum of the values of function *f* on *X*, where $a_1, a_2, \cdots,$ and a_n are weights. As we have seen in Example 5.5, this weighted sum is just a Lebesgue integral of *f*, i.e.,

$$a_1f(x_1) + a_2f(x_2) + \cdots + a_nf(x_n) = \int f\, d\mu,$$

where classical measure μ on $\mathscr{P}(X)$ is determined by $\mu(\{x_i\}) = a_i$, $i = 1, 2, \cdots, n$. Consequently, the linear multiregression model (9.1) can be expressed as

$$y = a_0 + \int f\, d\mu + N(0, \sigma^2) \tag{9.4}$$

or, equivalently,

$$y = \int f\, d\mu + N(a_0, \sigma^2).$$

Ignoring the probabilistic background, we can also understand this linear multiregression as a *linear data fitting* problem. That is, given the above data set in n+1 dimensional space, we want to find an n dimensional hyper plane

$$y = a_0 + a_1x_1 + a_2x_2 + \cdots + a_nx_n$$

such that the total squared vertical (i.e., parallel to the y-axis) distance from each data point to the hyper plane is minimized.

Expression (9.4) leads us to develop new multiregression models based on nonlinear integrals.

9.2 Nonlinear Multiregression Based on the Choquet Integral

We still consider n+1 attributes $x_1, x_2, \cdots, x_n$ and y in a database shown in the previous section. The linear multiregression model (9.1) works well only when the interaction among the contributions from predictive attributes towards the target can be ignored. However, in many real systems, such an interaction is significant. We have seen from Chapter 4 that the interaction among the contributions from predictive attributes $x_1, x_2, \cdots, x_n$ towards the target y can be described by a signed efficiency measure $\mu: \mathscr{P}(X) \to (-\infty, \infty)$ and, therefore, the aggregation tool should be a nonlinear integral, such as the Choquet integral, with respect to μ. Thus, a new nonlinear multiregression model now is expressed as

$$y = c + (\mathrm{C})\int (a + bf)\, d\mu + N(0, \sigma^2)\,, \tag{9.5}$$

where c is a constant, both a and b are real-valued functions defined on $X = \{x_1, x_2, \cdots, x_n\}$, f is an observation of $x_1, x_2, \cdots, x_n$, μ is a signed efficiency measure, and $N(0, \sigma^2)$ is a normally distributed random perturbation with expectation 0 and variance σ^2. Functions a and b can be expressed as vectors or n-tupes, i.e., $a = (a_1, a_2, \cdots, a_n)$ and $b = (b_1, b_2, \cdots, b_n)$. They should satisfy the following constraints:

$$\min_{1 \le i \le n} a_i = 0\,;$$

$$\max_{1 \le i \le n} |b_i| = 1\,.$$

Under these constraints, of course, we have $a_i \geq 0$ and $-1 \leq b_i \leq 1$ for $i = 1, 2, \cdots, n$. In the Choquet integral, integrand $(a + bf)$ is called a *linear core* of the integral, where shifting vector a is used to balance the phases, i.e., the starting point of each predictive attribute to make interaction with the other predictive attributes, while scaling vector b is used to balance the measuring units of each predictive attribute. These two vectors are necessary in the multiregression since the various predictive attributes may have different measurement systems (such as the Celsius degree and the Fahrenheit degree of the temperature) and may have various dimensions (such as the length and the weight) in the database. All elements in these two vectors are unknown and should be optimally determined in a learning procedure with the other unknown parameters based on the given data set.

Thus, in this multiregression model, the regression coefficients are constant c, all elements of vectors a and b, and $\mu(A)$ for every $A \in \mathscr{P}(X) - \{\emptyset\}$. Totally, there are $1 + 2n - 2 + 2^n - 1 = 2^n + 2n - 2$ independent unknown parameters. So, the data size l should be much larger than $2^n + 2n - 2$. Unlike the linear multiregression discussed in Section 9.1, this regression model now is not linear with respect to the regression coefficients since a (or b) and μ have a form of product and, therefore, these regression coefficients cannot be optimally determined by using only an algebraic method. We have to ask for some soft computing technique, such as the genetic algorithm, to obtain an approximate numerical solution. Fortunately, we may still partially use an algebraic method to reduce the complexity of the genetic algorithm, that is, once the parameters a and b have been created in the genetic algorithm, the other parameters c and μ may be optimally determined by using the least square method based on the above-mentioned data. As for σ^2, similar to the linear regression, it may be estimated by the regression residual

$$\hat{\sigma}^2 = \frac{1}{l + 2 - 2^n - 2n} \sum_{j=1}^{l} [y_j - c - (\mathrm{C}) \int (a + bf_j)\, d\mu]^2 .$$

The relevant algorithm is presented as follows.

A. Preparations

(1) For given *n*, express positive integer *k* in binary digits as bit string $k_n k_{n-1} \cdots k_1$ for every $k = 1, 2, \cdots, 2^n - 1$.

(2) Use μ_k to denote $\mu(A)$ where $A = \bigcup_{i|k_i=1}\{x_i\}$, $k = 1, 2, \cdots, 2^n - 1$.

B. Part 1

Use the least square method to determine c, μ_1, $\mu_2, \cdots$, μ_{2^n-1} when the values of all elements of a and b are specified in the genetic algorithm given in Part 2.

(1) Construct the $l \times (2^n + 1)$ augmented matrix $\mathbf{Z} = [z_{jk}]$ as follows.

$$z_{j0} = 1,$$

$$z_{jk} = \begin{cases} \min\limits_{k_i=1}(a_i + b_i f_{ji}) - \max\limits_{k_i=0}(a_i + b_i f_{ji}), & \text{if it is} > 0 \text{ or } k = 2^n - 1 \\ 0, & \text{otherwise} \end{cases},$$

for $k = 1, 2, \cdots, 2^n - 1$; $j = 1, 2, \cdots, l$ and $z_{j2^n} = y_j$.

(2) Find the least square solution of the system of linear equations having above augmented matrix for unknown variables c, $\mu_1, \mu_2, \cdots, \mu_{2^n-1}$.

(3) Calculate the regression residual error $\hat{\sigma}^2$ by

$$\begin{aligned} \hat{\sigma}^2 &= \frac{1}{l - 2^n - 2n + 2} \sum_{j=1}^{l} [y_j - c - (\mathrm{C})\int (a + bf_j)\, d\mu]^2 \\ &= \frac{1}{l - 2^n - 2n + 2} \sum_{j=1}^{l} (y_j - c - \sum_{k=1}^{2^n-1} z_{jk}\mu_k)^2 . \end{aligned}$$

C. *Part 2 (main algorithm)*

Use a genetic algorithm to optimize the values of vectors a and b.

(1) Input integers n, l, and the data.

(2) Choose a large prime p as the seed for the random number generator, which generates random numbers obeying the uniform distribution on unit interval $[0, 1)$. Set the value for each parameter listed in the following.

- λ: The bit length of each gene, i.e., λ bits are used for expressing each gene. It depends on the required precision of the results. e.g., λ=10 means that the precision is almost 10^{-3}. Its default is 10.
- s: The population size. It should be a large positive even integer. Its default is 200.
- α and β: The probabilities used in a random switch to control the choice of genetic operators for producing offspring from selected parents. They should satisfy the condition that $\alpha \geq 0$, $\beta \geq 0$, and $\alpha + \beta \leq 1$. Their defaults are 0.2 and 0.5 respectively.
- ε and δ: Small positive numbers used in the stopping controller. Their defaults are 10^{-6} and 10^{-10} respectively.
- w: The limit number of generations that have not significant progression successively. Its default is 10.

(3) Calculate

$$\hat{\sigma}_y^2 = \frac{1}{l-1}\sum_{j=1}^{l}(y_j - \bar{y})^2 ,$$

where

$$\bar{y} = \frac{1}{l}\sum_{j=1}^{l} y_j .$$

(4) Randomly create the initial population that consists of s chromosomes. Each chromosome consists of $2n$ genes, denoted by

$g_1, g_2, \cdots, g_n, g_{n+1}, g_{n+2}, \cdots, g_{2n}$. The first n of them represent vector a, while the next n represent vector b. Each gene consists of λ bits and represents a real number in $[0, 1)$. Initialize counter GC by 0, counter WT by 0, and SE by $\hat{\sigma}_y^2$.

(5) Decode each chromosome to get vectors a and b by the following formulae:

$$a_i = \frac{g_i - m(g)}{(1 - g_i)(1 - m(g))},$$

$$b_i = \frac{2g_{n+i} - 1}{M(g)},$$

for $i = 1, 2, \cdots, n$, where $m(g) = \min_{1 \le k \le n} g_k$ and $M(g) = \max_{1 \le k \le n} |2g_{n+k} - 1|$.

(6) For each chromosome in the population, through algorithm Part 1, use a and b obtained above and the data to determine the corresponding optimal values of c and μ, and find the residual $\hat{\sigma}^2$.

(7) The residual error of the q-th chromosome in the current population is denoted by $\hat{\sigma}_q^2$. Let

$$m(\hat{\sigma}^2) = \min_{1 \le q \le s} \hat{\sigma}_q^2 \text{ and } Q = \{q \mid \hat{\sigma}_q^2 = m(\hat{\sigma}^2)\}.$$

Erasing the record saved for the last generation if any, save $m(\hat{\sigma}^2)$ and a, b, c, μ of q-th chromosomes for all $q \in Q$ in the current population. Display GC, WT, and $m(\hat{\sigma}^2)$.

(8) If $m(\hat{\sigma}^2) < \varepsilon\hat{\sigma}_y^2$, then go to (16); otherwise, take the next step.

(9) If $SE - m(\hat{\sigma}^2) < \delta\hat{\sigma}_y^2$, then $WT + 1 \Rightarrow WT$ and take the next step; otherwise, $0 \Rightarrow WT$ and go to (11).

(10) If $WT > w$, then go to (16); otherwise, take the next step.

(11) The relative goodness of the q-th chromosome in the current population is defined by

$$G_q = \frac{m(\hat{\sigma}^2)}{\hat{\sigma}_q^2}, \quad q = 1, 2, \cdots, s, \quad \text{if } m(\hat{\sigma}^2) > 0.$$

(12) Let

$$p_q = \frac{G_q}{\sum_{q=1}^{s} G_q}, \quad q = 1, 2, \cdots, s.$$

(13) According to the probability distribution $\{p_q \mid q = 1, 2, \cdots, s\}$ (using a random switch), select two different chromosomes in the current population as the parents. Randomly select one operator among the three-bit mutation (with probability α), the two-point crossover (with probability β), and one of the equally likely 48 two-point realignments (with probability $1-\alpha-\beta$) to produce two new chromosomes as the offspring.

(14) Repeat step (13) for $s/2$ times totally to get s new chromosomes. $GC+1 \Rightarrow GC$ Save $m(\hat{\sigma}^2)$ in *SE*.

(15) For each new chromosome, take steps (5) and (6) to find the corresponding values of a, b, c, μ, and $\hat{\sigma}^2$. Add these new chromosomes into the current population. According to the magnitude of $\hat{\sigma}^2$ (the smaller the better), select s best chromosomes among these $2s$ chromosomes to form the new population. Then go to (7).

(16) Check the sign of μ_{2^n-1} corresponding the q-th chromosome for all $q \in Q$. In case $\mu_{2^n-1} < 0$, replace c by $c + \mu_{2^n-1} \max_{1 \le i \le n} a_i$, then replace a_i by $\max_{1 \le i \le n} a_i - a_i$ and switch the sign of vector b and set function μ such that $\mu_{2^n-1} > 0$. Display p, s, λ, α, β, ε, δ, and w. After deleting any duplicates, display a, b, c, and μ of q-th chromosomes for all $q \in Q$.

(17) Stop.

9.3 A Nonlinear Multiregression Model Accommodating Both Categorical and Numerical Predictive Attributes

Sometimes, we may find databases involving both numerical and categorical predictive attributes. Let $X = \{x_1, x_2, \cdots, x_n\}$ be the set of all considered predictive attributes, in which $x_1, x_2, \cdots, x_m$ are numerical and $x_{m+1}, x_{m+2}, \cdots, x_n$ are categorical where $1 \le m \le n-1$. The set of all possible states of categorical attribute x_i is denoted by

$$S_i = \{s_{i1}, s_{i2}, \cdots, s_{iN_i}\}$$

and is called the *range* of x_i, where N_i is the number of possible states of attribute x_i and is called the *potential* of x_i, $i = m+1, m+2, \cdots, n$. In each S_i, $i = m+1, m+2, \cdots, n$, each state s_{ik}, $k = 1, 2, \cdots, N_i$, may be or may not be a real number. The same as before, f is a function defined on *X*. It has a real value at each attributes x_i for $i = 1, 2, \cdots, m$, but has a value in S_i at attribute x_i for $i = m+1, m+2, \cdots, n$. Each f_j, $j = 1, 2, \cdots, l$, in the data set is an observation of such a function. In order to use the nonlinear multiregression model discussed in Section 9.2, we must numericalize attributes $x_{m+1}, \cdots, x_n$. Our numericalization strategy is, for each $i = m+1, m+2, \cdots, n$, optimally assigning a real value to each state s_{ik}, $k = 1, 2, \cdots, N_i$. The optimization is in the sense that, after replacing these states with corresponding real-valued assignments respectively, the regression

$$y = c + (\mathrm{C})\int (a + bf)\, d\mu$$

fits the data as well as possible. This optimization procedure takes place with optimizing regression coefficients together in a genetic algorithm that is similar to the one mentioned in the previous section. Thus, for each s_{ik}, we use a gene to represent it and align all of them in the chromosome. The corresponding value of this gene is denoted by d_{ik} that is the value we want to assign to s_{ik}. For a fixed i, different values d_{ik} indicate the different influences of s_{ik} to the target attribute, and they should be regularized when being used as the value of function f at attribute x_i to calculate the integral of f to avoid the indeterminacy of coefficients $b_1, b_2, \cdots, b_n$. The regularization is made as follows:

$$d_{ik}^* = \frac{d_{ik}}{\sum_{k=1}^{N_i} d_{ik}}$$

for $i = m+1, m+2, \cdots, n$. In comparison with the model for pure numerical attributes given in Section 9.2, now the number of unknown parameters is increased. In the current model, each chromosome consists of

$$2n + \sum_{i=m+1}^{n} N_i$$

genes ($2n$ genes for a and b, and N_i genes for each attribute x_i, $i = m+1, m+2, \cdots, n$), and there are

$$1 + (2^n - 1) + 2(n-1) + \sum_{i=m+1}^{n} (N_i - 1) = 2^n + n + m - 2 + \sum_{i=m+1}^{n} N_i$$

independent unknown parameters to be determined from data by minimizing the error

$$\hat{\sigma}^2 = \frac{1}{l + 2 - 2^n - n - m - \sum_{i=m+1}^{n} N_i} \sum_{j=1}^{l} [y_j - c - (\mathrm{C})\int (a + bf_j)\, d\mu]^2 .$$

Of course, using this model require that the data size must be much larger than the number of the independent unknown parameters.

An improved model can be also established by replacing a one-dimensional value optimally assigned to each state s_{ik} with an $(N_i - 1)$-dimensional value. A successful example using this improved model can be found in [Hui and Wang 2005].

9.4 Advanced Consideration on the Multiregression Involving Nonlinear Integrals

Using some various types of nonlinear integrals with various types of integrands as the aggregation tool, we may develop more multiregression models to handle data set to obtain valuable information on the relation between one attribute and the others. Some of them are briefly shown in the following.

9.4.1 *Nonlinear multiregressions based on the Choquet integral with quadratic core*

The multiregression model (9.5) is nonlinear with respect to the regression parameters. It can capture only the linear interaction among the contributions from predictive attributes towards the target. In some real-world problems, the above-mentioned interaction may not be linear. In this case, we may try to use quadratic core in the Choquet integral. Thus, the multiregression model (9.5) is changed to be

$$y = c + (\mathrm{C})\int (a + bf + df^2)\, d\mu + N(0, \sigma^2), \tag{9.6}$$

where vectors *a*, *b*, *d*, constant *c*, and signed efficiency measure, μ, are unknown regression coefficients satisfying $\min_{1 \le i \le n} a_i = 0$ and $\max_{1 \le i \le n} |b_i| = 1$. Similar to the way shown in Section 9.2, based on given data set (8.10), the values of these parameters can be optimally determined via a combination of algebraic method and genetic algorithm.

The multiregression model (9.6), which has a quadratic core in the Choquet integral, is a real generalization of model (9.5). The new model has more unknown parameters. So the program needs longer running time.

9.4.2 *Nonlinear multiregressions based on the Choquet integral involving unknown periodic variation*

Assume that the data set (8.10) is recorded according to the time uniformly. If there is a periodic affection by the time to the target, we may add an artificial predictive attribute, x_{n+1}, in the multiregression model (9.5) to capture it.

Thus, a new column consists of

$$f_{j,n+1} = f_j(x_{n+1}) = \cos[2\pi(\frac{j-1}{t} + d)], \quad j = 1, 2, \cdots, l,$$

where t is the period and d is the phase, is added into the data set (8.10) as the $(n+1)$-th column. Both t and d are unknown and they will be optimally determined from data with the other unknown regression coefficients together.

Now, let $X' = X \cup \{x_{n+1}\}$. The interaction among the contributions from predictive attributes towards the objective attribute is described by a signed efficiency measure defined on the power set of X'. The multiregression model has still a form as (9.5), but now $a = (a_1, a_2, \cdots, a_{n+1})$ and $b = (b_1, b_2, \cdots, b_{n+1})$ are (n+1)-vectors. They should satisfy the following constraints

$$\min_{1 \le i \le n+1} a_i = 0$$

and

$$\max_{1 \le i \le n+1} |b_i| = 1$$

as well.

After all regression coefficients have been optimally determined via a combination of algebraic method and genetic algorithm, once a new observation of n original predictive attributes is recorded as $(f(x_1), f(x_2), \cdots, f(x_n))$ at time t', we may add

$$f(x_{n+1}) = \cos[2\pi(\frac{t'-1}{t} + d)]$$

to obtain a function f on X' and then calculate

$$y = c + (\mathrm{C})\int(a + bf)\, d\mu$$

as the predicted value of objective attribute Y.

9.4.3 *Nonlinear multiregressions based on upper and lower integrals*

In multiregression model (9.5), the aggregation is performed by the Choquet integral. From Chapter 5, we know that the Choquet integral is just one of the nonlinear integrals and it has the maximal coordination manner. Usually, people cannot know what coordination manner among predictive attributes exactly exists in a given real regression problem. That is to say, there is no sufficient reason to choose the Choquet integral as the aggregation tool in the regression. Thus, a new idea is to use the upper and the lower integrals, which are also discussed in Chapter 5, to dominate any possible nonlinear integral if the coordination manner is unknown. Then, regarding any real number as a special interval number, an interval-valued multiregression model can be established as follows:

$$Y = c + [(\mathrm{L})\int(a + bf)\, d\mu, (\mathrm{U})\int(a + bf)\, d\mu] + N(0, \sigma^2),$$

where regression coefficients a, b, c, and μ have the same meaning as in model (9.5) but with a little different restriction $\min_{1\le i\le n} a_i = 0$,

$\min_{1\le i\le n} b_i \ge 0$, and $\max_{1\le i\le n} b_i = 1$. Based on given data set (8.10) that should now be restricted to be all nonnegative, these unknown parameters can be optimally determined by minimizing the total squared error

$$e^2 = \sum_{j=i}^{l} (e_{1j}^2 + e_{2j}^2),$$

where

$$e_{1j} = \begin{cases} 0 & \text{if } y_j - c \in [(\mathrm{L})\int (a+bf_j)\, d\mu,\ (\mathrm{U})\int (a+bf_j)\, d\mu] \\ \min(\left|y_j - c - (\mathrm{L})\int (a+bf_j)\, d\mu\right|,\ \left|y_j - c - (\mathrm{U})\int (a+bf_j)\, d\mu\right|) & \text{otherwise} \end{cases}$$

and

$$e_{2j} = (\mathrm{U})\int (a+bf_j)\, d\mu - (\mathrm{L})\int (a+bf_j)\, d\mu \ .$$

Error e_{1j} describes the random error while error e_{2j} describes the uncertainty carried by the signed efficiency measure μ for the j-th observation.

After determining all regression coefficients, once a new observation f is available, the prediction for the objective attribute Y is an interval number

$$\hat{Y} = [c + (\mathrm{L})\int (a+bf)\, d\mu,\ c + (\mathrm{U})\int (a+bf)\, d\mu].$$

Chapter 10

Classifications Based on Nonlinear Integrals

In Section 9.3, we allow the predictive attributes to be either numerical or categorical, but required the objective attribute (the target) to be only numerical. Instead, if we allow the objective attribute to be categorical, then it becomes a classification problem discussed in the following sections. In this case, the predictive attributes are called the *feature attributes*, while the objective attribute is called the *classifying attribute*. The number of possible states to the classifying attribute is just the number of classes in the classification problem. Let the number of the states of the classifying attribute be m. Then the corresponding classification is called *m-classification*. It is easy to see that any m-classification problem can be separated as $(m-1)$ 2-classification problems. So, in this chapter, we only discuss the 2-classification problems, unless a special statement is given.

The classification is an essential component of pattern recognition problem. It is, the same as the multiregression, one of the major techniques used in data mining.

10.1 Classification by an Integral Projection

Now we consider a classical 2-classification problem and express it via an abstract integral, the Lebesgue integral, which is discussed in Section 5.3.

Let a complete data set

$$\begin{array}{ccccc} x_1 & x_2 & \cdots & x_n & Y \\ \hline f_{11} & f_{12} & \cdots & f_{1n} & y_1 \\ f_{21} & f_{22} & \cdots & f_{2n} & y_2 \\ \vdots & \vdots & & \vdots & \vdots \\ f_{l1} & f_{l2} & \cdots & f_{ln} & y_l \end{array} \tag{10.1}$$

be available, where $x_1, x_2, \cdots, x_n$ are feature attributes, Y is the classifying attribute, and l is the number of samples in the data set. The set of all feature attributes, $X = \{x_1, x_2, \cdots, x_n\}$, is considered as the universal set. The range of feature attributes is called the *feature space*. It is a subset of n-dimensional Euclidean space. Unlike the multiregression problem, now Y is categorical and has only two possible states, denoted by s_1 and s_2. Set $S = \{s_1, s_2\}$ is called the *state set* of attribute Y. Each row f_{j1}, f_{j2}, $\cdots, f_{jn}$ ($j = 1, 2, \cdots, l$) in the data set is a sample of the feature attributes and is a real-valued function on X; while y_j is the corresponding state that indicates a specified class. A 2-classification problem is, based on the given data set, to find a classification model that divides the feature space into 2 disjoint pieces. Each of them corresponds to a class. Then once a new sample of the feature attributes is available, we may use the model to determine to which class the sample belongs. The classification model is usually called a *classifier*.

The simplest classification model is linear, that is, the two pieces of the feature space corresponding to two classes is divided by an ($n-1$)-dimensional hyper-plane that can be expressed by a linear equation of n variables $x_1, x_2, \cdots, x_n$:

$$a_1 x_1 + a_2 x_2 + \cdots + a_n x_n = c \tag{10.2}$$

in the n-dimensional Euclidean space. In expression (10.2), a_i, $i = 1, 2, \cdots, n$, and c are unknown parameters that we want to determine based on the given data. This ($n-1$)-dimensional hyper-plane is called the *classifying boundary*. Essentially, a linear classification model is just a linear projection $y = a_1 x_1 + a_2 x_2 + \cdots + a_n x_n$ from the n-dimensional feature space onto a one-dimensional real line, on which a point c is selected as the critical value for optimally separating the projections of the two-class samples in the data set. Point c corresponds to the classifying boundary. In fact, the projection of the whole $(n-1)$-dimensional classifying boundary onto the one-dimensional real line is just the critical point c. In most linear models, the criterion of the optimization is to minimize the misclassification rate when a nonempty subset of the data set is used as the training set. In this case, there are infinitely many optimal classifying boundaries generally and, usually, they are close to each other. Sometimes, the optimization criterion can also be formed by a certain function of the distance from sample points to the classifying boundary in the feature space. The values of the parameters a_i, $i = 1, 2, \cdots, n$, and c corresponding to one of the optimal classifyingboundaries can be calculated via an algebraic and analytical method precisely or be found via a numerical method approximately. Based on the found classifying boundary, once a new sample, i.e., a new observation of the feature attributes, $(f(x_1), f(x_2), \cdots, f(x_n))$ is available, we may conclude that this sample belongs to the first or the second class according to whether inequality

$$a_1 f(x_1) + a_2 f(x_2) + \cdots + a_n f(x_n) \le c \tag{10.3}$$

holds or not.

Inequality (10.3) can be rewritten in terms of the classical Lebesgue integral as follows.

$$\int f \, d\mu \le c ,$$

where f is a real-valued function defined on X and μ is an additive measure on measurable space $(X, \mathscr{P}(X))$ satisfying $\mu(\{x_i\}) = a_i$ for $i = 1, 2, \cdots, n$.

Example 10.1 Let $X = \{x_1, x_2\}$. There are 14 samples shown in Table 10.1 for (x_1, x_2) with their corresponding classes. Based on this training data, one of the optimal classifying boundaries may be the straight line $x_1 + 2x_2 = 1.4$. It separates the data very well with misclassification rate zero. If a new sample (0.3, 0.7) is obtained, we may immediately conclude that this sample belongs to class 2 since $0.3 + 2 \times 0.7 = 1.7 > 1.4$. This is shown in Figure 10.1. In this example, in fact, there are infinitely many classifying boundaries that can separate the given data with zero misclassification rate. For instance, straight line $0.99x_1 + 1.99x_2 = 1.41$ is also one of the optimal solutions of this linear classification problem. This classifying boundary is very close to the first one.

Table 10.1 Data for linear classification in Example 10.1.

x_1	x_2	class	x_1	x_2	class	x_1	x_2	class
0.9	0.2	1	0.8	0.2	1	0.3	0.6	2
0.5	0.4	1	0.7	0.3	1	0.9	0.3	2
0.2	0.5	1	0.1	0.6	1	0.1	0.7	2
0.1	0.2	1	0.6	0.6	2	0.5	0.8	2
0.6	0.3	1	0.5	0.5	2			

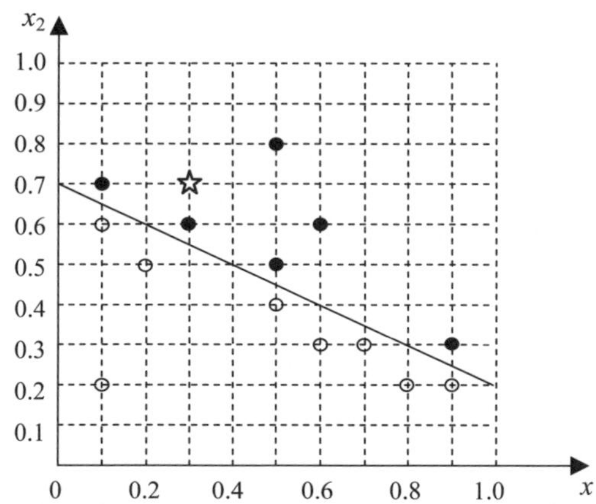

Fig. 10.1 The training data and one optimal classifying boundaries x_1+$2x_2$ =1.4 with a new sample (0.3, 0.7) in Example 10.1.

Similar to the multiregression problems, using the above-mentioned linear classification model needs a basic assumption that the interaction among the contributions from feature attributes towards the classification can be ignored. Unfortunately, in many real-world classification problems, the samples in the data are not linearly separatable, that is, the optimal classifying boundary is not approximately linear, since the above-mentioned interaction cannot be ignored. In this case, similar to the multiregression, we should adopt a nonlinear integral, such as the Choquet integral with respect to a signed efficiency measure μ, which describes the above-mentioned interaction, to express the classifying boundary. Such a nonlinear classification model is discussed in the remaining part of this chapter.

10.2 Nonlinear Classification by Weighted Choquet Integrals

If the interaction among the contributions from the feature attributes towards the classification cannot be ignored, then a signed efficiency measure should be used and, therefore, a relative nonlinear integral should be involved in the classifier, where the nonadditivity of the signed efficiency measure describes the interaction.

The following Example illustrates the interaction existing in nonlinear classification.

Example 10.2 A mail box is assumed to be large enough, but its slot is only 5 inches long. Thus, envelops are classified into two classes according to their size as follows.

(1) *small*: Those can be inserted into the mail box;
(2) *large*: Those cannot be inserted into the mail box.

This means that, to a given envelop, if only its length or width is large, it is not really "large"; only when both the length and the width are large, it then is "large". This shows a strong interaction between the contributions from the two dimensions of envelops towards the "size". Due to the strong interaction, such a classification is not linear. In fact, a good classifying boundary should be a segment of a broken line, but not a straight line as shown in Figure 10.2. In the discussion below, we can see that such a broken line can be exactly expressed as a contour of the function expressed by the Choquet integral.

In Chapters 5, 6, and 9, we have seen that the Choquet integral with a signed efficiency measure is nonlinear with respect to its integrand and can be used as an aggregation tool in multiregression. Now let us see how the Choquet integral can be regarded as a projection from a high-dimensional Euclidean space onto a one-dimensional Euclidean space and can be used in nonlinear classification. To illustrate it easily, we consider only two feature attributes.

Let x_1 and x_2 be two feature attributes. Denote $X = \{x_1, x_2\}$. Furthermore, let $\mu : \mathscr{P}(X) \to [0, 1]$ be an efficiency measure and $f : X \to (-\infty, \infty)$ be a real valued function. The Choquet integral $(\mathrm{C})\int f\, d\mu$ is a function of $f(x_1)$ and $f(x_2)$, or say, a functional of function *f*. For any specified constant *c*, the contour $(\mathrm{C})\int f\, d\mu = c$ is a broken line, but not a straight line if μ is not additive. This can be seen in Example 10.3.

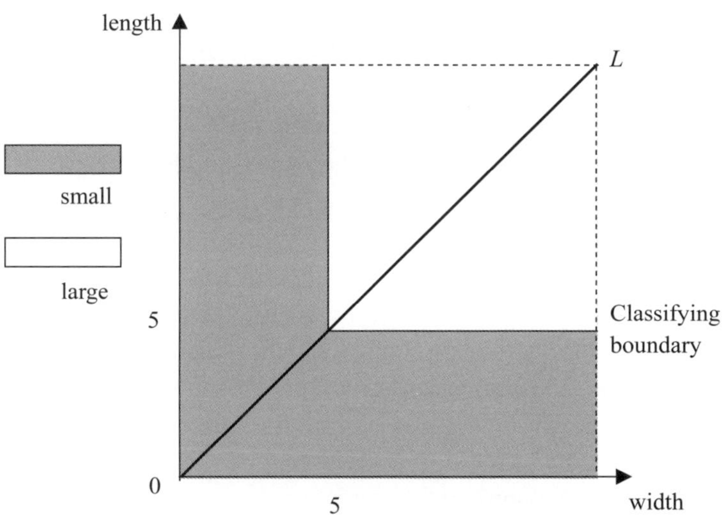

Fig. 10.2 Interaction between length and width of envelops in Example 10.2.

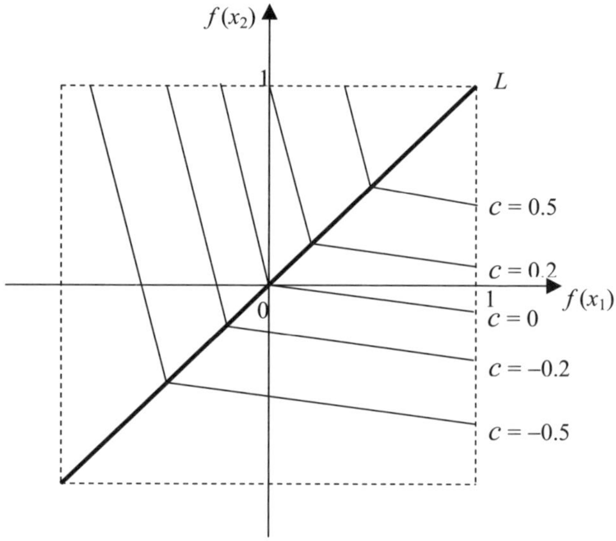

Fig. 10.3 The contours of the Choquet integral in Example 10.3.

Example 10.3 Let $X = \{x_1, x_2\}$ and efficiency measure μ have values $\mu(\{x_1\}) = 0.2$, $\mu(\{x_2\}) = 0.5$, $\mu(X) = 1$, and $\mu(\varnothing) = 0$. Then the contours of the Choquet integral $(\mathrm{C})\int f\, d\mu$ on the plane are shown in Figure 10.3.

From Example 10.3, we can see that the Choquet integral projects every point $(f(x_1), f(x_2))$ on the plane onto line L with the projection value $c = (\mathrm{C})\int f\, d\mu$. Straight line L is formed by points satisfying $f(x_1) = f(x_2)$. It passes through the origin and has angle 45° with the x-axis. The projection is not along with straight lines, but with broken lines, that is, the projection directions on the two sides of line L are different. However, the projection directions on the same side of line L are parallel, as shown in Figure 10.4 if the same attributes and the efficiency measure in Example 10.3 are used.

Thus, replacing the straight line used as the classifying boundary in linear classification, now each contour of the Choquet integral, $(\mathrm{C})\int f\, d\mu = c$, can be chosen as the classifying boundary of two classes.

Such a classifier is nonlinear. Similar to the roll being a linear function of the observations of feature attributes, the projection by the Choquet integral converts a high-dimensional classification problem into a one-dimensional classification problem. This means that, by using the Choquet integral, we just need to select an appropriate signed efficiency measure and an appropriate real value c on line L, and then can classify a new observation of feature attributes, f, to one class if $(C)\int f\,d\mu \le c$ and to another class if $(C)\int f\,d\mu > c$. This is also shown in Figure 10.4.

In such a way, the classifying boundary in Example 10.2 for the small size and the large size of envelops can be expressed as $(C)\int f\,d\mu = 5$,where signed efficiency measure μ has values $\mu(\{x_1\}) = \mu(\{x_2\})$ $= \mu(\emptyset) = 0$ and $\mu(X) = 1$. Of course, to form a classifier, the essential mission is to optimally determine the values of the signed efficiency measure and constant c based on given training data. It is discussed later in this section.

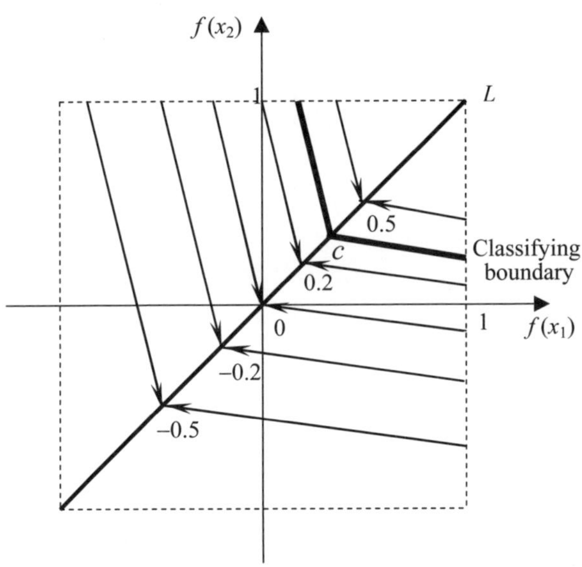

Fig. 10.4 The projection by the Choquet integral in Example 10.3.

From Figure 10.3, we can also see that the angle between two branches of each broken line is obtuse. In fact, when the value of μ at a singleton of attribute is positive, then the angle between the corresponding branch of the broken line and line L is greater than 45°. It can be seen from Figure 10.2 that the angle between any branch of the broken line and line L is equal to 45° if and only if the value of μ at the corresponding singleton of attribute is zero. Furthermore, if a signed efficiency measure μ is considered with a negative value at a singleton of attribute, then the angle between the corresponding branch of the broken line and line L is less than 45°. An example is given in Example 10.4 and illustrated in Figure 10.5.

Example 10.4 Let $X=\{x_1, x_2\}$ and signed efficiency measure μ have values $\mu(\{x_1\})=-0.2$, $\mu(\{x_2\})=-0.5$, $\mu(X)=1$, and $\mu(\varnothing)=0$. Then the contour of the Choquet integral $(\mathrm{C})\int f\,d\mu$ with value $c=-0.6$ on the plane is shown in Figure 10.5. The angle between two branches of the broken line is acute.

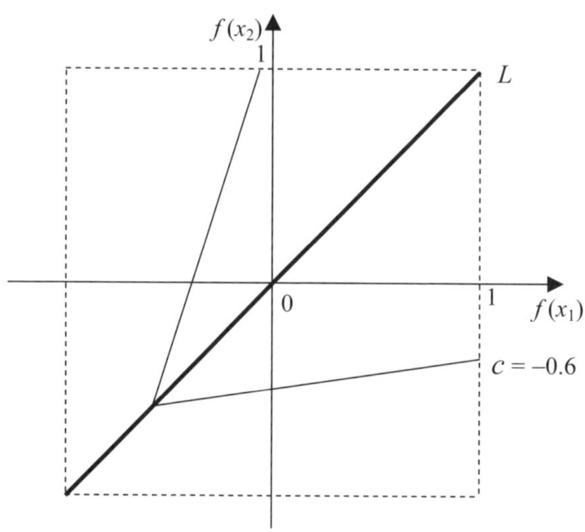

Fig. 10.5 A contour of the Choquet integral with respect to a signed efficiency measure in Example 10.4.

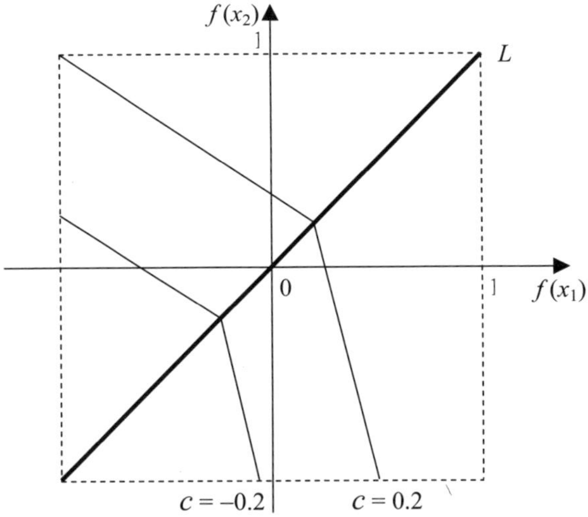

Fig. 10.6 Contours of the Choquet integral with respect to a subadditive efficiency measure in Example 10.5.

In Figures 10.2 to 10.5, all broken lines are concave towards the right and up direction, i.e., the positive direction on axis L. This is due to the superadditivity of signed efficiency measure μ. If μ is subadditive, then the contours of the Choquet integral are concave towards left and down direction, i.e., the negative direction on axis L. This is shown in Example 10.5 and illustrated in Figure 10.6.

Example 10.5 Let $X = \{x_1, x_2\}$ and signed efficiency measure μ has values $\mu(\{x_1\}) = 0.8$, $\mu(\{x_2\}) = 0.6$, $\mu(X) = 1$, and $\mu(\emptyset) = 0$. μ is subadditive. Then the contours of the Choquet integral $(C)\int f\, d\mu$ with value $c = -0.2$ and $c = 0.2$ on the plane are shown in Figure 10.6. These broken lines are concave towards left and down direction.

In the above examples, signed efficiency measure μ is used to describe the interaction between the contributions from attributes x_1 and x_2 towards the target, the classification. This interaction is considered based on "one unit of x_1 with one unit of x_2". However, to a given real problem, people do not know what the actual units, based on which the interaction of x_1 and x_2 is expressed, are adopted. Hence, it is necessary to use weights at the front of the integrand of the Choquet integral to balance the units, unless the units are naturally concord in some special problem such as the workers' products in Example 5.21. The weights, of course, are unknown generally and will be optimally determined in data mining problems.

Thus, the weighted Choquet integral should have a form of $(C)\int bf\, d\mu$, where $b: X \to [-1, 1]$ is the weights. It balances the scaling of various attributes. The geometric meaning of the weights is to adjust the direction of projection line *L*. Unlike the Choquet integral without the weights, now the projection line *L* is allowed to be any straight line passing through the origin. An example of the weighted Choquet integral with some contour is given in Example 10.6 and illustrated in Figure 10.7.

Example 10.6 Let $X = \{x_1, x_2\}$, weights b have values $b_1 = b(x_1) = 1$ and $b_2 = b(x_2) = -0.5$, and signed efficiency measure μ have values $\mu(\{x_1\}) = 0.1$, $\mu(\{x_2\}) = 0.6$, $\mu(X) = 1$, and $\mu(\emptyset) = 0$. μ is superadditive. The projection line *L*, at which the contours change their direction, can be determined as follows. Since the contours change their direction at points $(f(x_1), f(x_2))$ satisfying $b_1 f(x_1) = b_2 f(x_2)$, we may obtain the equation of line *L* as

$$f(x_2) = \frac{b_1}{b_2} f(x_1)$$

directly when $b_2 \neq 0$, or as $f(x_1) = 0$ when $b_2 = 0$. Hence, in this example, the slop of line *L* is –2. The contours of the Choquet integral $(C)\int bf\, d\mu$ with value $c = -0.2$ and $c = 0.2$ respectively on the plane are shown in Figure 10.7. These broken lines are concave towards down direction.

From the above discussion, we may see that, replacing the linear function of the feature attributes, the weighted Choquet integral can be used for classification. The classical linear classifier is just a special case

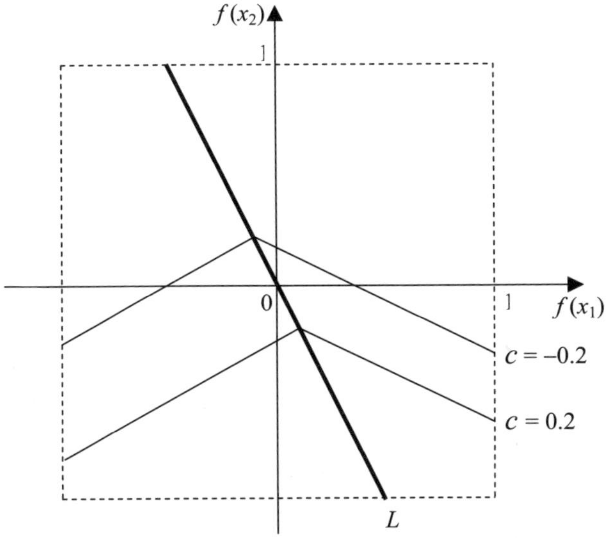

Fig. 10.7 Projection line and contours of the weighted Choquet integral in Example 10.6.

of the classifier based on the weighted Choquet integral. This new model is nonlinear, in which the unknown parameters are the weights and the values of the signed efficiency measure. When a training data set is available, we may use a soft computing technique, such as a genetic algorithm, to search for the optimal values of the parameters and then establish a classifier. The optimization criterion may be chosen as minimizing the misclassification rate. After the values of the unknown parameters, containing vector b, set function μ, and constant c, are optimally determined based on the training data, once a new observation f is available, we just need to calculate the value of $(C)\int bf\, d\mu$, and then classify f into one class if $(C)\int bf\, d\mu \le c$, into another class if $(C)\int bf\, d\mu > c$.

10.3 An Example of Nonlinear Classification in a Three-Dimensional Sample Space

The following is an example of using the weighted Choquet integral for classification in a three-dimensional sample space. Though the involved set function is monotone, this restriction is not essential.

Example 10.7 ([Xu et al 2003]) An artificial data set now is used to test the effectiveness, including the convergence and the running time, of the algorithm and the program. Regarding the values of monotone measure μ and weight function b as parameters, we preset them together with a value for the magnitude of the weighted Choquet integral as the classifying boundary, and then construct the training data possessing the form of $(f_j(x_1), f_j(x_2), f_j(x_3), Y_j)$ with categorical $Y_j = C_1$ or C_2, $j = 1, 2, \cdots, l$. Using these constructed training data, we run the program to check whether the preset values of the parameters and the classifying boundary can be recovered approximately with a low misclassification rate. The following is the detailed procedure for constructing the training data.

There are three feature attributes, two classes, and 200 records in the data set, that is, $X = \{x_1, x_2, x_3\}$, $C = \{C_1, C_2\}$, and $l = 200$.

(1) Preset the values of normalized monotone measure μ and weight function b by assigning $\mu(\{x_1\}) = 0.1$, $\mu(\{x_2\}) = 0.2$, $\mu(\{x_1, x_2\}) = 0.6$, $\mu(\{x_3\}) = 0.05$, $\mu(\{x_1, x_3\}) = 0.8$, $\mu(\{x_2, x_3\}) = 0.9$, $b_1 = 0.1$, $b_2 = 0.3$, and $b_3 = 0.6$.

(2) Use a random number generator with the uniform distribution on $[0, 1)$ to create a sequence of the values of feature attributes, $f(x_i)$ for i = 1, 2, 3, independently. Each $f = (f(x_1), f(x_2), f(x_3))$ is the left part of a record.

(3) For each $f = (f(x_1), f(x_2), f(x_3))$, calculate the corresponding value of the weighted Choquet integral with respect to μ:

$$\hat{Y} = (\mathrm{C})\int bf \, d\mu ,$$

where μ and b are given in (1).

(4) Create a random number, ξ, with the uniform distribution. Take 0.08 as the preset classification boundary. In case $\hat{Y} \le 0.08$, if $\xi \le e^{-(\hat{Y}-0.05)^2/0.0018}$, then assign class C_1 to the right part of the record; otherwise, abandon this record. While in case $\hat{Y} > 0.08$, if $\xi \le e^{-(\hat{Y}-0.12)^2/0.0032}$, then, assign class C_2 to the right part of the record; otherwise, abandon this record.

(5) Collect the first 80 records with class C_1 and the first 120 records with class C_2 in the sequence of records to form the sample data.

In Step (4), random number ξ is used to construct a random switch and the normal distributions $N(0.05, 0.03^2)$ and $N(0.12, 0.04^2)$ are used for controlling the distribution of the data in classes C_1 and C_2 respectively. In fact, the probability density of $N(0.05, 0.03^2)$ is

$$p_1(t) = \frac{1}{\sqrt{2\pi} \times 0.03} e^{-\frac{(t-0.05)^2}{2\times 0.03^2}},$$

and the probability density of $N(0.12, 0.04^2)$ is

$$p_2(t) = \frac{1}{\sqrt{2\pi} \times 0.04} e^{-\frac{(t-0.12)^2}{2\times 0.04^2}}.$$

Inequality $\xi \le e^{-(\hat{Y}-0.05)^2/0.0018}$ means that $\xi \le 0.03\sqrt{2\pi} \cdot p_1(\hat{Y})$. Similarly, Inequality $\xi \le e^{-(\hat{Y}-0.12)^2/0.0032}$ means that $\xi \le 0.04\sqrt{2\pi} \cdot p_2(\hat{Y})$. Thus, the remaining data in C_1 have a right truncated unimodal distribution with mode 0.05, while those in C_2 have a left truncated unimodal distribution with mode 0.12. For both of them, the truncating point is 0.08. The entire sample data are listed in Table 10.2.

Table 10.2 Artificial training data in Example 10.7.

No.	f_1	f_2	f_3	Class
1	0.001648	0.061432	0.303497	C_1
2	0.647797	0.342316	0.060577	C_1
3	0.581604	0.059906	0.809631	C_1
4	0.328979	0.151184	0.850067	C_1
5	0.517639	0.209778	0.404083	C_1
6	0.149719	0.112335	0.727692	C_1
7	0.419647	0.104828	0.659882	C_1
8	0.461670	0.132233	0.529663	C_1
9	0.581879	0.339691	0.115265	C_1
10	0.122192	0.008789	0.257477	C_1
11	0.372955	0.061401	0.098785	C_1
12	0.382751	0.148621	0.882111	C_1
13	0.037994	0.623016	0.071930	C_1
14	0.211914	0.182770	0.075897	C_1
15	0.304382	0.105347	0.886597	C_1
16	0.473602	0.307281	0.124573	C_1
17	0.439056	0.024261	0.440338	C_1
18	0.378296	0.058411	0.727631	C_1
19	0.617828	0.136444	0.404449	C_1
20	0.126465	0.270142	0.034119	C_1
21	0.097778	0.592224	0.027618	C_1
22	0.449707	0.147278	0.723419	C_1
23	0.988495	0.292572	0.102325	C_1
24	0.184052	0.285339	0.086853	C_1
25	0.028931	0.155975	0.116486	C_1
26	0.117859	0.119293	0.569458	C_1
27	0.166626	0.404388	0.027344	C_1
28	0.523834	0.107117	0.574585	C_1
29	0.564758	0.217438	0.108917	C_1

30	0.802551	0.125397	0.077576	C_1
31	0.668335	0.107056	0.251007	C_1
32	0.191589	0.977539	0.008331	C_1
33	0.116821	0.036713	0.201721	C_1
34	0.221222	0.478790	0.107666	C_1
35	0.392151	0.021454	0.819183	C_1
36	0.039001	0.099060	0.707642	C_1
37	0.102264	0.169525	0.826904	C_1
38	0.200653	0.059357	0.244843	C_1
39	0.044403	0.135010	0.757813	C_1
40	0.182373	0.155670	0.595337	C_1
41	0.193451	0.497986	0.107544	C_1
42	0.490753	0.101624	0.757355	C_1
43	0.464813	0.198517	0.011169	C_1
44	0.722382	0.332397	0.081055	C_1
45	0.419800	0.047302	0.729675	C_1
46	0.412628	0.217896	0.110535	C_1
47	0.409851	0.061707	0.613770	C_1
48	0.169891	0.024048	0.821594	C_1
49	0.266144	0.057281	0.363220	C_1
50	0.338989	0.126190	0.932922	C_1
51	0.841187	0.217224	0.070190	C_1
52	0.624512	0.034515	0.633820	C_1
53	0.726349	0.190857	0.328186	C_1
54	0.000305	0.165833	0.114258	C_1
55	0.963348	0.098694	0.088104	C_1
56	0.273499	0.012939	0.852173	C_1
57	0.815430	0.061737	0.105927	C_1
58	0.680023	0.095703	0.075867	C_1
59	0.119324	0.034668	0.122925	C_1
60	0.232697	0.951843	0.015808	C_1
61	0.099854	0.254822	0.090729	C_1

62	0.128143	0.092590	0.194061	C_1
63	0.884338	0.474182	0.052155	C_1
64	0.157898	0.316803	0.008850	C_1
65	0.752625	0.025055	0.085144	C_1
66	0.558441	0.029999	0.181854	C_1
67	0.726807	0.041962	0.665619	C_1
68	0.246704	0.221497	0.296570	C_1
69	0.913483	0.375244	0.062500	C_1
70	0.155670	0.202271	0.121826	C_1
71	0.205597	0.631683	0.035675	C_1
72	0.135254	0.056976	0.718323	C_1
73	0.207214	0.400482	0.107391	C_1
74	0.093140	0.113251	0.580200	C_1
75	0.934906	0.153015	0.085785	C_1
76	0.111206	0.181915	0.838623	C_1
77	0.462616	0.131317	0.362183	C_1
78	0.144043	0.181641	0.189270	C_1
79	0.097687	0.415833	0.087921	C_1
80	0.330933	0.047821	0.374481	C_1
81	0.001862	0.531677	0.464325	C_2
82	0.473663	0.198853	0.920166	C_2
83	0.846161	0.620850	0.147034	C_2
84	0.764221	0.543243	0.367493	C_2
85	0.078735	0.280304	0.868378	C_2
86	0.682800	0.402771	0.433380	C_2
87	0.431519	0.339752	0.715729	C_2
88	0.989838	0.227264	0.998505	C_2
89	0.090240	0.302216	0.281830	C_2
90	0.536987	0.378998	0.411957	C_2
91	0.422363	0.727570	0.859802	C_2
92	0.327423	0.299530	0.425232	C_2
93	0.607300	0.406372	0.269135	C_2

94	0.990082	0.410522	0.660370	C_2
95	0.098785	0.461487	0.317657	C_2
96	0.950226	0.734314	0.098022	C_2
97	0.915192	0.183929	0.201874	C_2
98	0.128204	0.930908	0.195618	C_2
99	0.050323	0.995270	0.279694	C_2
100	0.941650	0.084015	0.990997	C_2
101	0.891846	0.050201	0.771179	C_2
102	0.630951	0.493530	0.449402	C_2
103	0.296295	0.367554	0.359619	C_2
104	0.047760	0.350159	0.490356	C_2
105	0.460999	0.449219	0.909332	C_2
106	0.960876	0.03418	0.914154	C_2
107	0.891266	0.483276	0.641266	C_2
108	0.697449	0.490234	0.338043	C_2
109	0.727905	0.497223	0.547302	C_2
110	0.940948	0.084869	0.660492	C_2
111	0.249176	0.491241	0.733459	C_2
112	0.847290	0.489594	0.149536	C_2
113	0.822815	0.697052	0.150482	C_2
114	0.320435	0.660126	0.157043	C_2
115	0.289978	0.431396	0.868164	C_2
116	0.655975	0.601501	0.361847	C_2
117	0.974792	0.313782	0.213165	C_2
118	0.478058	0.329315	0.671051	C_2
119	0.963257	0.599457	0.503632	C_2
120	0.858795	0.501892	0.624878	C_2
121	0.011414	0.770996	0.297058	C_2
122	0.809235	0.749512	0.407593	C_2
123	0.652588	0.705353	0.115295	C_2
124	0.273987	0.618317	0.734528	C_2
125	0.907318	0.205109	0.359558	C_2

126	0.699860	0.111511	0.948425	C_2
127	0.291290	0.770294	0.457947	C_2
128	0.931915	0.136658	0.843903	C_2
129	0.647522	0.655518	0.385864	C_2
130	0.493195	0.604858	0.303436	C_2
131	0.436737	0.262299	0.964539	C_2
132	0.975586	0.380249	0.940430	C_2
133	0.002869	0.918579	0.160156	C_2
134	0.866180	0.758240	0.166809	C_2
135	0.936798	0.302490	0.863312	C_2
136	0.305878	0.621948	0.847595	C_2
137	0.630493	0.436707	0.885223	C_2
138	0.446014	0.399323	0.178009	C_2
139	0.743713	0.650726	0.152466	C_2
140	0.145752	0.607574	0.361450	C_2
141	0.031372	0.437317	0.357635	C_2
142	0.502228	0.622620	0.135010	C_2
143	0.926453	0.066620	0.936218	C_2
144	0.263367	0.315155	0.770172	C_2
145	0.768768	0.405579	0.212433	C_2
146	0.029358	0.949219	0.140411	C_2
147	0.850098	0.269318	0.835114	C_2
148	0.945038	0.141418	0.906036	C_2
149	0.877502	0.026184	0.990540	C_2
150	0.484436	0.606445	0.673431	C_2
151	0.190460	0.320526	0.853210	C_2
152	0.788513	0.460297	0.292267	C_2
153	0.919617	0.449951	0.238831	C_2
154	0.658691	0.292084	0.755005	C_2
155	0.263184	0.73587	0.251648	C_2
156	0.591827	0.543274	0.294861	C_2
157	0.713135	0.170441	0.600342	C_2

158	0.632996	0.328278	0.689148	C_2
159	0.043579	0.444183	0.768951	C_2
160	0.325195	0.266998	0.494843	C_2
161	0.188660	0.263062	0.940857	C_2
162	0.854584	0.709229	0.180634	C_2
163	0.715637	0.809875	0.016541	C_2
164	0.835144	0.383942	0.842346	C_2
165	0.931824	0.386749	0.115662	C_2
166	0.320648	0.550262	0.449554	C_2
167	0.332031	0.666809	0.245636	C_2
168	0.554504	0.407043	0.280457	C_2
169	0.916504	0.429352	0.173584	C_2
170	0.324127	0.374847	0.779175	C_2
171	0.758820	0.184753	0.980347	C_2
172	0.263794	0.544067	0.877136	C_2
173	0.992462	0.444916	0.666656	C_2
174	0.376190	0.683777	0.258362	C_2
175	0.445465	0.632935	0.240784	C_2
176	0.469543	0.926727	0.237762	C_2
177	0.096771	0.918213	0.319611	C_2
178	0.170715	0.420593	0.366394	C_2
179	0.225739	0.399689	0.131470	C_2
180	0.872681	0.096710	0.945313	C_2
181	0.837341	0.936005	0.225616	C_2
182	0.938477	0.269531	0.542755	C_2
183	0.910889	0.466827	0.980377	C_2
184	0.223846	0.311432	0.449524	C_2
185	0.656830	0.562958	0.791687	C_2
186	0.298309	0.557129	0.291565	C_2
187	0.756744	0.717316	0.171234	C_2
188	0.535095	0.373199	0.183929	C_2
189	0.133331	0.419434	0.770355	C_2

190	0.507507	0.645264	0.327209	C_2
191	0.819916	0.283051	0.665192	C_2
192	0.410004	0.290100	0.759583	C_2
193	0.414276	0.930817	0.105347	C_2
194	0.519745	0.869232	0.035309	C_2
195	0.066223	0.818970	0.709808	C_2
196	0.177429	0.393524	0.935272	C_2
197	0.388336	0.701660	0.278198	C_2
198	0.052399	0.445374	0.505188	C_2
199	0.323578	0.315887	0.788910	C_2
200	0.567902	0.682190	0.120605	C_2

Setting $s = 200$ as the population size and running the program with the whole sample data (such a test is called reclassification), after a number (thousands, depending on the seed chosen for generating random numbers) of individuals were produced, a resulting weighted Choquet integral projection and a classification with misclassification rate 0 are obtained. The values of the monotone measure and the weights in the weighted Choquet integral projection are rather close to their preset values, that is to say, the algorithm retrieves the values of parameters very well. To be convenient to compare them, all preset and retrieved values of parameters are listed in Table 10.3, where μ_1, μ_2, μ_{12}, μ_3, μ_{13}, μ_{23}, and μ_{123} represent $\mu(\{x_1\})$, $\mu(\{x_2\})$, $\mu(\{x_1, x_2\})$, $\mu(\{x_3\})$, $\mu(\{x_1, x_3\})$, $\mu(\{x_2, x_3\})$, and $\mu(X)$ respectively. In the result, the centers of classes C_1 and C_2 are numericalized as 0.059467 and 0.125833 respectively with a classifying boundary 0.08333, which is close to the preset value 0.08.

From three different view directions, Figures 10.8(a)-(f) illustrate the distribution of the data in a three-dimensional feature space, $[0, 1)^3$. The red balls are of class C_1, while the green balls are of class C_2. Figures 10.8(a), 10.8(c), and 10.8(e) show the data set without the classifying boundary, while Figures 10.8(b), 10.8(d), and 10.8(f) are added with the resulting classifying boundary. The classifying boundary is a broken plane with six pieces that divides the feature space into two parts—one

Table 10.3 The preset and retrieved values of monotone measure μ and weights b.

Parameters	Preset	Recovered
μ_1	0.1	0.015625
μ_2	0.2	0.209961
μ_{12}	0.6	0.606445
μ_3	0.05	0.057617
μ_{13}	0.8	0.718750
μ_{23}	0.9	0.982422
μ_{123}	1.0	1.0
b_1	0.1	0.113074
b_2	0.3	0.296820
b_3	0.6	0.590106

contains all red balls and another contains all green balls. These pieces of the broken plane have a common vertex (0.737024, 0.280771, 0.141226) on axis L that passes through the origin and has equations $b_1 f_1 = b_2 f_2 = b_3 f_3$. The weighted Choquet integral $\hat{Y} = (\mathrm{C})\int bf\,d\mu$ projects each point from the feature space onto axis L along one of the six pieces of a broken plane that is parallel to the broken plane shown in Figure 10.8.

The distribution of the resulting (final) projection $\hat{Y}$ on axis L is presented by a histogram in Figure 10.9(c). Also, Figures 10.9(a) and 10.9(b) present a histogram of the distribution of $\hat{Y}$ under the weighted Choquet integral projection with deferent parameters' values at the beginning and in the middle of the pursuit process performed by the genetic algorithm respectively. In these figures, several small black triangles indicate the numerical center of classes on axis L, while the yellow bars illustrate the location of classifying boundaries on L. From Figure 10.9, we can see how the weighted Choquet integral projection works and how the classifying boundary divides the feature space to classify the data.

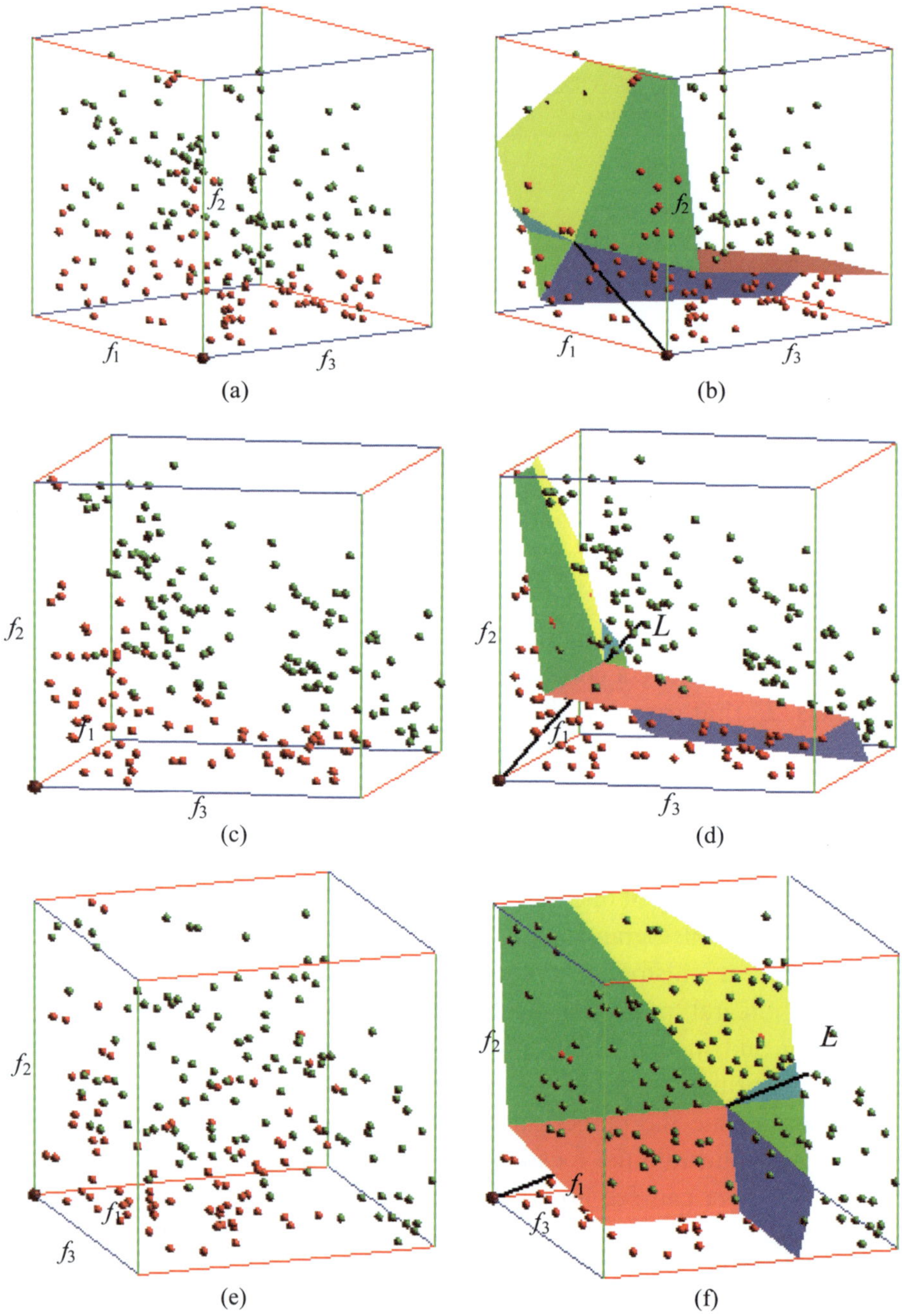

Fig. 10.8 View classification in Example 10.7 from three different directions.

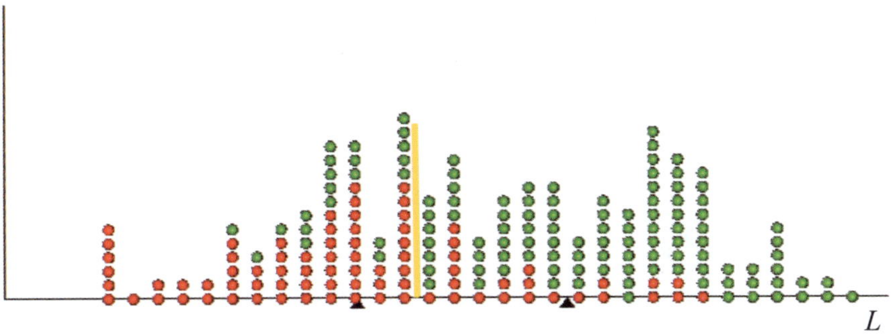

(a) Beginning: 43 misclassified points with $\mu_1 = 0.450484$, $\mu_2 = 0.740959$, $\mu_{12} = 0.872860$, $\mu_3 = 0.574908$, $\mu_{13} = 0.672933$, $\mu_{23} = 0.793390$, $b_1 = 0.472160$, $b_2 = 0.314903$, and $b_3 = 0.212937$.

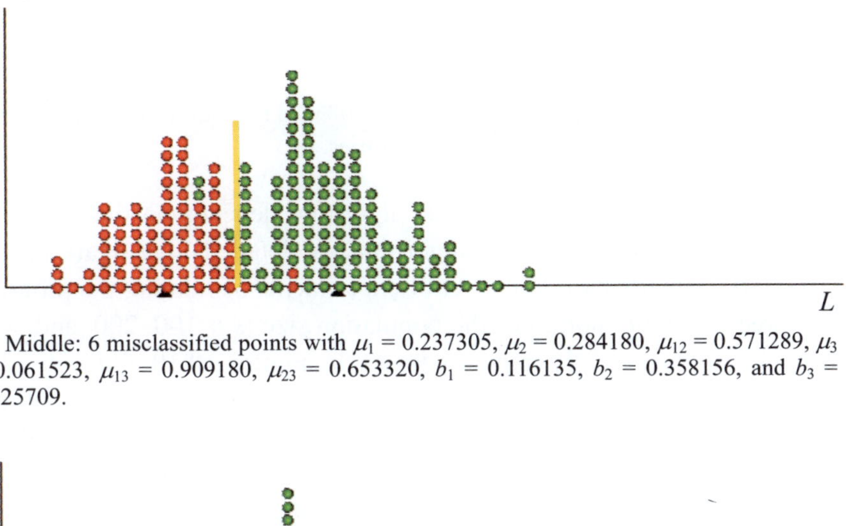

(b) Middle: 6 misclassified points with $\mu_1 = 0.237305$, $\mu_2 = 0.284180$, $\mu_{12} = 0.571289$, $\mu_3 = 0.061523$, $\mu_{13} = 0.909180$, $\mu_{23} = 0.653320$, $b_1 = 0.116135$, $b_2 = 0.358156$, and $b_3 = 0.525709$.

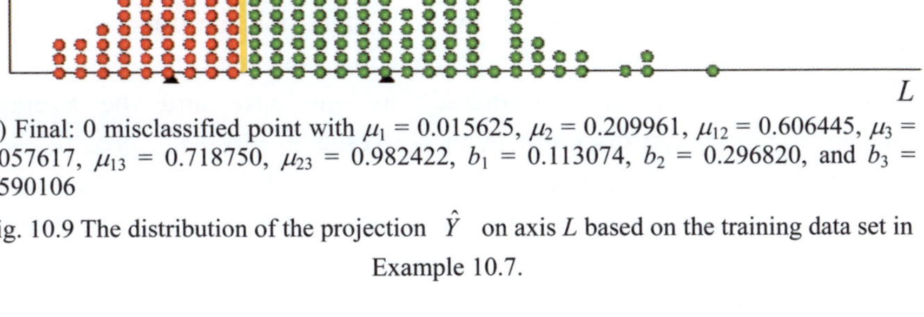

(c) Final: 0 misclassified point with $\mu_1 = 0.015625$, $\mu_2 = 0.209961$, $\mu_{12} = 0.606445$, $\mu_3 = 0.057617$, $\mu_{13} = 0.718750$, $\mu_{23} = 0.982422$, $b_1 = 0.113074$, $b_2 = 0.296820$, and $b_3 = 0.590106$

Fig. 10.9 The distribution of the projection $\hat{Y}$ on axis L based on the training data set in Example 10.7.

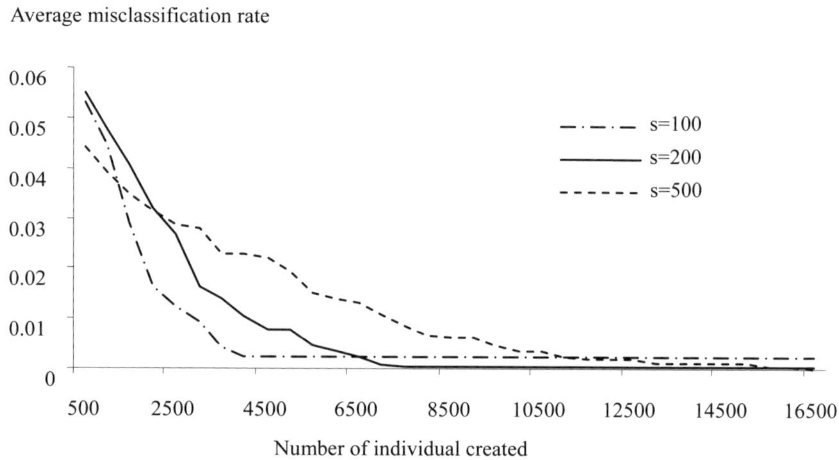

Fig. 10.10 The convergence of the genetic algorithm in Example 10.7 with different population sizes.

The convergence of the genetic algorithm depends on the choice of population size s used in the algorithm. A detailed investigate on the effect of s is made based on a set of experiments of running the program for three different choices of the population size (s = 100, 200, and 500) with ten trails (ten different seeds) each. Figure 10.10 illustrates the average convergence rate under these three different choices of the population size. The horizontal axis indicates the number of individuals created up to some moment during the running of the program. The vertical axis indicates the average best misclassification rate of these ten trails at that moment. Here the best misclassification rate is that of the best-so-far individual with the minimum misclassification rate. From Figure 10.10, we can see that all three curves for s = 100, 200, and 500 are decreasing. The curve for s = 100 decreases fastest at the beginning of the optimization process. However, there is a serious premature convergence (the population converging to identity before the misclassification rate is minimized) in this case and the average misclassification rate does not even converge to close to zero. In fact, five out of ten trails converged prematurely. The curve for s = 500 decreases too slowly though no premature among ten trails is found. Choosing s = 200 seems to be the best among these three choices. It decreases rather fast and there is only one premature convergence among the ten trails. Generally, too small a population size will lead to serious premature convergence in the optimization process, while too large a population size slows down the convergence speed and prolongs the running time of the program unnecessarily.

More examples of classification based on the weighted Choquet integral for real-world data sets can also be found in [Xu et al 2003].

In general, when the weighted Choquet integral is adopted in the classifier, the nonadditivity of signed efficiency measure μ describes the interaction among the contributions from feature attributes towards the classification. Thus, the classifying boundary is not an $(n-1)$-dimensional hyper-plane generally, but an $(n-1)$-dimensional broken hyper-plane with $n!$ pieces. The parameters, vector b and constant c as well as signed efficiency measure μ, can be optimally determined by the training samples in the given data set via a soft computing technique such as the genetic algorithm approximately. Such a nonlinear classification model is a real generalization of the classical linear classification model. After determining the parameters, b, c, and μ based on the training data, if a new individual f is obtained, we only need to calculate the value of $y(f) = (\mathrm{C})\int bf\, d\mu$. Then, we can classify f into one of the two classes according to whether $y(f) \le c$.

10.4 The Uniqueness Problem of the Classification by the Choquet Integral with a Linear Core

A natural idea to generalize the classification model presented in Section 10.2 is to replace the weighted integrand of the Choquet integral with a linear core similar to the nonlinear multiregression model discussed in Section 9.2, that is, the classifying boundary is identified by equation

$$(\mathrm{C})\int (a + bf)\, d\mu = c\,,$$

where a, b, c, f, and μ have the same meaning shown before. Unfortunately, it will violate the uniqueness of the Choquet integral expression of the classifying boundary. Though this does not affect the effectiveness of the classification, explaining the importance of the various attributes making contributions towards the classification becomes difficult.

Example 10.8 ([Zhang et al 2009]) An artificial data set now is used to show the problem on the uniqueness of the Choquet integral expression with linear core in the classification model. There are two feature attributes, two classes, and 26 records in the data set, that is, $X=\{x_1, x_2\}$, $S=$ {I , II} and $l=26$. The data are listed in Table 10.4 and shown in Figure 10.11 by white dots for class I and black dots for class II respectively, where f_1 and f_2 are two coordinates on the plane.

Geometrically, the classification by the Choquet integral with linear core can be described as follows. There is a projection axis L located by equation $a_1+b_1f_1=a_2+b_2f_2$. For each sample point (f_1, f_2), along with the directions presented by the two branches of each contour of the Choquet integral, is projected onto axis L as a point with corresponding value $(\mathrm{C})\int(a+bf)\,d\mu$. Thus, the 2-dimensional classification problem is converted to be a one-dimensional classification problem on L and, therefore, can be solved by using only one critical value, c, as the boundary of two classes on line L. The two classes in the given data on the plane can be actually well separated (with misclassification rate 0) by a contour of the Choquet integral with linear core. The contour is a broken line indicated, for example, by

$$f_2=-\frac{20}{9}f_1+\frac{80}{9} \quad \text{when} \quad a_1+b_1f_1\le a_2+b_2f_2$$

and

$$f_2=-\frac{40}{3}f_1+20 \quad \text{when} \quad a_1+b_1f_1> a_2+b_2f_2\,.$$

The vertex of the broken line is (1, 20/3), which is on axis L_1 possessing equation

$$f_2 = \frac{10}{3} f_1 + \frac{10}{3}.$$

These are shown in Figure 10.11. The corresponding values of parameters in the Choquet integral are $a = (1, 0)$, $b = (1, 3/10)$, $\mu(\emptyset) = 0$, $\mu(\{x_1\}) = 4/5$, $\mu(\{x_2\}) = 3/5$, $\mu(X) = 1$, and $c = 2$.

However, for the same boundary, it is not difficult to find other values of parameters in the expression of Choquet integral's contour. For example, $a = (0, 0)$, $b = (1, 3/20)$, $\mu(\emptyset) = 0$, $\mu(\{x_1\}) = 2/3$, $\mu(\{x_2\}) = 3/4$, $\mu(X) = 1$, and $c = 1$. The corresponding contour of the Choquet integral with these parameters' values coincides with the previous one, though the projection axis L_2 is different from L_1 and their projection directions are different in some area. These are also illustrated in Figure 10.11.

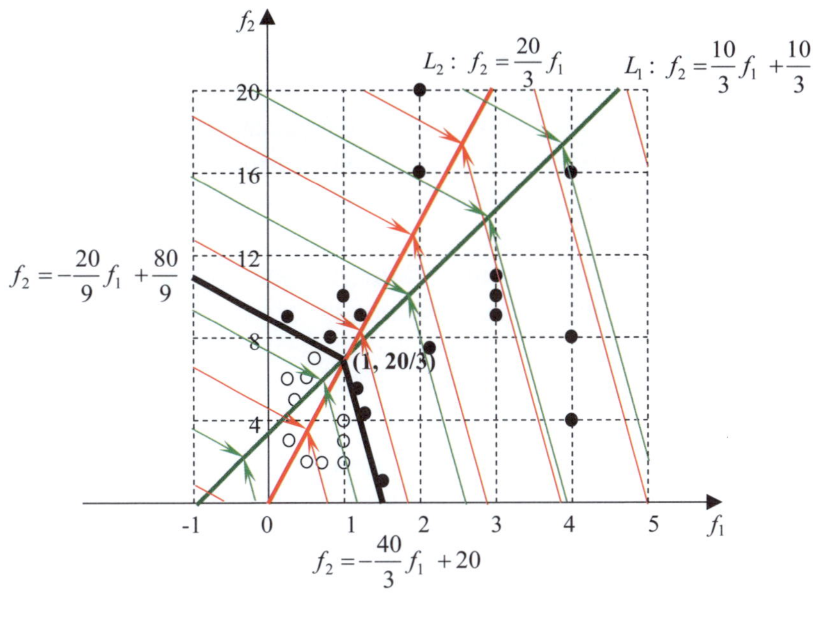

Fig. 10.11 Different projections share the same classifying boundary in Example 11.8.

Table 10.4 Data and their projections in Example 10.8.

Given data			Projected to line L_1			Projected to line L_2		
f_{j1}	f_{j2}	class	$a_1+b_1f_{j1}$ $=1+f_{j1}$	$a_2+b_2f_{j2}$ $=\frac{3}{10}f_{j2}$	$c=2$	$a_1+b_1f_{j1}$ $=f_{j1}$	$a_2+b_2f_{j2}$ $=\frac{3}{20}f_{j2}$	$c=1$
1/5	6	I	6/5	9/5	24/25	1/5	9/10	29/40
1/4	3	I	5/4	9/10	59/50	1/4	9/20	2/5
1/3	5	I	4/3	3/2	43/30	1/3	3/4	31/48
1/2	2	I	3/2	6/10	33/25	1/2	3/10	13/30
1/2	6	I	3/2	9/5	42/25	1/2	9/10	8/10
3/5	7	I	5/8	21/10	19/10	3/5	21/20	15/16
2/3	2	I	5/3	3/5	109/75	2/3	3/10	49/90
1	2	I	2	2/10	83/55	1	3/10	23/30
1	3	I	2	9/10	89/50	1	9/20	49/60
1	4	I	2	6/5	46/25	1	3/5	13/15
1	10	II	2	3	13/5	1	3/2	11/8
1/5	9	II	6/5	27/10	21/10	1/5	27/20	17/16
4/5	8	II	9/5	12/5	54/25	4/5	6/5	11/10
21/20	15/2	II	41/20	9/4	211/100	21/20	9/8	177/160
11/10	17/3	II	21/10	17/10	101/50	11/10	17/20	61/10
6/5	21/5	II	11/5	63/50	251/50	6/5	63/100	101/100
6/5	9	II	11/5	27/10	5/2	6/5	27/20	21/16
3/2	1	II	5/2	3/10	103/50	3/2	3/20	21/20
2	16	II	3	24/5	102/25	2	12/5	23/10
2	20	II	3	6	24/5	2	3	11/4
3	9	II	4	27/10	187/50	3	9/20	43/20
3	10	II	4	3	19/5	3	3/2	5/2
3	11	II	4	33/10	193/50	3	33/20	51/20
4	4	II	5	6/5	106/25	4	3/5	43/15
4	8	II	5	12/5	86/25	4	6/5	46/15
4	16	II	5	24/5	56/5	4	12/5	49/10

From Example 10.8, we may see that the classifying boundary can be expressed by more than one (in fact, infinitely many) different Choquet integrals, which have different values of the signed efficiency measure, different values of the parameters a and b, or different critical value c.

10.5 Advanced Consideration on the Nonlinear Classification Involving the Choquet Integral

Based on the basic classification model involving the Choquet integral discussed in Section 10.2, there are several improvements or generalizations. Some of them are briefly shown in the following.

10.5.1 *Classification by the Choquet integral with the widest gap between classes*

We have seen the uniqueness problem on the expression of the Choquet integral with a linear core in the classification. To avoid the trouble, an additional optimization criterion should be added to the original model discussed in the previous sections. One of the possible additional optimization criteria may be chosen as follows.

For a given data set (10.1), by using the model of the weighted Choquet integral, assume that the minimal misclassification rate is η. Each optimal solution (a classifier) with the minimal misclassification rate η is denoted as (a, b, c, μ). Let $\mathbf{Q}$ be the set of all optimal solutions, i.e.,

$$\begin{aligned}\mathbf{Q} &= \{(a, b, c, \mu) \mid (a, b, c, \mu) \text{ has misclassification rate } \eta\} \\ &= \{e^{(t)} = (a^{(t)}, b^{(t)}, c^{(t)}, \mu^{(t)}) \mid t \in T\},\end{aligned}$$

where T is a certain index set. For each $e^{(t)}$, find the interval $(c_1^{(t)}, c_2^{(t)})$ such that $(a^{(t)}, b^{(t)}, c\ , \mu^{(t)}) \in \mathbf{Q}$ for every $c \in (c_1^{(t)}, c_2^{(t)})$. Let $t_0 = \arg\max_{t\in T}(c_2^{(t)} - c_1^{(t)})$. Then take

$$e^{(t_0)} = (a^{(t_0)}, b^{(t_0)}, \frac{c_2^{(t_0)} + c_1^{(t_0)}}{2}, \mu^{(t_0)})$$

as the final unique optimal solution. This solution can be understood as the middle boundary of the widest gap between two classes.

10.5.2 *Classification by cross-oriented projection pursuit*

From Section 10.4, we may see that, even the optimal broken line as a classifying boundary of two classes has been found, expressing it as a contour of the Choquet integral with a linear core still has infinitely many different ways with respective projection axis L. This suggests us to restrict the location of the axis. Setting a by the zero vector, i.e., reducing the linear core to weighted integrand as we have done in Section 10.2 is just one of the possible ways. In this way, the projection axis is restricted to pass through the origin, of course, pass through the vertex of the broken line as well. Thus, an alternative way for reducing the number of optimal solutions is to fix the projection axis by two vertices of broken lines that are expressed as the contours of two different Choquet integrals, i.e., let the two Choquet integrals share one common projection axis. This method is called the *cross-oriented projection pursuit* based on the Choquet integral.

Let data set (10.1) be given. Based on the data set, now the classifier is a mapping $\mathcal{M}$: $R^n \to R^2$ with a boundary that separates R^2 to form a partition $\{S_1, S_2\}$ of R^2.

To reflect the complex interaction among the feature attributes towards the classification, two signed efficiency measures, μ and ν, defined on the power set of $X = \{x_1, x_2, \cdots, x_n\}$ are used for measuring the strength of contributions from each individual feature attribute as well as the strength of the joint contributions from each possible combination of feature attributes in two different points of view. Regarding each observation of the feature attributes as a function f defined on X, the Choquet integrals of $(af + b)$ with respect to μ or ν, in symbol $\int(af + b)d\mu$ and $\int(af + b)d\nu$, are used to project f from the feature space to R respectively, where $a = (a_1, a_2, \cdots, a_n)$ and $b = (b_1, b_2, \cdots, b_n)$ are n-dimensional vectors satisfying $\min_i a_i = 0$, and $\max_i |b_i| = 1$, for $i = 1, 2, \cdots, n$. Thus, ordered pair

$$\left(\int (af+b)d\mu, \ \int (af+b)d\nu \right)$$

forms mapping $\mathcal{M}$ from R^n to R^2. Vector a and b are called *oriented coefficients* of the projection axis. The partition of R^2 is formed by $S_1 = (c_1, \infty) \times (-\infty, c_2]$ and $S_2 = R^2 - S_1$, where c_1 and c_2 are boundary points on the projection axis L for the classification. That is, the corresponding regions of the classes in the sample space are: one class is indicated by

$$\int (af+b)d\mu > c_1 \ \text{ and } \ \int (af+b)d\nu \le c_2;$$

while another by

$$\int (af+b)d\mu \le c_1 \ \text{ or } \ \int (af+b)d\nu > c_2.$$

In this model, the values of signed efficiency measures μ and ν (except $\mu(X) = \nu(X) = 1$), vectors a and b, and numbers c_1 and c_2 are unknown. All of them should be determined based on given data of the feature attributes and the classifying attribute optimally, that is, such that the misclassification rate for the given training data is minimized.

The learning procedure of the determination of these unknown parameters is a cross-oriented projection pursuit. It may be performed via a two-layer adaptive genetic algorithm. In the first layer of the algorithm, each pair of c_1 and c_2 is a chromosome. This layer is devoted to determine the boundary points c_1 and c_2 optimally when the values of signed efficiency measures μ, ν, and vectors a, b are generated as a chromosome in the second layer. While the second layer is used to determine the values of signed efficiency measures μ, ν and vectors a, b optimally based on the optimized values of c_1 and c_2 to each chromosome.

Some typical distributions of two-class data set that can be well classified by the cross-oriented projection pursuit based on the Choquet integral are shown in Figures 10.12-13.

10.5.3 *Classification by the Choquet integral with quadratic core*

When the linear core is replaced by a quadratic core, the classifier based on the Choquet integral is more powerful. Such a classifier can capture the quadratic interaction among the contributions from feature attributes towards the classification. The classifying boundary has a form of $(C)\int(a+bf+cf^2)d\mu=d$, where $a=(a_1,a_2,\cdots,a_n)$, $b=(b_1,b_2,\cdots,b_n)$, $c=(c_1,c_2,\cdots,c_n)$ are n-vectors satisfying

$$\min_i a_i=0\text{, and }\max_i|b_i|=1\text{ for }i=1,2,\cdots,n,$$

μ is a signed efficiency measure with $\mu(X)=1$, d is a constant. All of them are unknown parameters, whose values should be optimally determined via a learning procedure based on data set (10.1).

Some typical two-class data distributions that can be well classified by a classifier based on the Choquet integral with quadratic core can be found in [Liu and Wang 2005].

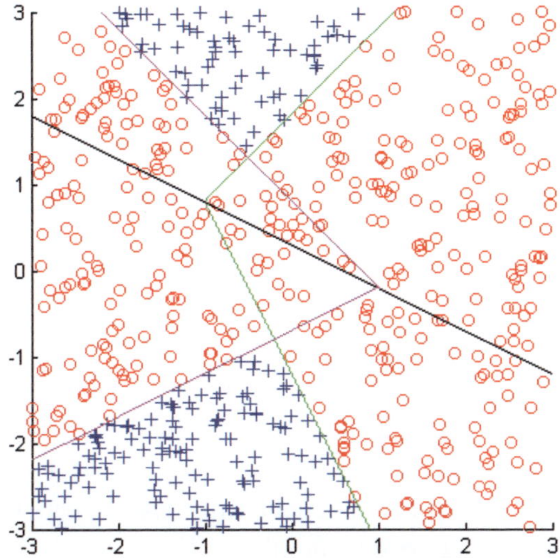

Fig. 10.12 Two-class two-dimensional data set that can be well classified by cross-oriented projection pursuit.

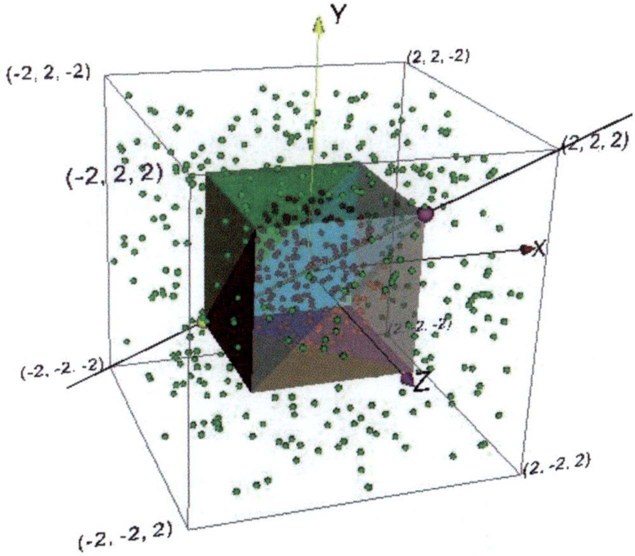

Fig. 10.13 Two-class three-dimensional data set that can be well classified by cross-oriented projection pursuit.

Chapter 11

Data Mining with Fuzzy Data

The Choquet integral discussed in Chapter 5 only supports real-valued integrand. It means that both the integrand and the integration result of the Choquet integral are real-valued. Thus, the data mining based on Choquet integral described in Chapters 8 and 9 can only handle the problems concerning crisp real numbers. However, in many databases, some attributes may not be numerical, but categorical, or may have linguistic words (or fuzzy numbers directly) as their values. Thus, to extend the advantages of Choquet integral to fuzzy domain such that it can manage fuzzy information, the original Choquet integral needs to be generalized (or say, fuzzified) such that it can be used to deal with fuzzy or linguistic data.

There is more than one way to fuzzify the Choquet integral. For a given signed efficiency measure whose values are crisp real numbers, when the integrand is allowed to be fuzzy-valued, the integration result of its Choquet integral may be defined as either a crisp real number or a fuzzy number. The former is called the *Defuzzified Choquet Integral with Fuzzy-valued Integrand* (DCIFI) which is named by its defuzzified (real-valued) integration result, while the latter is called the *Fuzzified Choquet Integral with Fuzzy-valued Integrand* (FCIFI) due to its fuzzy-valued integration result.

Both fuzzifications of the Choquet integral are applicable to different problems in the data mining area. The non-fuzzy integral result in the DCIFI facilitates to solve the classification or clustering problems where crisp boundaries are pursued. On the other hand, the FCIFI is more suitable to the regression problems where the objective attribute is also fuzzy-valued.

11.1 Defuzzified Choquet Integral with Fuzzy-Valued Integrand (DCIFI)

Definition 11.1 Let $\tilde{f}: X \to \mathcal{N}_F$ be a fuzzy-valued function defined on a finite universal set $X = \{x_1, x_2, \cdots, x_n\}$ and μ be a signed efficiency measure defined on $\mathcal{P}(X)$, the power set of X, where $\mathcal{N}_F$ is the set of all fuzzy numbers. The *defuzzified Choquet integral with fuzzy-valued integrand* (DCIFI) of $\tilde{f}$ is defined as

$$\int \tilde{f}\, d\mu = (\mathrm{C})\int_{-\infty}^{0} [\mu(\tilde{F}_\alpha) - \mu(X)]d\alpha + (\mathrm{C})\int_{0}^{\infty} \mu(\tilde{F}_\alpha)d\alpha ,$$

where $\tilde{F}_\alpha$ is the α-level set of the fuzzy-valued function $\tilde{f}$.

Obviously, the way to compute the value of the Choquet integral given in Section 5.4 cannot be directly applied for computing the DCIFI since the range of the fuzzy-valued function is not full-ordered, and therefore, the values of function $\tilde{f}$ at variant attributes cannot be rearranged in a nondecreasing order. However, we still can derive a calculation scheme of the DCIFI according to the fuzzy set theory and relevant properties of the Choquet integral. Actually, from the definition of the DCIFI, we can see that the calculation of the DCIFI can be rendered down into two subproblems:

(1) How to get $\tilde{F}_\alpha$ for a fuzzy-valued function $\tilde{f}$?
(2) How to get the value of $\mu(\tilde{F}_\alpha)$?

The following subsections aim to answer these questions respectively.

11.1.1 *The α-level set of a fuzzy-valued function*

Let $\mathscr{F}(X)$ be the class of all fuzzy subsets of X, the fuzzy power set of X. Any fuzzy subset of X, $\tilde{A}$, can be expressed as

$$\tilde{A} = \left\{ \frac{d_1}{x_1}, \frac{d_2}{x_2}, \cdots, \frac{d_n}{x_n} \right\},$$

where d_i is the degree of the membership of $\tilde{A}$ at x_i, $i = 1, 2, \cdots, n$.

Let $\tilde{f}$ be a fuzzy-valued function defined on X. Function $\tilde{f}$ can be expressed as $(m_1, m_2, ..., m_n)$, where m_i is the membership function of $\tilde{f}(x_i)$ at x_i, $i = 1, 2, \cdots, n$.

Definition 11.2 For any given $\alpha \in (-\infty, \infty)$, the *α-level set* of a fuzzy-valued function $\tilde{f} = (m_1, m_2, ..., m_n)$, denoted by $\tilde{F}_\alpha$, is a fuzzy subset of X, whose membership function $m_{\tilde{F}_\alpha}$ has a degree of membership

$$m_{\tilde{F}_\alpha}(x_i) = \frac{\int_\alpha^\infty m_i(t)dt}{\int_{-\infty}^\infty m_i(t)dt} \tag{11.1}$$

at attribute x_i if $\int_{-\infty}^\infty m_i(t)dt \neq 0$, $i = 1, 2, \cdots, n$. When $\tilde{f}(x_i)$ is a crisp number, then $\int_{-\infty}^\infty m_i(t)\, dt = 0$. In this case, the degree of membership at x_i, denoted by $m_{\tilde{F}_\alpha}(x_i)$, is equal to 1 if $\tilde{f}(x_i) \geq \alpha$, or 0 if $\tilde{f}(x_i) < \alpha$.

Example 11.1 Let $X = \{x_1, x_2, x_3\}$ and let a fuzzy-valued function $\tilde{f}$ assign each element of X a trapezoidal fuzzy number, denoted by four parameters $[a_l\ a_b\ a_c\ a_r]$, that is, $\tilde{f}(x_1) = [1.0\ \ 1.5\ \ 2.0\ \ 2.5]$, $\tilde{f}(x_2) = [4.0\ \ 4.5\ \ 5.0\ \ 5.0]$, and $\tilde{f}(x_3) = [3.0\ \ 3.5\ \ 4.0\ \ 4.5]$. Then, we have

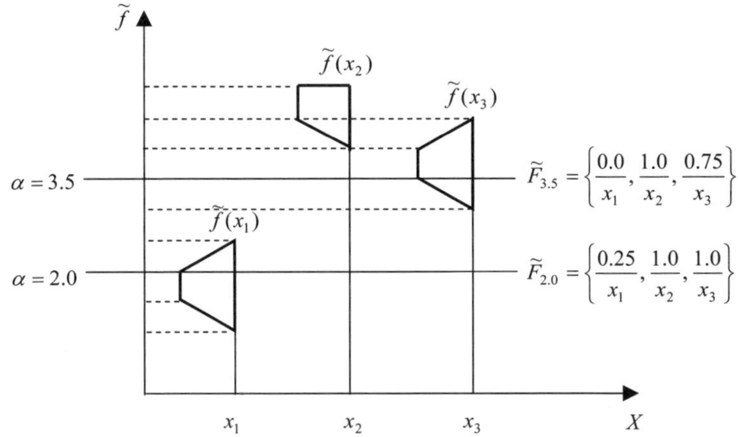

Fig. 11.1 The α-level set of a fuzzy-valued function in Example 11.1.

$$\tilde{F}_\alpha = \left\{ \frac{0.25}{x_1}, \frac{1.0}{x_2}, \frac{1.0}{x_3} \right\} \quad \text{when} \quad \alpha = 2.0,$$

while

$$\tilde{F}_\alpha = \left\{ \frac{0.0}{x_1}, \frac{1.0}{x_2}, \frac{0.75}{x_3} \right\} \quad \text{when} \quad \alpha = 3.5,$$

as shown in Fig. 11.1.

11.1.2 *The Choquet extension of* μ

We can derive the signed efficiency measure $\tilde{\mu}$ defined on $\mathscr{F}(X)$ based on the signed efficiency measure μ defined on $\mathscr{P}(X)$.

Definition 11.3 Let μ be a signed efficiency measure defined on $\mathscr{P}(X)$, the signed efficiency measure $\tilde{\mu}$ is a set function mapping from the fuzzy power set of X, $\mathscr{F}(X)$, to $(-\infty, \infty)$. For any fuzzy set $\tilde{A} \in \mathscr{F}(X)$ with membership function $m_{\tilde{A}}(x): X \to [0,1]$, we have

$$\widetilde{\mu}(\widetilde{A}) = \int m_{\widetilde{A}} d\mu \,, \tag{11.2}$$

where the integral is the Choquet integral with real-valued function, i.e., the membership function $m_{\widetilde{A}}$ of $\widetilde{A}$.

Here, for any crisp subset $A \in \mathscr{P}(X)$, we have $\widetilde{\mu}(A) = \int \chi_A d\mu = \mu(A)$, where

$$\chi_A(x) = \begin{cases} 1 & \text{if } x \in A \\ 0 & \text{if } x \notin A \end{cases}$$

is the characteristic function of A. Thus, $\widetilde{\mu}$ coincides with μ on $\mathscr{P}(X)$, that is, $\widetilde{\mu}$ is an extension of μ from $\mathscr{P}(X)$ onto $\mathscr{F}(X)$ and called the *Choquet extension* of μ.

For simplification, we use μ to replace $\widetilde{\mu}$ on $\mathscr{F}(X)$ without any confusion in the following context.

Example 11.2 Let $X = \{x_1, x_2, x_3\}$ and a signed efficiency measure μ be given as

$$\mu_0 = \mu(\varnothing) = 0 \,,$$

$$\mu_1 = \mu(\{x_1\}) = 1 \,,$$

$$\mu_2 = \mu(\{x_2\}) = -1 \,,$$

$$\mu_3 = \mu(\{x_1, x_2\}) = 3 \,,$$

$$\mu_4 = \mu(\{x_3\}) = 2 \,,$$

$$\mu_5 = \mu(\{x_1, x_3\}) = -1 \,,$$

$$\mu_6 = \mu(\{x_2, x_3\}) = 4 \,,$$

$$\mu_7 = \mu(\{x_1, x_2, x_3\}) = 5 .$$

For fuzzy set

$$\widetilde{A} = \widetilde{F}_{2.0} = \left\{\frac{0.25}{x_1}, \frac{1.0}{x_2}, \frac{1.0}{x_3}\right\} \text{ and } \widetilde{B} = \widetilde{F}_{3.5} = \left\{\frac{0.0}{x_1}, \frac{1.0}{x_2}, \frac{0.75}{x_3}\right\}$$

in Example 11.1, we have

$$\begin{aligned}\mu(\widetilde{F}_{2.0}) &= \int m_{\widetilde{F}_{2.0}} d\mu \\ &= m_{\widetilde{A}}(x_1) \cdot \mu_7 + [m_{\widetilde{A}}(x_2) - m_{\widetilde{A}}(x_1)] \cdot \mu_6 \\ &= 0.25 \cdot 5 + (1.0 - 0.25) \cdot 4 \\ &= 4.25\end{aligned}$$

and

$$\begin{aligned}\mu(\widetilde{F}_{3.5}) &= \int m_{\widetilde{F}_{3.5}} d\mu \\ &= m_{\widetilde{A}}(x_3) \cdot \mu_6 + [m_{\widetilde{A}}(x_2) - m_{\widetilde{A}}(x_3)] \cdot \mu_2 \\ &= 0.75 \cdot 4 + (1.0 - 0.75) \cdot (-1) \\ &= 2.75.\end{aligned}$$

11.1.3 *Calculation of DCIFI*

Obviously, it is rather difficult to express $\mu(\widetilde{F}_\alpha)$ in an explicit form involving only fundamental functions of α, and by which, to compute the precise value of the DCIFI. However, we can numerically calculate it approximately. Before illustrating the algorithm, some concepts and properties are reviewed.

The *support set* of a fuzzy number $\widetilde{a}$, denoted by $\widetilde{a}_{0+}$, is defined by $\widetilde{a}_{0+} = \{t | m_{\widetilde{a}}(t) > 0\}$, which is a crisp subset of the domain of the membership function of $\widetilde{a}$. We denote the left and the right terminals of the support set of $\widetilde{a}$ by $\widetilde{a}_l$ and $\widetilde{a}_r$, respectively. For example, a trapezoidal fuzzy number $\widetilde{a} = [1.0\ 1.5\ 2.0\ 2.5]$ has $\widetilde{a}_l = 1.0$ and $\widetilde{a}_r = 2.5$; a normal fuzzy number $\widetilde{b}$ has $\widetilde{b}_l = -\infty$ and $\widetilde{b}_r = \infty$.

A fuzzy-valued function $\widetilde{f}$ assigns each element x_i in the universal set a fuzzy number $\widetilde{f}(x_i)$, represented by its membership function $m_{\widetilde{f}(x_i)}(t)$, $i = 1, 2, \cdots, n$. Now, we denote the left and the right terminals of the support set of $\widetilde{f}(x_i)$ as $\widetilde{f}(x_i)_l$ and $\widetilde{f}(x_i)_r$, respectively.

Theorem 11.1 For a universal set X, let μ be a signed efficiency measure on $\mathscr{F}(X)$ and $\widetilde{f}$ be a fuzzy-valued function on X. Then,

$$(\mathrm{C})\int \widetilde{f} d\mu = (\mathrm{C})\int (\widetilde{f} - q) d\mu + q \cdot \mu(X),$$

where q is a crisp value and $(\widetilde{f} - q)$ is also a fuzzy-valued function with its values represented by $\widetilde{f}(x_i) - q$, $i = 1, 2, \cdots, n$.

Proof. Let $\widetilde{g} = \widetilde{f} - q$. Then $\widetilde{g}$ is also a fuzzy-valued function and its α-level set, $\widetilde{G}_\alpha$, satisfies $\widetilde{G}_\alpha = \widetilde{F}_{\alpha+q}$ or, equivalently, $\widetilde{G}_{\alpha-q} = \widetilde{F}_\alpha$, for any real number α. Thus, denoting $\alpha - q$ by β, we have

$$\begin{aligned}
(\mathrm{C})\int \widetilde{f} d\mu &= \int_{-\infty}^{0} [\mu(\widetilde{F}_\alpha) - \mu(X)] d\alpha + \int_0^\infty \mu(\widetilde{F}_\alpha) d\alpha \\
&= \int_{-\infty}^{0} [\mu(\widetilde{G}_{\alpha-q}) - \mu(X)] d\alpha + \int_0^\infty \mu(\widetilde{G}_{\alpha-q}) d\alpha \\
&= \int_{-\infty}^{0} [\mu(\widetilde{G}_{\alpha-q}) - \mu(X)] d(\alpha - q) + \int_0^\infty \mu(\widetilde{G}_{\alpha-q}) d(\alpha - q) \\
&= \int_{-\infty}^{-q} [\mu(\widetilde{G}_\beta) - \mu(X)] d\beta + \int_{-q}^{\infty} \mu(\widetilde{G}_\beta) d\beta \\
&= \int_{-\infty}^{-q} [\mu(\widetilde{G}_\beta) - \mu(X)] d\beta + \int_{-q}^{0} \mu(\widetilde{G}_\beta) d\beta + \int_0^\infty \mu(\widetilde{G}_\beta) d\beta \\
&\quad - \int_{-q}^{0} \mu(X) d\beta + \int_{-q}^{0} \mu(X) d\beta \\
&= \int_{-\infty}^{0} [\mu(\widetilde{G}_\beta) - \mu(X)] d\beta + \int_0^\infty \mu(\widetilde{G}_\beta) d\beta + \int_{-q}^{0} \mu(X) d\beta \\
&= (\mathrm{C})\int \widetilde{g} d\mu + q \cdot \mu(X) \\
&= (\mathrm{C})\int (\widetilde{f} - q) d\mu + q \cdot \mu(X) .
\end{aligned}$$

□

Using Theorem 11.1, we can write

$$(C)\int \tilde{f}d\mu = \int_0^{\infty} \mu(\tilde{G}_\alpha)d\alpha + q\cdot\mu(X)$$
$$= \int_0^{r-q} \mu(\tilde{G}_\alpha)d\alpha + q\cdot\mu(X),$$

where $\tilde{G}_\alpha$ is the α-level set of function $\tilde{g} = \tilde{f} - q$, $q = \min_{1\le i\le n} \tilde{f}(x_i)_l$ and $r = \max_{1\le i\le n} \tilde{f}(x_i)_r$.

Now, we can numerically calculate the approximate value of the DCIFI through the following algorithm.

(1) Input attributes' number n in X, subintervals' number K (with default value K=100) required in the approximate computing, function's values $\tilde{f}(x_i)$ for $i = 1, 2, \cdots, n$, and the values of the signed efficiency measure μ_j, $j = 1, 2, \cdots, 2^n - 1$.
(2) Find $q = \min_{1\le i\le n} \tilde{f}(x_i)_l$, $r = \max_{1\le i\le n} \tilde{f}(x_i)_r$. If $q = -\infty$ or $r = \infty$, then take $\tilde{f}(x_i)_l$ and $\tilde{f}(x_i)_r$ as the left and right terminal of $\tilde{f}(x_i)|_{\alpha=\varepsilon}$, $i = 1, 2, \cdots, n$, respectively. Here, ε is a very small positive real value defined by user with default value 10^{-3}. Then reset $q = \min_{1\le i\le n} \tilde{f}(x_i)_l$, $r = \max_{1\le i\le n} \tilde{f}(x_i)_r$, and set $\delta = (r-q)/K$.
(3) Replace $\tilde{f}(x_i)$ by $\tilde{f}(x_i) - q$.
(4) Initialize $\alpha = 0$ and $S = \mu_{2^n-1}/2$.
(5) $\alpha + \delta \Rightarrow \alpha$.
(6) Whether $\alpha > (r-q)$? If yes,

$$\delta\cdot(S - \frac{\Delta S}{2}) + q\cdot\mu_{2^n-1} \Rightarrow S,$$

output S as an approximate value of $\int \tilde{f}d\mu$, and stop; otherwise, continue.
(7) Find $c_i = m_{\tilde{F}_\alpha}(x_i)$ by Equation (11.1), $i = 1, 2, \cdots, n$.
(8) Regarding $h = (c_1, c_2, \cdots, c_n)$ as a function on X, calculate $\Delta S = \int hd\mu$ by scheme of calculation of classical Choquet integral with real-valued integrand.

(9) $S+\Delta S \Rightarrow S$ and go to (5).

We can see now, given a signed efficiency measure, the value of the DCIFI is a crisp real number. Though the information on the fuzziness is compressed, applying such an aggregation tool in data mining is usually more convenient than giving a fuzzy number.

Example 11.3 Suppose that the evaluation of submitted papers is based on three criteria: originality, significance, and presentation. They are denoted by x_1, x_2, and x_3 respectively. The importance of each individual criterion and their joint importance are described by a signed efficiency measure, μ, defined on $\mathscr{P}(X)$, where $X=\{x_1, x_2, x_3\}$. Also suppose that the values of μ are $\mu_1=0.2$, $\mu_2=0.3$, $\mu_3=0.8$, $\mu_4=0.1$, $\mu_5=0.4$, $\mu_6=0.4$, and $\mu_7=1$.

The range of the evaluation to each criterion for submitted papers by a journal editor is in the interval I = [0, 5]. However, the reviewers, usually, are only required to rate the criteria by the following words: "bad", "weak", "fair", "good", and "excellent". These are fuzzy concepts and can be described by fuzzy subsets of I, $\tilde{a}_b$, $\tilde{a}_w$, $\tilde{a}_f$, $\tilde{a}_g$, and $\tilde{a}_e$, with membership functions

$$m_b(t)=\begin{cases}1 & \text{if } t\in[0,1]\\ 3-2t & \text{if } t\in(1,1.5]\\ 0 & \text{otherwise}\end{cases},$$

$$m_w(t)=\begin{cases}1 & \text{if } t\in[1.5,2]\\ 2t-2 & \text{if } t\in[1,1.5)\\ 5-2t & \text{if } t\in(2,2.5]\\ 0 & \text{otherwise}\end{cases},$$

$$m_f(t) = \begin{cases} 1 & \text{if } t \in [2.5, 3] \\ 2t - 4 & \text{if } t \in [2, 2.5) \\ 7 - 2t & \text{if } t \in (3, 3.5] \\ 0 & \text{otherwise} \end{cases},$$

$$m_g(t) = \begin{cases} 1 & \text{if } t \in [3.5, 4] \\ 2t - 6 & \text{if } t \in [3, 3.5) \\ 9 - 2t & \text{if } t \in (4, 4.5] \\ 0 & \text{otherwise} \end{cases},$$

and

$$m_e(t) = \begin{cases} 1 & \text{if } t \in [4.5, 5] \\ 2t - 8 & \text{if } t \in [4, 4.5) \\ 0 & \text{otherwise} \end{cases},$$

respectively, then $\{\widetilde{a}_b, \widetilde{a}_w, \widetilde{a}_f, \widetilde{a}_g, \widetilde{a}_e\}$ is a fuzzy partition of *I*. Here, $\widetilde{a}_b$, $\widetilde{a}_w$, $\widetilde{a}_f$, $\widetilde{a}_g$, and $\widetilde{a}_e$ are trapezoidal fuzzy numbers (see Figure 3.5), and we can write $\widetilde{a}_b = [0\ 0\ 1\ 1.5]$, $\widetilde{a}_w = [1\ 1.5\ 2\ 2.5]$, $\widetilde{a}_f = [2\ 2.5\ 3\ 3.5]$, $\widetilde{a}_g = [3\ 3.5\ 4\ 4.5]$, and $\widetilde{a}_e = [4\ 4.5\ 5\ 5]$. Now, a paper is evaluated as "excellent" for originality, "fair" for significance, and "weak" for presentation by a reviewer. This reviewer's evaluation can be represented as a fuzzy-valued function $\widetilde{f} = (\widetilde{a}_e, \widetilde{a}_f, \widetilde{a}_w)$ on $X = \{x_1, x_2, x_3\}$. Thus, a global evaluation for the quality of the paper is given by the Choquet integral of $\widetilde{f}$ with respect to μ, $(\mathrm{C})\int \widetilde{f} d\mu$. Using the algorithm above, a rather precise approximate value of $(\mathrm{C})\int \widetilde{f} d\mu$ can be obtained:

$$(\mathrm{C})\int \widetilde{f} d\mu \approx 2.92176 \quad \text{when} \quad K = 100,$$

$$(\mathrm{C})\int \tilde{f} d\mu \approx 2.92222 \qquad \text{when} \quad K = 1000 .$$

For another paper evaluated as "bad" for originality, "good" for significance, and "excellent" for presentation, denoting $\tilde{g} = (\tilde{a}_b, \tilde{a}_g, \tilde{a}_e)$, we have

$$(\mathrm{C})\int \tilde{g} d\mu \approx 1.96618 \qquad \text{when} \quad K = 100 ,$$

$$(\mathrm{C})\int \tilde{g} d\mu \approx 1.96611 \qquad \text{when} \quad K = 1000 .$$

It means that the paper represented by function $\tilde{f}$ is more suitable than the one represented by function $\tilde{g}$ for publishing in the journal.

Since the procedure of calculating the value of the Choquet integral with fuzzy integrand will be repeated for a large number in multiregression or classification problems, we should reduce its running time as much as possible. For most real problems in decision-making, the precision of the relevant results reaching three or four decimal digits is sufficient. So, this example also suggests us to use 100 as the default value of K in the algorithm.

In Example 11.3, all attributes have the same dimension. This is a rather special case in data analysis. Generally, the attributes may have variant dimensions. Thus, for a given function $\tilde{f}$ on X, we should usually use $a + b\tilde{f}$ as the integrand in the Choquet integral to balance the scales of the variant dimensions, where both $a = (a_1, a_2, \cdots, a_n)$ and $b = (b_1, b_2, \cdots, b_n)$ are functions defined on X and their values are optimally determined from given data via genetic algorithms.

11.2 Classification Model Based on the DCIFI

In classification, an observation is denoted by an n-dimensional vector $(f(x_1), f(x_2), \cdots, f(x_n))$, whose components $f(x_i)$ are measurements of the feature attributes x_i, $i = 1, 2, \cdots, n$. We assume that there exist

m groups or classes in the n-dimensional space, denoted by $C_1, C_2, \cdots, C_m$, and associated with each observation is a categorical attribute Y that denotes the class or group membership. For example, if $Y = j$, then the observation belongs to C_j, $j \in \{1, 2, \cdots, m\}$. To design the classifier, we are usually given a set of training data with observations of known classes, represented as

x_1	x_2	$\cdots$	x_n	Y
$f_1(x_1)$	$f_1(x_2)$	$\cdots$	$f_1(x_n)$	y_1
$f_2(x_1)$	$f_2(x_2)$	$\cdots$	$f_2(x_n)$	y_2
$\vdots$	$\vdots$		$\vdots$	$\vdots$
$f_l(x_1)$	$f_l(x_2)$	$\cdots$	$f_l(x_n)$	y_l

The training data set is used to set up internal parameters of the classifier. Here, the positive integer l is the number of samples in the training data set. Once a classifier has been devised, we may estimate the class belongingness for any new observation.

11.2.1 *Fuzzy data classification by the DCIFI*

When the measurements of feature attributes of an observation are heterogeneous fuzzy data, such as crisp data, fuzzy data, interval values, or linguistic variables, they are denoted by an n-dimensional fuzzy data vector $(\tilde{f}(x_1), \tilde{f}(x_2), \cdots, \tilde{f}(x_n))$. Such an n-dimensional fuzzy data vector can be visualized as a fuzzy point, which is not a single point but a special fuzzy subset in the n-dimensional space. Each coordinate value of a fuzzy point is a fuzzy number. A typical 2-dimensional heterogeneous fuzzy data $(\tilde{f}(x_1), \tilde{f}(x_2))$ is shown in Fig. 11.2. It is depicted as a frustum of a prism with height as 1. It has two coordinates which are represented by two different trapezoidal fuzzy numbers with their membership functions shown on the $m(t_1)$ - t_1 and the $m(t_2)$ - t_2 planes in Fig. 11.2, respectively.

Remember that the DCIFI takes a fuzzy-valued function as its integrand and gives a crisp value as its integration result. It can be regarded as a projection from the feature space onto the real axis. Under

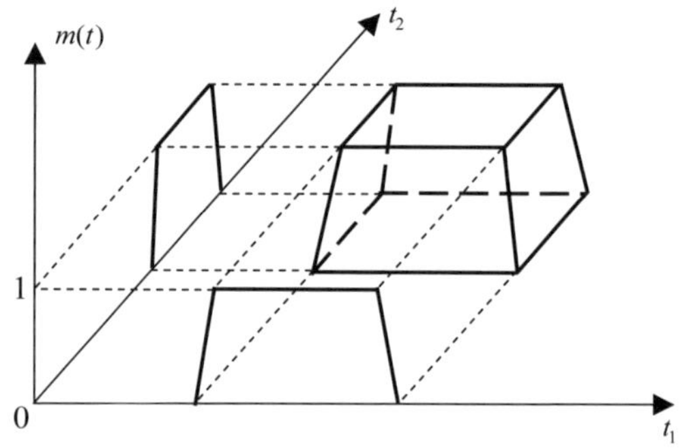

Fig. 11.2 A typical 2-dimensional heterogeneous fuzzy data.

such a scheme, any fuzzy point $(\tilde{f}(x_1), \tilde{f}(x_2), \cdots, \tilde{f}(x_n))$, denoted simply by $(\tilde{f}_1, \tilde{f}_2, \cdots, \tilde{f}_n)$ in the feature space, is regarded as a fuzzy-valued function $\tilde{f}$ defined on $X = \{x_1, x_2, \cdots, x_n\}$, and furthermore, projected onto a virtual variable, denoted by $\hat{Y}$, on the real axis through a DCIFI defined by

$$\hat{Y} = (\mathrm{C})\int \tilde{f} d\mu . \tag{11.3}$$

Figure 11.3 illustrates the DCIFI projection of some heterogeneous fuzzy data in the 2-dimensional space. Here, all heterogeneous fuzzy data are distributed into two classes. Each class has three observations. Each observation is identified by its fuzzy-valued coordinates $\tilde{f}(x_1)$ and $\tilde{f}(x_2)$. By certain DCIFI projection, each observation has been projected onto a virtual point (denoted by the black dots in Fig. 11.3) on the real axis L. It is natural to assume that there exists a boundary in the 2-dimensional space, on which each point can be projected onto an identical virtual point (denoted by the white dot in Fig. 11.3), called the virtual boundary, on the real axis by the same DCIFI projection. According to this assumption, a classification problem of n-dimensional heterogeneous fuzzy data can be simplified to that of one-dimensional real data.

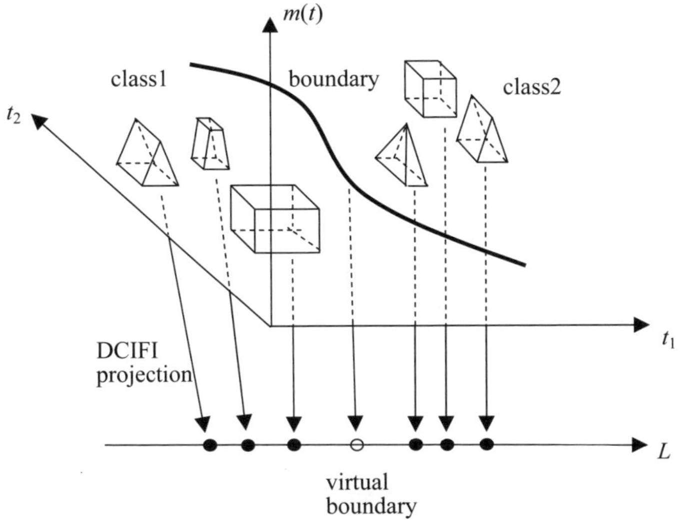

Fig. 11.3 The DCIFI projection for 2-dimensional heterogeneous fuzzy data.

Good performance of the DCIFI projection classifier is expected due to the use of the signed efficiency measure and the relevant nonlinear integral which can handle heterogeneous fuzzy data, since the nonadditivity of the signed efficiency measure reflects the importance of feature attributes, as well as their inherent interaction, toward the discrimination of the fuzzy points. In fact, the global contribution of several feature attributes to the decision of classification is not just the simple sum of the contributions of each feature to the decision. A combination of the feature attributes may have a mutually restraining or a complementary synergy effect on their contributions toward the classification decision. So, the signed efficiency measure defined on the power set of all feature attributes is a proper representation of the respective importance of the feature attributes and the interaction among them, and a relevant DCIFI is a good fusion tool to aggregate information in different forms coming from the feature attributes for the classification.

11.2.2 *GA-based adaptive classifier-learning algorithm via DCIFI projection pursuit*

Now, based on the DCIFI, we want to find an appropriate aggregation formula that projects the n-dimensional feature space onto the real axis, L, such that each fuzzy point $\tilde{f} = (\tilde{f}_1, \tilde{f}_2, \cdots, \tilde{f}_n)$ becomes a value of the virtual variable that is optimal with respect to classification. In such a way, each classifying boundary is just a point on the real axis L.

The classification task by the DCIFI projection classifier can be divided into two parts:

(1) The DCIFI projection classifier depends on the signed efficiency measure μ, so how to determine the values of μ is the first problem we are facing with.
(2) Once the values of μ are retrieved, the DCIFI projection classifier is established. To classify new data, boundaries on the real axis L should be determined.

The following two parts focus on the above two problems respectively.

A. Boundaries determination

A DCIFI projection classifier is described by a signed efficiency measure μ. Once the values of μ are given, the n-dimensional classification problem of heterogeneous fuzzy data is reduced to a one-dimensional classification problem of crisp data on the axis L of the virtual variable. The m classes of records in the original training data set are now projected to be m classes on the projection axis L. We can still use symbol C_k, $k = 1, 2, \cdots, m$, to denote these classes. The center, c_k, of each class C_k on L is the medium of the values of the virtual variables corresponding to the points in class C_k. The center c_k, expressed as a real number, is a numericalization of class C_k. After arranging $\{c_k \mid k = 1, 2, \cdots, m\}$, and therefore, $\{C_k \mid k = 1, 2, \cdots, m\}$, in an increasing order as $(c_{k_1}, c_{k_2}, \cdots, c_{k_m})$ and $(C_{k_1}, C_{k_2}, \cdots, C_{k_m})$, where $(k_1, k_2, \cdots, k_m)$ is a permutation of $\{1, 2, \cdots, m\}$, we carry out a point-wise search for the best classifying boundary between each pair of successive classes one by one under the criterion of minimizing the misclassification rate which is defined as the number of misclassified records (points) in the training set divided by data size l. The following algorithm is devoted to determining the boundaries of successive classes which have been rearranged according to the ascending order of their centers:

(1) Initialize $i = 1$.
(2) Find $\hat{Y}^*(k_i)$, the farthest right (largest) point of C_{k_i}, and $\hat{Y}^*(k_{i+1})$, the farthest left (smallest) point of $C_{k_{i+1}}$.
(3) If $\hat{Y}^*(k_i) \le \hat{Y}^*(k_{i+1})$ (as shown in Fig. 11.4(a))

$$b_i = \frac{(\hat{Y}^*(k_i) + \hat{Y}^*(k_{i+1}))}{2},$$

where b_i is the boundary between class C_{k_i} and $C_{k_{i+1}}$.
(4) Else if $\hat{Y}^*(k_i) > \hat{Y}^*(k_{i+1})$ (as shown in Fig. 11.4(b))
b_i is the average of the collection points which satisfy three conditions:

(a) are members of class C_{k_i} and $C_{k_{i+1}}$;
(b) are between $\hat{Y}^*(k_i)$ and $\hat{Y}^*(k_{i+1})$; and
(c) have property "possessing the lowest number of misclassified points if being a classifying boundary".

(5) Check whether $i = m$, if not, $i+1 \Rightarrow i$, and go to (2); if yes, go to (6).
(6) End.

Thus, $b_1, b_2, \cdots, b_{m-1}$ are the best classification boundaries for the DCIFI projection classifier with respect to the given signed efficiency measure μ. The corresponding global misclassification rate is the sum of the numbers of misclassified points in these $(m-1)$ pairs of successive classes divided by l.

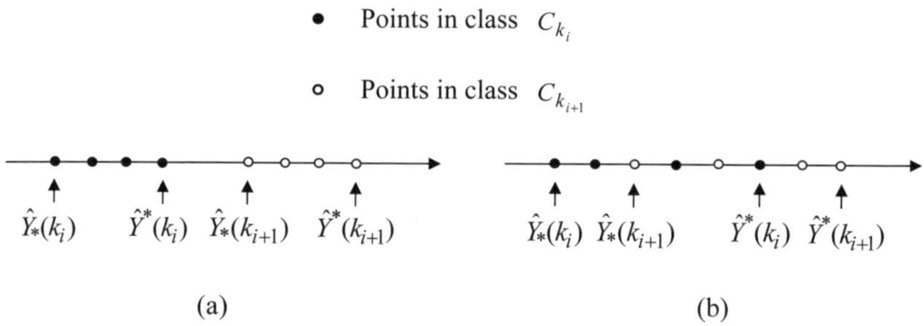

Fig. 11.4 Illustration of virtual projection axis L when determining the boundary of a pair of successive classes C_{k_i} and $C_{k_{i+1}}$: (a) when $\hat{Y}^*(k_i) \le \hat{Y}_*(k_{i+1})$; (b) when $\hat{Y}^*(k_i) > \hat{Y}^*(k_{i+1})$.

B. GA-based adaptive classifier-learning algorithm

In this part, we discuss the optimization of the signed efficiency measure μ under the criterion of minimizing the corresponding global misclassification rate, and then obtain an optimal DCIFI projection classifier. The optimizing process is just a "pursuit" for searching an appropriate projection direction. It is performed by the GA-based adaptive classifier-learning algorithm (GACA). The optimization is also a data-driven process, where a training data set in the form of

x_1	x_2	$\cdots$	x_n	Y
$\tilde{f}_{11}$	$\tilde{f}_{12}$	$\cdots$	$\tilde{f}_{1n}$	$\hat{Y}_1$
$\tilde{f}_{21}$	$\tilde{f}_{22}$	$\cdots$	$\tilde{f}_{2n}$	$\hat{Y}_2$
$\vdots$			$\vdots$	$\vdots$
$\tilde{f}_{l1}$	$\tilde{f}_{l2}$	$\cdots$	$\tilde{f}_{ln}$	$\hat{Y}_l$

is needed. Here, $\tilde{f}_{ji}$ denotes the fuzzy value of the i-th feature at the j-th observation and $\hat{Y}_j$ denotes the class tag of the j-th observation, $i = 1, 2, \cdots, n;\ j = 1, 2, \cdots, l$.

In the GACA, each individual of chromosome represents a DCIFI projection which is identified by the values of a signed efficiency measure μ. Since real coding method is employed, each individual of chromosome consists of $(2^n - 1)$ genes. Each gene represents a real value between 0 and 1. The population in the GACA consists of s individuals of chromosome. The misclassification rate is adopted for estimating the fitness value of each individual of chromosome (i.e., the DCIFI projection). The probability of an individual of chromosome in the population being chosen as a parent to produce offspring depends on its fitness value. The optimization in the GACA is performed under the criterion of minimizing the misclassification rate. Fig. 11.5 shows the flow chart of the GACA.

It starts off with an initialized population. Individuals of chromosome in the population are decoded into their corresponding signed efficiency measures to further determine their corresponding DCIFI projections. For a DCIFI projection, each observation in the training data set can be projected onto its virtual point on the real axis. According to the class tags provided by the training data, we can pursue the best virtual boundaries of the DCIFI projection being considered using the boundaries determination approach presented in Subsection 11.2.2-A. Then, cooperated with the training set, we can derive the misclassification

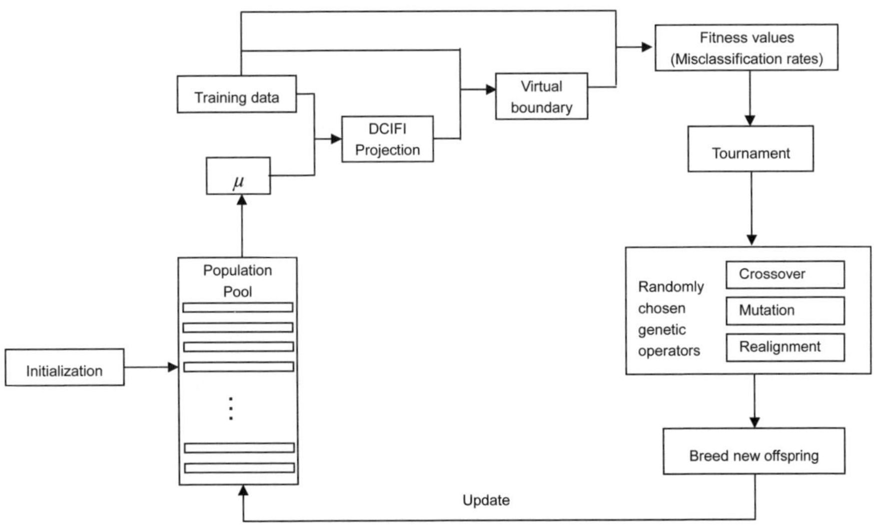

Fig. 11.5 Flowchart of the GACA.

rate of the current DCIFI projection, which also represents the fitness value of the corresponding individual in the population. After that, a tournament selection is performed. Better individuals have more chance to produce offspring by some randomly chosen genetic operators. The newly created offspring update the population. This process repeats until we get zero misclassification rate or the generation number exceeds the preset maximum number of generations.

To maintain the diversity of the searching space of our genetic algorithm, a special set of operations is used when the best fitness value remains unchanged for several consecutive generations (default value is 20). At that time, original population is divided into three parts by ascending order on fitness values. The individuals of chromosome in the first part are kept, while those in the second part create new offspring by random mutation, and those in the third part are replaced by new randomly created individuals of chromosome. Then, the population is updated and the iteration is continued.

After determining the signed efficiency measure μ and the respective classification boundaries $b_1, b_2, \cdots, b_{m-1}$ from the training data, any new observation of the feature attributes $\tilde{f} = (\tilde{f}_1, \tilde{f}_2, \cdots, \tilde{f}_n)$ can be classified by calculating its corresponding value of the virtual variable

$$\hat{Y} = (\mathrm{C}) \int \tilde{f} d\mu$$

and checking its location relative to the classification boundaries in the order of $b_1, b_2, \cdots, b_{m-1}$ one by one. If $\hat{Y} \le b_1$, then $\tilde{f}$ is classified into class C_{k_1}; if $\hat{Y} \in (b_{j-1}, b_j]$, then $\tilde{f}$ is classified into class C_{k_j}, $j = 2, 3, \cdots, m-1$; otherwise, $\tilde{f}$ is classified into class C_{k_m}.

11.2.3 *Examples of the classification problems solved by the DCIFI projection classifier*

To evaluate the performance of the DCIFI projection classifier, a series of examples both on synthetic and real data sets have been conducted.

A. Examples on synthetic data

Two synthetic data sets, one containing 2-dimensional heterogeneous fuzzy data distributed in 3 classes, and the other containing 3-dimensional heterogeneous fuzzy data distributed in 2 classes, are generated and used to verify the efficiency and the effectiveness of the DCIFI and the GACA. To evaluate the performance of the GACA on recovering the classifier parameters, the classifier parameters, including the values of the signed efficiency measure and the virtual boundaries, are preset. The preset DCIFI projection constructs normally distributed heterogeneous fuzzy data for each class which is separated by the preset virtual boundaries. Then, using the created training data sets, the GACA should recover the preset values of the parameters and obtain a low misclassification rate. The procedure to construct the synthetic training data sets is detailed as follows.

Assume that the data set has n feature attributes $\{x_1, x_2, \cdots, x_n\}$, m classes $\{C_1, C_2, \cdots, C_m\}$, and l records with l_j records for class C_j, $j = 1, 2, \cdots, m$. Here, $l = \sum_{j=1}^{m} l_j$. Each sample in the created data sets has the form of

$$\{(\tilde{f}(x_1), \tilde{f}(x_2), \cdots, \tilde{f}(x_n)), \quad class \ \ tag\}.$$

The following algorithm creates the heterogeneous fuzzy data (with trapezoidal fuzzy number in each dimension) which are distributed in a unit hypercube in the n-dimensional space and classified into m classes.

(1) Preset the values of the signed efficiency measure μ by assigning $\mu_1, \mu_2, \cdots, \mu_{2^n-1}$ and the virtual boundaries $b_1, b_2, \cdots, b_{m-1}$.

(2) Create the center of a fuzzy point in the n-dimensional space, represented as a vector $(c_1, c_2, \cdots, c_n)$. Each coordinate c_i, $i = 1, 2, \cdots, n$, of the center is a real number generated by a random number generator with the uniform distribution in $[0, 1)$. Create a fuzzy point $(\tilde{f}(x_1), \tilde{f}(x_2), \cdots, \tilde{f}(x_n))$, where $\tilde{f}(x_i)$ is a randomly generated trapezoidal fuzzy number with its support set

$[c_i - r_i, c_i + r_i]$, $i = 1, 2, \cdots, n$. Here, r_i is a random value between 0.0 and 0.05.

(3) For each observation $(\tilde{f}(x_1), \tilde{f}(x_2), \cdots, \tilde{f}(x_n))$, calculate the corresponding value of the DCIFI, denoted by $\hat{Y}$, with respect to the preset μ.

(4) Create a random number, $\xi \in [0, 1)$, with the uniform distribution. In case $\hat{Y} \le b_1$, if $\xi \le e^{-(\hat{Y}-a_1)^2/2\sigma_1^2}$, then assign class C_1 to the right part of record, otherwise, abandon this record. In case $\hat{Y} \in (b_{j-1}, b_j]$, if $\xi \le e^{-(\hat{Y}-a_j)^2/2\sigma_j^2}$, then assign class C_j to the right part of record, $j = 1, 2, \cdots, m-1$; otherwise, abandon this record. In case $\hat{Y} > b_{m-1}$, if $\xi \le e^{-(\hat{Y}-a_m)^2/2\sigma_m^2}$, then assign class C_m to the right part of record; otherwise, abandon this record. Here, the normal distribution $N(a_j, \sigma_j^2)$ are used to control the distribution of data in class C_j, $j = 1, 2, \cdots, m$.

(5) Repeat step (2) to step (4) until l_j records of class C_j, $j = 1, 2, \cdots, m$, have been created.

Example 11.4 Consider a classification problem of 2 feature attributes and 3 classes, that is, $X = \{x_1, x_2\}$, $C = \{C_1, C_2, C_3\}$. Totally 100 records are provided in the training data set, where 20 records for C_1, 50 records for C_2, and 30 records for C_3. The preset parameters to generate the training data are as follows: $\mu(\{x_1\}) = -0.1$, $\mu(\{x_2\}) = 0.2$, $\mu(\{x_1, x_2\}) = 1.0$, $b_1 = 0.2$, and $b_2 = 0.6$. Each record in the training data set presents a fuzzy point in the 2-dimensional space. Here, the fuzzy point is described by a 2-tuple vector whose elements are trapezoidal fuzzy number represented by their membership functions. Fig. 11.6 shows the sample data, where each frustum of a prism denotes a 2-dimensional fuzzy point (with dashed contours for data of C_1, solid contours for data of C_2, and dash dotted contours for data of C_3). Setting $s = 20$ as the population size and running the GACA with the whole sample data, after 3 generations, zero misclassification rate is achieved, and we obtain a trained DCIFI projection classifier with the classifying boundaries (thick broken lines in Fig. 11.6). Here, the straight line starting from the origin shows the virtual real axis to which the 2-dimensional heterogeneous fuzzy data are projected by the DCIFI. The values of the signed efficiency measure and boundaries in the retrieved

DCIFI projection classifier are rather close to the preset ones. That is to say, the GACA can retrieve the values of parameters well and perform the classification task successfully. The comparison of the preset and the retrieved values of parameters is listed in Table 11.1.

Example 11.5 Consider a classification problem of 3 feature attributes and 2 classes, that is, $X=\{x_1,x_2,x_3\}$, $C=\{C_1,C_2\}$. 200 records are generated by the preset DCIFI parameters as: $\mu(\{x_1\})=0.1$, $\mu(\{x_2\})=0.2$, $\mu(\{x_1,x_2\})=0.3$, $\mu(\{x_3\})=0.05$, $\mu(\{x_1,x_3\})=0.25$,

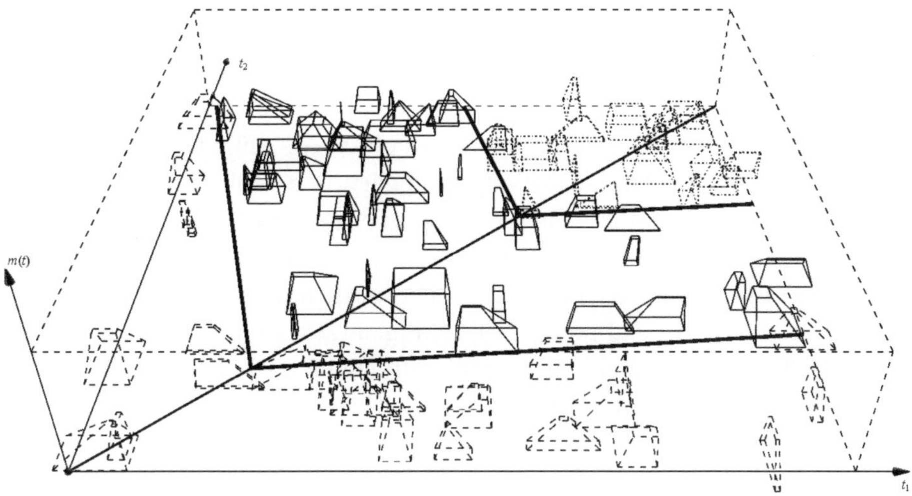

Fig. 11.6 The training data and the trained classifying boundaries in Example 11.4.

Table 11.1 Preset and retrieved values of the signed efficiency measure and boundaries in Example 11.4.

Parameters	Preset	Retrieved
$\mu(\{x_1\})$	-0.1	-0.105981
$\mu(\{x_2\})$	0.2	0.189793
$\mu(\{x_1,x_2\})$	1.0	1.000000
b_1	0.2	0.201631
b_2	0.6	0.598042

$\mu(\{x_2, x_3\}) = 0.9$, $\mu(\{x_1, x_2, x_3\}) = 1.0$ and $b_1 = 0.23$, where 80 records are for C_1 and 120 records are for C_2. Setting $s = 30$ as the population size and running the GACA with the whole sample data, after 50 generations, we obtain the trained DCIFI projection classifier with misclassification rate 0. The values of the signed efficiency measure in the retrieved DCIFI projection are rather close to their corresponding preset values. This experiment also confirms that our GACA can retrieve the values of the classifier parameters accurately. The comparison of the preset and the retrieved values of parameters are listed in Table 11.2. Fig. 11.7 illustrates the distribution of the training data and the classifying boundary in 3-dimensional feature space from two different viewing directions. The 3-dimensional fuzzy data are represented by cubes in the graph. The lengths on three dimensions of a cube denote the ranges of support sets of the membership functions of three feature attributes in each observation. The blue cubes are of class C_1, while the yellow cubes are of class C_2. The classifying boundary is a broken plane with six pieces that divide the feature space into two parts. These pieces of broken planes have a common vertex $(0.239537, 0.239537, 0.239537)$ on the virtual axis L (denoted by the black line in graph) that passes through the origin and point $(1, 1, 1)$. Fig. 11.7 also reveals the ability of the DCIFI projection classifier on classifying data which are separated by boundaries with irregular shape.

Table 11.2 Preset and retrieved values of the signed efficiency measure and boundaries in Example 11.5.

Parameters	Preset	Retrieved
$\mu(\{x_1\})$	0.10	0.105585
$\mu(\{x_2\})$	0.20	0.181064
$\mu(\{x_1, x_2\})$	0.30	0.318546
$\mu(\{x_3\})$	0.05	0.053967
$\mu(\{x_1, x_3\})$	0.25	0.246499
$\mu(\{x_2, x_3\})$	0.90	0.907981
$\mu(\{x_1, x_2, x_3\})$	1.00	1.000000
b_1	0.23	0.239537

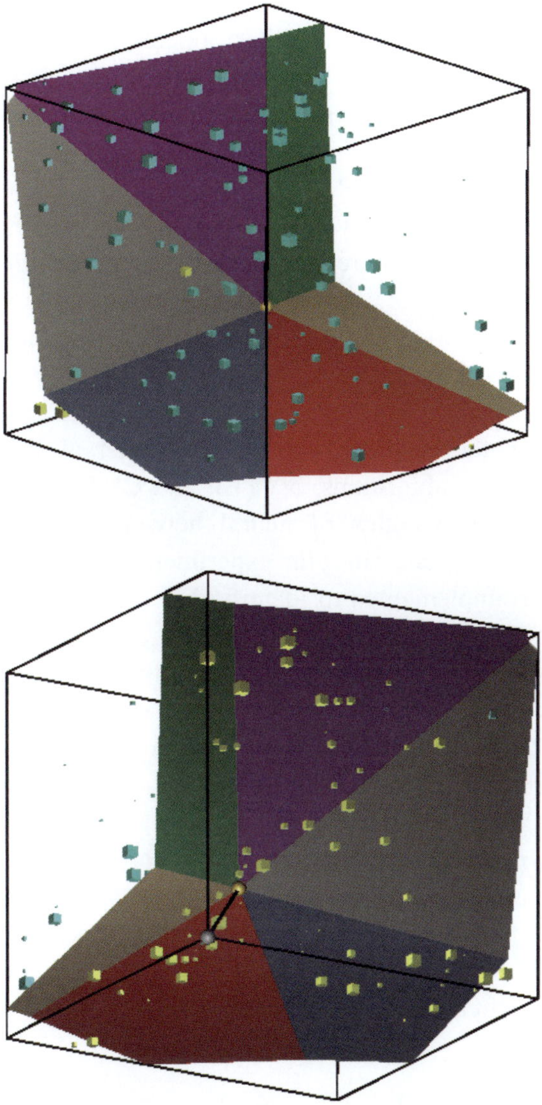

Fig. 11.7 Artificial data and the classifying boundaries in Example 11.5 — from two view directions.

B. Application on emitters identification

It is a high-priority problem in military operation to identify and track unique mobile transmitters for targeting. A powerful emitter identification function is necessary to warn of immediate threat with enough information to take evasive action. In military operation, such identification is accessed by Radio Frequency (RF), Pulse Width (PW), and Pulse Repetition Interval (PRI), of the collected pulse descriptor words. They form the feature attributes of an observation recognition problem, denoted by x_1, x_2, and x_3, respectively. The values of these features vary in interval ranges in accordance with a specific radar emitter. Shieh et al proposed a fuzzy-neuro network to identify the emitters in [Shieh et al 2002], where an interval activation function is applied so that the network can process interval data. Two back propagation learning algorithms, NVTBP and CVTBP algorithms, were derived to tune the weights of neural network, and furthermore, to classify the observations. In our experiments, the DCIFI projection classifier is also implemented to identify different types of emitters, and its performance is compared to that of the fuzzy-neuro network.

We use both the two-emitters and the three-emitter identification problems to test and compare the performance of the DCIFI projection classifier and those of the neural network approaches [Shieh et al 2002]. The training and testing data sets are the same as those in [Shieh et al 2002], where the data in training set are interval values while the data in testing set are crisp values. To evaluate the robustness of the proposed methods, a measurement distortion is also used as in [Shieh et al 2002] to simulate the adding of noise to the testing data. To perform the testing at different levels of adding noise, an Error Deviation Level (EDL) is defined in [Shieh et al 2002] by

$$EDL_i(\%) = \frac{\xi_{ji}}{x_{ji}} \times 100\%,$$

for $i = 1, 2, 3$, and $j = 1, 2, \cdots, l$, where l is the number of observations. Here, x_{ji} denotes the values of attribute x_i of j-th observation in the testing data set, and ξ_{ji} is a small alteration added to the values of x_{ji}. The noisy testing data are obtained by adding random noise ξ_{ji} to each original testing observation, denoted by

$$(x_{j1} \pm \xi_{j1}, x_{j2} \pm \xi_{j2}, x_{j3} \pm \xi_{j3})$$

with different *EDL*'s (from 0% to 15%).

First, we consider the two-emitter identification problem with the input data corrupted by adding noise. For the DCIFI projection classifier, it is a 3 attributes and 2 classes problem. We set the population size s as 30, and the maximum number of generations as 1000. 10 training samples are used to train the DCIFI projection classifier and the neural network approaches respectively. The estimated values of the signed efficiency measure and the virtual boundary are listed in Table 11.3.

9 sets of 80 testing samples with different EDLs (from 0% to 15%) are generated and used to test the performance of the considered identification approaches. The experimental results on average accuracy are compared in Table 11.4.

Table 11.3 The estimated values of the signed efficiency measure and the virtual boundary in two-emitter identification problem.

Parameters	Estimated Values
$\mu(\{x_1\})$	0.504087
$\mu(\{x_2\})$	0.476912
$\mu(\{x_1, x_2\})$	0.568434
$\mu(\{x_3\})$	0.394032
$\mu(\{x_1, x_3\})$	0.487458
$\mu(\{x_2, x_3\})$	0.503144
$\mu(\{x_1, x_2, x_3\})$	1.000000
boundary	6.885570

Table 11.4 Testing results on two-emitter identification problem with/without noise.

Error Deviation Level (%)	Total Average Accuracy (%)		
	NN by NVTBP Algorithm	NN by CVTBP Algorithm	DCIFI Projection
15	99.71	91.04	100
13	99.90	93.75	100
11	99.91	94.85	100
9	99.91	95.49	100
7	99.91	95.83	100
5	99.91	96.03	100
3	99.91	96.15	100
1	99.91	96.23	100
0	99.91	96.26	100

Secondly, we consider the three-emitter identification problem with the input data corrupted by adding noise. For the DCIFI projection classifier, it is a 3 attribute and 3 classes problem. We set the population size s as 30, and the maximum number of generations as 1000. 15 training samples are used to train the DCIFI projection classifier and the neural network approaches respectively. The estimated values of the signed efficiency measure and the virtual boundary are listed in Table 11.5.

120 testing samples with different EDLs (from 0% to 15%) are used to train and test the performance of DCIFI projection classifier and the neural network approaches, respectively. The comparison results on average accuracy are shown in Table 11.6.

The comparison results shown in Tables 11.5 and 11.6 indicate that the proposed DCIFI projection not only has higher identification capability, but also relatively more robust to noise than the neural network approaches.

Table 11.5 The estimated values of the signed efficiency measure and the virtual boundary in three-emitter identification problem.

Parameters	Estimated Values
$\mu(\{x_1\})$	0.488003
$\mu(\{x_2\})$	0.434324
$\mu(\{x_1, x_2\})$	0.479056
$\mu(\{x_3\})$	0.490667
$\mu(\{x_1, x_3\})$	0.454789
$\mu(\{x_2, x_3\})$	0.507754
$\mu(\{x_1, x_2, x_3\})$	1.000000
Boundary 0	6.481580
Boundary 1	10.237300

Table 11.6 Testing results on three-emitter identification problem with/without noise.

Error Deviation Level (%)	Total Average Accuracy (%)		
	NN by NVTBP Algorithm	NN by CVTBP Algorithm	DCIFI Projection
15	75.75	72.21	80.83
13	79.16	73.10	85.83
11	80.49	73.76	85.00
9	84.09	76.17	91.67
7	89.44	80.25	94.17
5	96.04	85.96	99.17
3	99.44	89.19	100
1	99.80	90.63	100
0	99.84	91.08	100

11.3 Fuzzified Choquet Integral with Fuzzy-Valued Integrand (FCIFI)

Let $\tilde{f}: X \to \mathcal{N}_F$ be a fuzzy-valued function and μ be a signed efficiency measure on $\mathscr{P}(X)$. The defuzzified Choquet integral of $\tilde{f}$ with respect to μ has been defined and discussed in Section 11.1. As an aggregation tool, the DCIFI ignores the fuzziness in the integration result, that is, the result of the integration is a crisp number. Though it is convenient in many real data mining problems, missing the knowledge on the fuzziness will bring some error in optimization problems, such as the network optimizations. In this section, keeping the fuzzy knowledge in the integration result, we concentrate on another approach for fuzzifying the Choquet integral with fuzzy-valued integrand, called the *Fuzzified Choquet Integral with Fuzzy-valued Integrand* (FCIFI), which is named due to its fuzzy integrand and fuzzy integration result as well.

11.3.1 *Definition of the FCIFI*

First, we use the extension principle to define the Choquet integral with a measurable interval-valued integrand, that is, an integrand being a function whose range is a subset of $\mathcal{N}_I$ with the measurability in the following sense. Here, $\mathcal{N}_I$ denotes the set of all rectangular fuzzy numbers, which are identical to interval numbers.

Definition 11.4 An interval-valued function $\bar{f}: X \to \mathcal{N}_I$ is measurable if both $\bar{f}_l(x) = [\bar{f}(x)]_l$, the left end point of interval $\bar{f}(x)$, and $\bar{f}_r(x) = [\bar{f}(x)]_r$, the right end point of interval $\bar{f}(x)$, are measurable functions of x.

Definition 11.5 Let $\bar{f}: X \to \mathcal{N}_I$ be a measurable interval-valued function on X and μ be a signed efficiency measure on $\mathscr{P}(X)$. The Choquet Integral of $\bar{f}$ with respect to μ is defined by

$$(\mathrm{C})\int \bar{f}\, d\mu = \left\{ \int g\ d\mu \,\middle|\, g(x) \in \bar{f}(x) \quad \forall x \in X, g: X \to R \text{ is measurable} \right\}. \tag{11.4}$$

According to the representation theorem and the extension principle in fuzzy set theory, we can define the measurability of the fuzzy-valued function and the FCIFI as follows.

Definition 11.6 A fuzzy-valued function $\widetilde{f}: X \to \mathscr{N}_F$ is measurable if its α-cut function,

$$\bar{f}_\alpha(x) = M_\alpha^{\widetilde{f}(x)} = \{t \mid m_{\widetilde{f}(x)}(t) \geq \alpha\},$$

is a measurable interval-valued function for every $\alpha \in [0,1]$, where $m_{\widetilde{f}(x)}$ is the membership function of the value of $\widetilde{f}$ at x.

Definition 11.7 (FCIFI) Let $\widetilde{f}: X \to \mathscr{N}_F$ be a measurable fuzzy-valued function on X and μ be a signed efficiency measure on $\mathscr{P}(X)$. The fuzzified Choquet integral of $\widetilde{f}$ with respect to μ is defined by

$$(\mathrm{C})\int \widetilde{f} d\mu = \bigcup_{0 \leq \alpha \leq 1} \alpha \cdot (\mathrm{C}) \int \bar{f}_\alpha d\mu \tag{11.5}$$

where $\bar{f}_\alpha(x)$ is given in Definition 11.6.

Note that, the integration value of the FCIFI is also a fuzzy subset of R (a fuzzy number). Fig. 11.8 is helpful for understanding the relationship between $\widetilde{f}$ and $\bar{f}_\alpha$ in Definition 11.7. For further illustration, let us refer to an example.

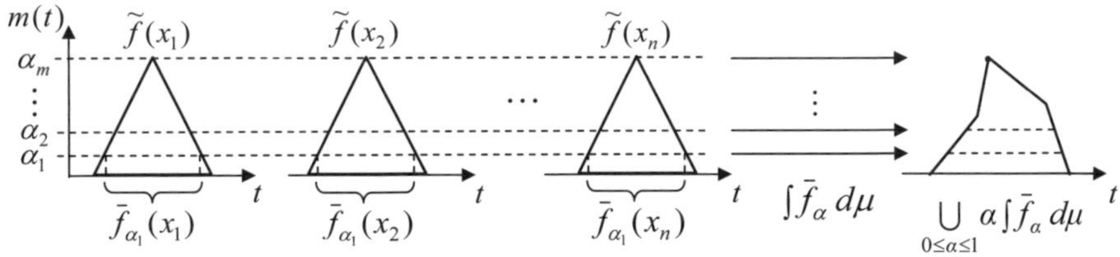

Fig. 11.8 Relationship between $\widetilde{f}$ and $\bar{f}_\alpha$.

Example 11.6 Let $\tilde{f}$ be a fuzzy-valued function defined on universal set $X = \{x_1, x_2, x_3\}$. Each element of X is mapped to a fuzzy number by function $\tilde{f}$, i.e., $\tilde{f}(x_1) = [1.0 \;\; 2.0 \;\; 3.0 \;\; 4.0]$, a trapezoidal fuzzy number; $\tilde{f}(x_2) = [5.0 \;\; 6.0]$, an interval number; and $\tilde{f}(x_3) = [7.0 \;\; 8.0 \;\; 9.0]$, a triangular fuzzy number. Their membership functions are depicted in Fig.11.9. Take $\alpha = 0.5$, the α-cut function of fuzzy-valued function $\tilde{f}$ is an interval-valued function, $\bar{f}_\alpha$, which maps each element of X to an interval number, i.e., $\bar{f}_\alpha(x_1) = [1.5 \;\; 3.5]$, $\bar{f}_\alpha(x_2) = [5.0 \;\; 6.0]$ and $\bar{f}_\alpha(x_3) = [7.5 \;\; 8.5]$, as shown in Fig. 11.9.

According to Definition 11.7, the calculation of the FCIFI is established on that of the Choquet Integral with Interval-valued Integrand (CIII). Due to the continuity of the Choquet integral, the integration value of the CIII is also an interval number. Now the problem we are facing with is how to determine the left and the right terminals of the interval-valued integration result. In the following subsections, we discuss two aspects of this problem, which are the CIII with respect to efficiency measures and signed efficiency measures, respectively.

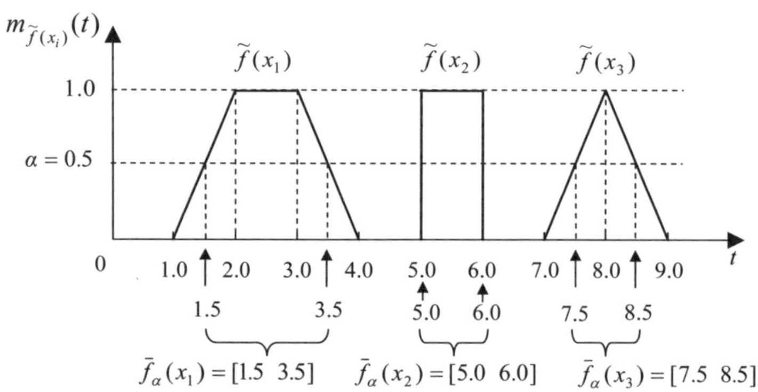

Fig. 11.9 The membership functions and α-cut function of $\tilde{f}$ in Example 11.6.

11.3.2 *The FCIFI with respect to monotone measures*

Using the continuity and the monotonicity of the Choquet integral with respect to monotone measures, we may prove the following theorem.

Theorem 11.2 Let $\bar{f}: X \to \mathcal{N}_I$ be a measurable interval-valued function on X and μ be a monotone measure on $\mathscr{P}(X)$. Then the Choquet integral of $\bar{f}$ with respect to μ is

$$(\mathrm{C})\int \bar{f}\, d\mu = [(\mathrm{C})\int \bar{f}_l\, d\mu, (\mathrm{C})\int \bar{f}_r\, d\mu] \qquad (11.6)$$

where $\bar{f}_l$ and $\bar{f}_r$ are two real-valued functions with $\bar{f}_l(x) = [\bar{f}(x)]_l$, the left end point of interval $\bar{f}(x)$, and $\bar{f}_r(x) = [\bar{f}(x)]_r$, the right end point of interval $\bar{f}(x)$, $\forall x \in X$.

As shown in Theorem 11.2, when the CIII is with respect to a monotone measure, terminals of the integration result can be directly calculated from the Choquet integrals of terminals of the integrand. Therefore, the FCIFI with respect to the monotone measure can be derived by (11.6) easily. Two examples are given as follows.

Example 11.7 Let $X = \{x_1, x_2\}$. Set function μ is a monotone measure with $\mu(\{x_1\}) = 0.1$, $\mu(\{x_2\}) = 0.2$, and $\mu(X) = 1$. $\tilde{f}$ is a triangular fuzzy-valued function with $\tilde{f}(x_1) = [0\ \ 1\ \ 1]$ and $\tilde{f}(x_2) = [0.5\ \ 0.5\ \ 1.5]$. The membership function of $\tilde{f}(x_1)$ and $\tilde{f}(x_2)$ are

$$m_1(t) = m_{\tilde{f}(x_1)}(t) = \begin{cases} t & \text{if } t \in [0,1] \\ 0 & \text{otherwise} \end{cases}$$

and

$$m_2(t) = m_{\tilde{f}(x_2)}(t) = \begin{cases} 1.5 - t & \text{if } t \in [0.5, 1.5] \\ 0 & \text{otherwise} \end{cases}$$

respectively. They are shown in Fig. 11.10. The α-cut function of $\tilde{f}$ is represented by intervals

$$\bar{f}_\alpha(x_1) = M_\alpha^{\tilde{f}(x_1)} = \{t \mid m_{\tilde{f}(x_1)}(t) \geq \alpha\} = [\alpha, 1]$$

and

$$\bar{f}_\alpha(x_2) = M_\alpha^{\tilde{f}(x_2)} = \{t \mid m_{\tilde{f}(x_2)}(t) \geq \alpha\} = [0.5, 1.5 - \alpha] .$$

When $0 \leq \alpha \leq 0.5$, we have $[\bar{f}_\alpha(x_1)]_l \leq [\bar{f}_\alpha(x_2)]_l$ and $[\bar{f}_\alpha(x_1)]_r \leq [\bar{f}_\alpha(x_2)]_r$. Therefore,

$$[\int \bar{f}_\alpha d\mu]_l = \alpha \cdot 1 + (0.5 - \alpha) \cdot 0.2 = 0.1 + 0.8\alpha$$

and

$$[\int \bar{f}_\alpha d\mu]_r = 1 \cdot 1 + (0.5 - \alpha) \cdot 0.2 = 1.1 - 0.2\alpha .$$

That is,

$$\int \bar{f}_\alpha d\mu = [0.1 + 0.8\alpha \;\; 1.1 - 0.2\alpha] .$$

Similarly, when $\alpha \in (0.5, 1]$, we have

$$\int \bar{f}_\alpha d\mu = [0.45 + 0.1\alpha \;\; 1.45 - 0.9\alpha] .$$

The membership function of $(C)\int \tilde{f}\, d\mu$, $m(t)$, is also shown in Fig. 11.10. We can see that $(C)\int \tilde{f}\, d\mu$ is not a triangular fuzzy numbers.

As for the Choquet integral with a normal fuzzy-valued integrand, its value may not be a normal fuzzy number either.

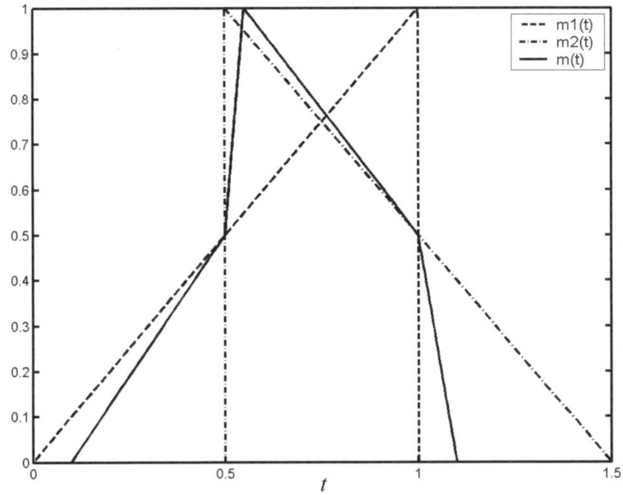

Fig. 11.10 The membership functions of the Choquet integral with triangular fuzzy-valued integrand in Example 11.7.

Example 11.8 Use the same X and μ given in Example 11.7. Let $\tilde{f}$ be a normal fuzzy-valued function having value $\tilde{n}(10, 1^2)$ at x_1 and $\tilde{n}(15, 10^2)$ at x_2. Fig. 11.11 shows the membership functions of $\tilde{n}(10, 1)$ and $\tilde{n}(15, 10)$, $m_1(t)$ and $m_2(t)$, respectively, as well as the membership function of $(\mathrm{C})\int \tilde{f}\, d\mu$, denoted by $m(t)$. We can see that $(\mathrm{C})\int \tilde{f}\, d\mu$ is not a normal fuzzy number. Its membership function $m(t)$ has a nondifferentiable point, also shown in Fig. 11.11.

We may also construct some examples to show similar conclusion for the Choquet integral with a trapezoidal fuzzy-valued or a cosine fuzzy-valued integrand. We can image that the membership function of the value of an FCIFI may have a large number of nondifferentiable points when n is not small.

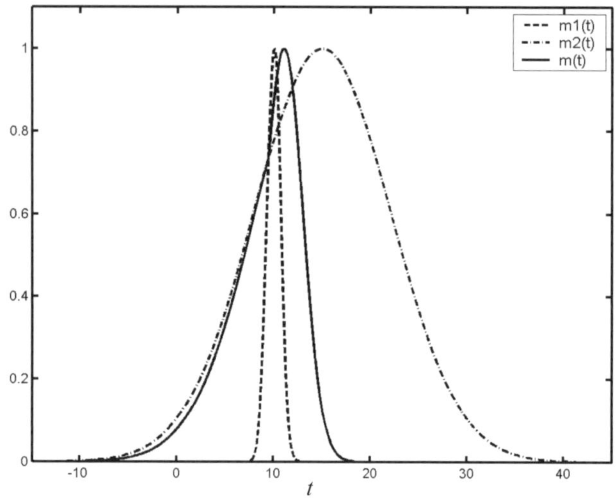

Fig. 11.11 The membership functions of the Choquet integral with normal fuzzy-valued integrand in Example 11.8.

11.3.3 *The FCIFI with respect to signed efficiency measures*

We should note that "μ be a monotone measure" cannot be replaced by "μ be a signed efficiency measure" in Theorem 11.2. The condition guaranteeing the nonnegativity of μ is essential. This can be verified by the following counterexample.

Example 11.9 Suppose that $X = \{x_1, x_2\}$, $\mu_1 = \mu(\{x_1\}) = 1$, $\mu_2 = \mu(\{x_2\}) = -2$, $\mu_3 = \mu(X) = 2$. Then, μ is a signed efficiency measure, but not a monotone measure. Taking interval-valued function $\bar{f}$ that has value [10 12] at x_1 and [8 14] at x_2, we have $(\mathrm{C})\int \bar{f} d\mu = [18\ \ 24]$. However, $\bar{f}_l(x_1) = 10$, $\bar{f}_r(x_1) = 12$, $\bar{f}_l(x_2) = 8$, $\bar{f}_r(x_2) = 14$, therefore, $(\mathrm{C})\int \bar{f}_l d\mu = 18$ and $(\mathrm{C})\int \bar{f}_r d\mu = 20 \neq 24$.

Furthermore, the decomposability described by

$$(\mathrm{C})\int f\, d\mu = (\mathrm{C})\int f\, d\mu^{+} - (\mathrm{C})\int f\, d\mu^{-}$$

in Section 5.4 is also violated by the CIII with respect to a signed efficiency measure. This can be shown in the following example.

Example 11.10 We still use the universal set X, the interval-valued function $\bar{f}$, and the signed efficiency measure μ given in Example 11.9. So $\mu_1^+ = 1$, $\mu_2^+ = 0$, $\mu_3^+ = 2$, $\mu_1^- = 0$, $\mu_2^- = 2$, and $\mu_3^- = 0$. Thus, $(\mathrm{C})\int \bar{f}\, d\mu^+ = [18\ \ 24]$ and $(\mathrm{C})\int \bar{f}\, d\mu^- = [0\ \ 8]$. Hence, $(\mathrm{C})\int \bar{f}\, d\mu^+ - (\mathrm{C})\int \bar{f}\, d\mu^- = [18\ \ 24] - [0\ \ 8] = [10\ \ 24]$. However, we have $(\mathrm{C})\int \bar{f}\, d\mu = [18\ \ 24]$. This violates the decomposability, that is,

$$(\mathrm{C})\int \bar{f}\, d\mu \neq (\mathrm{C})\int \bar{f}\, d\mu^{+} - (\mathrm{C})\int \bar{f}\, d\mu^{-}.$$

As shown above, with respect to a monotone measure μ, the left and the right terminals of $(\mathrm{C})\int \bar{f}\, d\mu$ can be directly calculated from the Choquet integrals of the integrand's left and right terminals, respectively. However, when the FCIFI is respect to a signed efficiency measure, Theorem 11.2 may not hold. In this case, terminals of $(\mathrm{C})\int \bar{f}\, d\mu$ may overstep the range which is restricted by $(\mathrm{C})\int \bar{f}_l d\mu$ and $(\mathrm{C})\int \bar{f}_r d\mu$. Hence, the exact membership function of the Choquet integral with respect to a signed efficiency measure for a fuzzy-valued integrand is rather difficult to be found.

In general case, we may give estimation on the integration result of the CIII with respect to signed efficiency measure through the following theorem.

Theorem 11.3 Let $\bar{f}: X \to \mathcal{N}_I$ be a measurable interval-valued function on X and μ be a signed efficiency measure on $\mathcal{P}(X)$. Then the Choquet integral of $\bar{f}$ with respect to μ, $(\mathrm{C})\int \bar{f}\, d\mu$, is still a rectangular fuzzy number (an interval number) and

$$\begin{aligned}(\mathrm{C})\int \bar{f}\, d\mu &\subseteq (\mathrm{C})\int \bar{f}\, d\mu^{+} - (\mathrm{C})\int \bar{f}\, d\mu^{-} \\ &= \Big[\, [(\mathrm{C})\int \bar{f}\, d\mu^{+}]_l - [(\mathrm{C})\int \bar{f}\, d\mu^{-}]_r,\ [(\mathrm{C})\int \bar{f}\, d\mu^{+}]_r - [(\mathrm{C})\int \bar{f}\, d\mu^{-}]_l \Big].\end{aligned}$$

Fig. 11.12 is helpful for understanding the above theorem.

In a simpler but common case where X is finite, we may obtain the valued of $(\mathrm{C})\int \bar{f}\, d\mu$ by solving two optimization problems with linear constraints and nonlinear objective functions

$$\min \sum_{j=1}^{2^n-1} z_j \mu_j \tag{11.7}$$

and

$$\max \sum_{j=1}^{2^n-1} z_j \mu_j \tag{11.8}$$

subject to $\bar{f}_l(x_i) \le f(x_i) \le \bar{f}_r(x_i)$, $i = 1, 2, \cdots, n$, where

$$z_j = \begin{cases} \min\limits_{i:\mathrm{frc}(j/2^i)\in[1/2,\,1)} f(x_i) - \max\limits_{i:\mathrm{frc}(j/2^i)\in[0,\,1/2)} f(x_i), & \text{if it is} > 0 \text{ or } j = 2^n - 1 \\ 0, & \text{otherwise} \end{cases}.$$

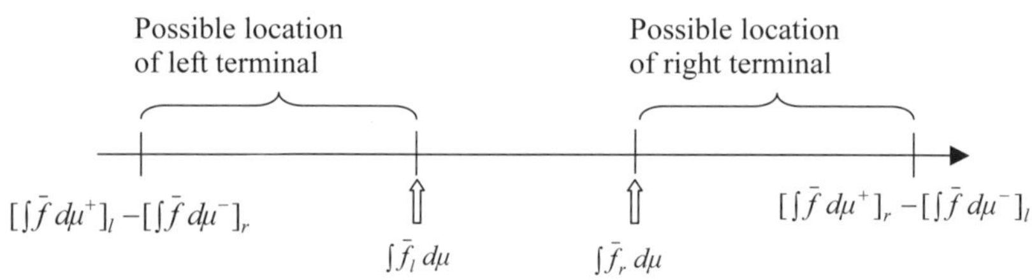

Fig. 11.12 Description of terminal ranges when μ is a signed efficiency measure.

We propose a numerical optimization method involving a genetic algorithm to approximately estimate the membership function of the CIII when μ is a signed efficiency measure in the following subsection.

11.3.4 *GA-based optimization algorithm for the FCIFI with respect to signed efficiency measures*

The core of the proposed numerical optimization is a genetic algorithm which is used to calculate the integration value of the CIII with respect to a signed efficiency measure. For clarification, we reintroduce the problem here. Let X be a finite set, i.e., $X = \{x_1, x_2, \cdots, x_n\}$. A signed efficiency measure $\mu : \mathscr{P}(X) \to (-\infty, \infty)$ is given. For an interval-valued function $\bar{f} : X \to \mathcal{N}_I$, where $\mathcal{N}_I$ denotes the set of all interval numbers, we are going to calculate the integration result of $(\mathrm{C})\int \bar{f}\, d\mu$. Since $(\mathrm{C})\int \bar{f}\, d\mu$ is also an interval number, only the left and the right terminals are required to be determined. These two terminals are calculated by the same GA approach, respectively.

A. Coding

In the GA-based optimization algorithm, real coding method is applied here. Each chromosome consists of n genes, denoted by $g_1, g_2, \cdots, g_n$, where n is the cardinality of the universal set X. Each gene takes a real number between zero and one. We introduce a real-valued function $v : X \to (-\infty, \infty)$, where $v(x_i)$ is a number in $\bar{f}(x_i)$, $i = 1, 2, \cdots, n$. For each i, g_i and $v(x_i)$ are one-to-one correspondence, and they can be coded and decoded by the following formula:

$$v(x_i) = \bar{f}_l(x_i) + (\bar{f}_r(x_i) - \bar{f}_l(x_i)) \cdot g_i ,$$

where $\bar{f}_l(x_i)$ and $\bar{f}_r(x_i)$ are the left and the right terminals of $\bar{f}(x_i)$ respectively. The correspondence among genes, function v and $\bar{f}$ are illustrated in Fig. 11.13.

We denote V as the set of all real-valued function v. Now, the problem can be summarized as follows:

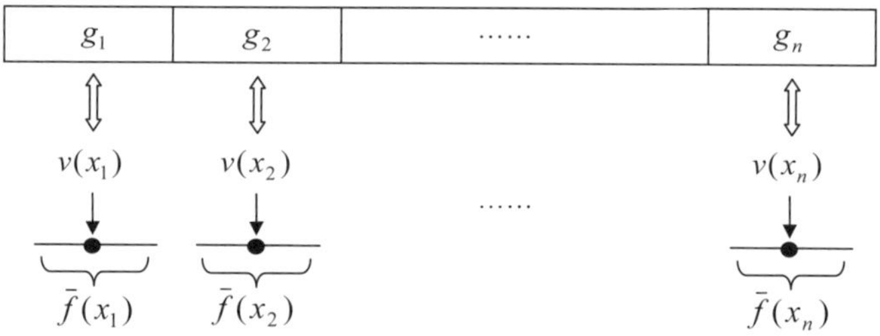

Fig. 11.13 Correspondence in coding method.

(1) Finding a real-valued function $v_l : X \to (-\infty, \infty)$, where $v_l(x_i) \in \bar{f}(x_i)$, $i = 1, 2, \cdots, n$, so that

$$(C)\int v_l d\mu = \min_{v \in V} (C)\int v\, d\mu, \tag{11.9}$$

in which the value of $(C)\int v_l d\mu$ is the left terminal of $(C)\int \bar{f}\, d\mu$.

(2) Finding a real-valued function $v_r : X \to (-\infty, \infty)$, where $v_r(x_i) \in \bar{f}(x_i)$, $i = 1, 2, \cdots, n$, so that

$$(C)\int v_r d\mu = \max_{v \in V} (C)\int v\, d\mu, \tag{11.10}$$

in which the value of $(C)\int v_r d\mu$ is the right terminal of $(C)\int \bar{f}\, d\mu$.

B. Evaluation criteria

To evaluate an individual of chromosome in the population, two reference values, $(C)\int \bar{f}_l d\mu$ and $(C)\int \bar{f}_r d\mu$, are pre-calculated. After decoding, each individual corresponds to a real-valued function v. When the left terminal of $(C)\int \bar{f} d\mu$ is calculated, we define the distance between $(C)\int v\, d\mu$ and $(C)\int \bar{f}_l d\mu$ as $\Delta_l = (C)\int \bar{f}_l d\mu - (C)\int v\, d\mu$. On the other hand, when the right terminal of $(C)\int \bar{f} d\mu$ is calculated, we

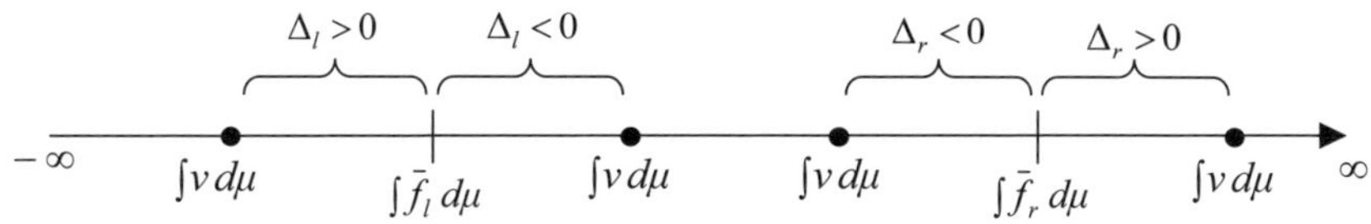

Fig. 11.14 Distance definition on calculation of the left and the right terminals of $(\mathrm{C})\int \bar{f}\, d\mu$.

define the distance between $(\mathrm{C})\int v\, d\mu$ and $(\mathrm{C})\int \bar{f}_r d\mu$ as $\Delta_r = (\mathrm{C})\int v\, d\mu - (\mathrm{C})\int \bar{f}_r\, d\mu$. Fig. 11.14 shows such a relationship.

Then, two fitness functions are defined to evaluate the performance of an individual in two situations respectively.

(1) When the left terminal of $(\mathrm{C})\int \bar{f} d\mu$ is calculated, the fitness valued of the individual in the population is derived by

$$\Delta_l = (\mathrm{C})\int \bar{f}_l d\mu - (\mathrm{C})\int v\, d\mu\,. \tag{11.11}$$

(2) When the right terminal of $(\mathrm{C})\int \bar{f} d\mu$ is calculated, the fitness valued of the individual in the population is derived by

$$\Delta_r = (\mathrm{C})\int v\, d\mu - (\mathrm{C})\int \bar{f}_r\, d\mu\,. \tag{11.12}$$

The positively larger the fitness value is, the better performance the individual has, and more chance it has to be selected to create new offspring.

C. GA-based optimization algorithm

The optimization algorithm used here is a GA-based algorithm. We take α-cuts of $\tilde{f}(x_i)$ from the bottom to the top in turn, i.e., $\alpha = \varepsilon \to 1$. If the α-cut of $\tilde{f}(x_i)$ is a closed interval when $\alpha = 0$, then $\varepsilon = 0$; otherwise, ε takes a small positive number to make the α-cut of $\tilde{f}(x_i)$ be a closed interval when $\alpha = \varepsilon$. For each α stage, calculate the left and the right terminals of $(\mathrm{C})\int \bar{f}_\alpha d\mu$ respectively by a genetic algorithm. Then, using the decomposition theorem, Eq. (11.5) is applied to reconstruct the final result of $(\mathrm{C})\int \tilde{f} d\mu$.

The main program is as follows.

(1) Input the following initial parameters:

n:	Cardinality of the universal set $X = \{x_1, x_2, \cdots, x_n\}$.
$\mu_1, \mu_2, \cdots, \mu_{2^n-1}$:	$(2^n - 1)$ real numbers representing the signed efficiency measure.
$\tilde{f}(x_1), \tilde{f}(x_2), \cdots, \tilde{f}(x_n)$:	n fuzzy numbers representing the integrand function $\tilde{f}$.
K:	Number of α-cuts with default value 100.
$step$:	$step = (1.0 - \varepsilon)/K$, the alteration of α value between two successive α stages.

(2) $i = 0$.

(3) $\alpha = step \cdot i$.

(4) For $j = 1 \to n$, calculate

$$\bar{f}_i(x_j) = \{t \mid m_{\tilde{f}(x_j)}(t) \geq \alpha\}.$$

(5) If $i = 0$, Go to Phase 1 to calculate $(\mathrm{C})\int \bar{f}_i \, d\mu$; otherwise, go to Phase 2 to calculate $(\mathrm{C})\int \bar{f}_i \, d\mu$.

(6) $i + 1 \Rightarrow i$. If $i = K$, go to (7); otherwise go to (3).

(7) Output the integration result $(\mathrm{C})\int \tilde{f} d\mu$.

Phase 1:

This part focuses on the calculation of $(\mathrm{C})\int \bar{f}_i d\mu$ when $i = 0$. Here, $\bar{f}_i|_{i=0}$ is the α-cut function of $\tilde{f}$ when $\alpha = \varepsilon$. In this phase, for lack of any information on integration result, a global search is required. The following genetic parameters have to be set before the iteration starts off:

s: The population size represented as a positive integer with default as 50.

δ: A small positive number with default as 10^{-10}.

$IC_{\max}$: The maximum number of the *Improvement Counter* (*IC*), which records the number of successive generations whose individuals are unimproved. It also acts as a marker to indicate that the optimal has been found. Its default is 20.

$GC_{\max}$: The maximum number of the *Generation Counter* (*GC*) with default as 100.

flag : A flag to determine which terminal (the left or the right) of $(C)\int \bar{f} d\mu$ is currently calculated, $flag = 0$ for the left terminal, while $flag = 1$ for the right terminal.

The program is summarized as follows:

(1) Randomly create an initial population that consists of s individuals of chromosome. Initialize both *GC* and *IC* as 0. Initialize $m_0 = 0.0$, where m_0 stores the fitness value of the best individual of the closest previous generation.
(2) Calculate $(C)\int(\bar{f}_i)_l\, d\mu$ and $(C)\int(\bar{f}_i)_r\, d\mu$.
(3) Decode and evaluate each individual in current population. The fitness value of the *k*-th individual is denoted by φ_k.
(4) Set $m_0 = \min_{1 \le k \le s} \varphi_k$.
(5) If $IC > IC_{\max}$ or $GC > GC_{\max}$, then go to (12).
(6) Do tournament selection (tournament size as 2). Randomly select one operator among the random mutation (with probability 0.4), the BLX-0.5 crossover (with probability 0.4), and the flat crossover (with probability 0.2) to produce new individuals of chromosome as offspring.
(7) Repeat (6) until totally getting s new offspring. Decode and evaluate each of the newly created individuals. Choose the best s individuals from the group of these s new created ones and the original s individuals in current generation to form the population for the next generation.
(8) Set $m = \min_{1 \le k \le s} \varphi_k$. If $|m_0 - m| < \delta$, then $IC + 1 \to IC$; otherwise, $0 \to IC$.

(9) Set $m_0 = m$.
(10) $GC + 1 \to GC$. Then go to (5).
(11) Output $(C)\int v\,d\mu$, where $v(x_i)$ is encoded from the genes of the best individual of current generation.
(12) Stop.

Phase 2:

In this phase, $(C)\int \bar{f}_i\,d\mu$, $i = 1, 2, \cdots, K$ are calculated. As $(C)\int \bar{f}_{i-1}\,d\mu$ has been obtained by the previous genetic process, according to the continuity and the monotonicity of the Choquet integral, we can find the left and the right terminals of $(C)\int \bar{f}_i\,d\mu$ nearby those of $(C)\int \bar{f}_{i-1}\,d\mu$. Thus, a relative local optimization is enough.

The genetic parameters are set as those in Phase 1. The program process follows the flowchart of Phase 1 except that some modifications are applied in steps (1), (6) and (7):

(1) Unlike Phase 1, here, the population is initialized the same as that of the last generation during the calculation on $(C)\int \bar{f}_{i-1}\,d\mu$.
(6) Do tournament selection (tournament size as 2). Produce new individual only by the random walk.
(7) Repeat (6) until totally getting s new offspring. To increase diversity of the searching space, randomly generate another s individuals of chromosome. Decode and evaluate these $2s$ individuals and select the best s ones to form the population for the next generation.

D. Examples

In this subsection, several examples of the FCIFI are shown, whose integration results are retrieved by the optimization algorithm presented above. It is shown that the proposed GA-based optimization algorithm is an effective algorithm to solve the calculation of the FCIFI.

Example 11.11 Suppose that $X = \{x_1, x_2\}$, μ is a signed efficiency measure valued by $\mu_1 = \mu(\{x_1\}) = 1$, $\mu_2 = \mu(\{x_2\}) = -2$, and $\mu_3 = \mu(X) = 2$. Taking fuzzy-valued function $\tilde{f}$ that assigns a triangular fuzzy number [0 1 1] at x_1 and [0.5 0.5 1] at x_2. Fig. 11.15 shows the membership function of $\tilde{f}(x_1)$ and $\tilde{f}(x_2)$. Set $K = 100$, the membership function of $(C)\int \tilde{f} d\mu$ is retrieved by the proposed genetic approaches and plotted in Fig.11.16.

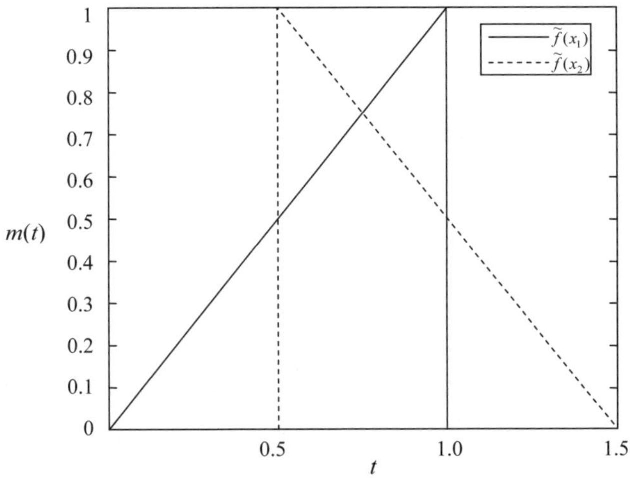

Fig. 11.15 Membership functions of $\tilde{f}(x_1)$ and $\tilde{f}(x_2)$ in Example 11.11.

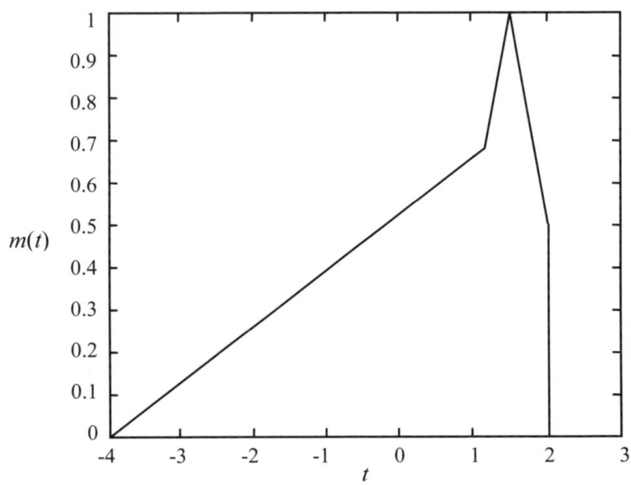

Fig. 11.16 The membership functions of $(C)\int \tilde{f} d\mu$ in Example 11.11.

Example 11.12 Let the universal set X and the signed efficiency measure μ be the same as those in Example 11.11. However, the fuzzy-valued function $\widetilde{f}$ assumes normal fuzzy numbers at x_1 and x_2. Their membership functions are

$$m_{\widetilde{f}(x_1)} = e^{-(\frac{t-10.0}{1.0})^2} \text{ and } m_{\widetilde{f}(x_2)} = e^{-(\frac{t-15.0}{10.0})^2},$$

respectively, where $t \in (-\infty, \infty)$, as shown in Fig. 11.17. Then, the membership function of $(C)\int \widetilde{f}\, d\mu$ is derived by the proposed GA-based optimization algorithm and its membership function is shown in Fig. 11.18.

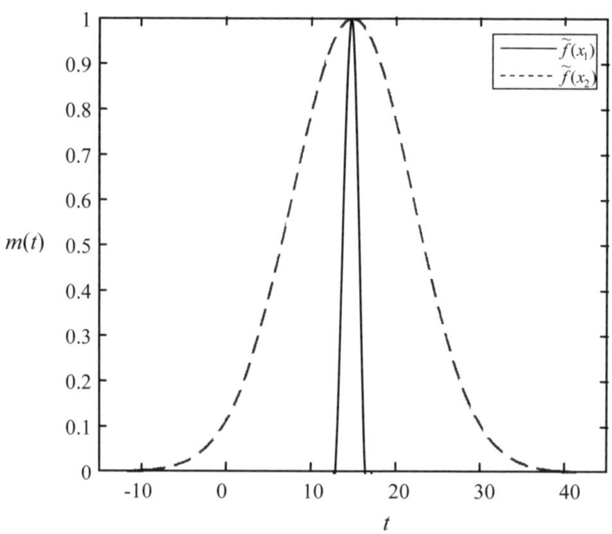

Fig. 11.17 Membership functions of $\widetilde{f}(x_1)$ and $\widetilde{f}(x_2)$ in Example 11.12.

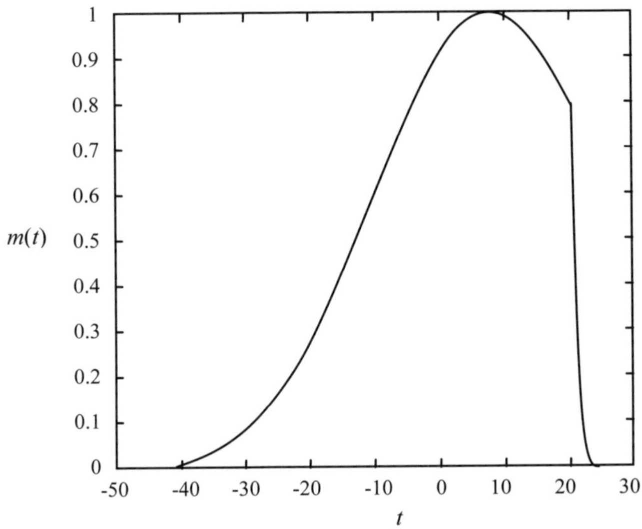

Fig. 11.18 The membership function of $(\mathrm{C})\int \tilde{f} d\mu$ in Example 11.12.

Example 11.13 We consider a more complex case. Here, the universal set X consists of 4 elements, x_1, x_2, x_3, and x_4. A signed efficiency measure is defined in Table 11.7. Take a fuzzy-valued function $\tilde{f}$ that assigns normal fuzzy numbers, which are the same as $\tilde{f}(x_1)$ and $\tilde{f}(x_2)$ in Example 11.12, to x_1 and x_2 respectively, and assigns triangular fuzzy numbers, which are the same as $\tilde{f}(x_1)$ and $\tilde{f}(x_2)$ in Example 11.11, to x_3 and x_4 respectively. The membership function of the value of $(\mathrm{C})\int \tilde{f}\, d\mu$ is derived by the proposed GA-based optimization algorithm and is shown in Fig. 11.19.

Table 11.7 Values of the signed efficiency measure μ in Example 11.13.

μ	value	μ	value
$\mu(\emptyset)$	0.0	$\mu(\{x_4\})$	2.0
$\mu(\{x_1\})$	1.0	$\mu(\{x_1, x_4\})$	7.0
$\mu(\{x_2\})$	-2.0	$\mu(\{x_2, x_4\})$	-9.0
$\mu(\{x_1, x_2\})$	2.0	$\mu(\{x_1, x_2, x_4\})$	1.0
$\mu(\{x_3\})$	3.0	$\mu(\{x_3, x_4\})$	2.0
$\mu(\{x_1, x_3\})$	11.0	$\mu(\{x_1, x_3, x_4\})$	-2.0
$\mu(\{x_2, x_3\})$	-1.0	$\mu(\{x_2, x_3, x_4\})$	2.0
$\mu(\{x_1, x_2, x_3\})$	4.0	$\mu(X)$	2.0

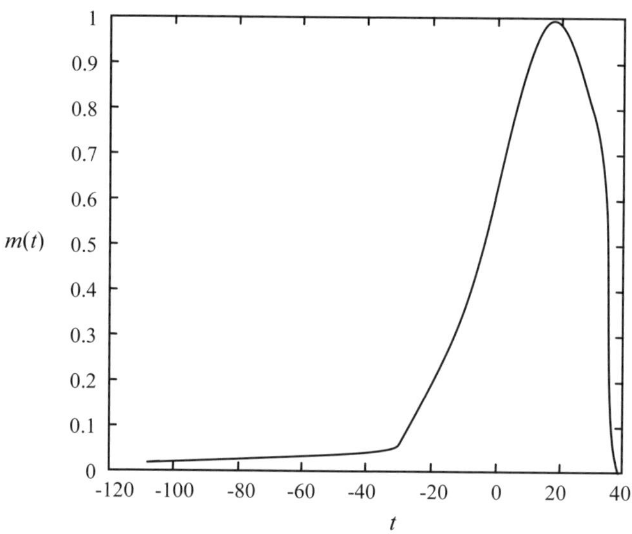

Fig. 11.19 Membership function of $(\mathrm{C})\int \tilde{f}\, d\mu$ in Example 11.13.

11.4 Regression Model Based on the CIII

Both the FCIFI and the CIII can be applied as regression tools. The former is a generalized model to the latter, since the FCIFI handles heterogeneous fuzzy data while the CIII manages interval data, and as we know, interval data are included in heterogeneous fuzzy data.

In this section, we focus our attention on the regression problems by the CIII because there are many practical cases where more complete information can be surely achieved by describing a set of variables in terms of interval data. For example, intervals may occur as transaction time and valid time ranges in temporal databases, as line segments on a space-filling curve in spatial applications, as inaccurate measurements with tolerances in engineering databases, as daily temperatures registered as the minimum and the maximum values, or for the minimum and the maximum transaction prices daily recorded for a set of stocks.

11.4.1 *CIII regression model*

From Chapter 9, we can see that the Choquet integral with a real-valued integrand is a very powerful regression tool because the nonadditivity of the signed efficiency measure can well capture the nonlinear relationship between the predictive attributes and the objective attribute. Similarly, the CIII can also be used as an aggregation tool in multiregression, which can represent the relationship among attributes with not only crisp data, but also interval data.

In the CIII regression model, let $x_1, x_2, \cdots, x_n$ be the predictive attributes and y be the objective attribute. Denote $X = \{x_1, x_2, \cdots, x_n\}$ as before. The provided training data set consists of l observations of $x_1, x_2, \cdots, x_n$ and y, and has a form as

x_1	x_2	$\cdots$	x_n	y
$\bar{f}_{11}$	$\bar{f}_{12}$	$\cdots$	$\bar{f}_{1n}$	$\bar{y}_1$
$\bar{f}_{21}$	$\bar{f}_{22}$	$\cdots$	$\bar{f}_{2n}$	$\bar{y}_2$
$\vdots$	$\vdots$		$\vdots$	$\vdots$
$\bar{f}_{l1}$	$\bar{f}_{l2}$	$\cdots$	$\bar{f}_{ln}$	$\bar{y}_l$

where each row

$$\bar{f}_{j1} \quad \bar{f}_{j2} \quad \cdots \quad \bar{f}_{jn} \quad \bar{y}_j$$

is the j-th observation of attributes $x_1, x_2, \cdots, x_n$, and y, $j=1,2,\cdots,l$. Note that, the values of observations in the training data set are all interval numbers, indicated by adding a top bar. Positive integer l is the size of the data set and should be much larger than 2^n. Usually, l is not less than 5 times of 2^n. Each observation of $x_1, x_2, \cdots, x_n$ can be regarded as an interval-valued function $\bar{f}: X \to \mathcal{N}_I$. Thus, the j-th observation of $x_1, x_2, \cdots, x_n$ is denoted by $\bar{f}_j$, and we write $\bar{f}_{ji}=\bar{f}_j(x_i)$, $i = 1, 2, \cdots, n$, for $j = 1, 2, \cdots, l$. Similarly, the j-th observation of y is denoted by $\bar{y}_j$, $j = 1, 2, \cdots, l$.

Hence, the CIII regression model (without showing the random perturbation) is expressed as

$$\bar{y} = \bar{c} + (\mathrm{C})\int(a + b\bar{f})\,d\mu,$$

where

- $\bar{y}$: value of the objective attribute y;
- $\bar{f}$: an interval-valued function on X with $\bar{f}(x_i)$ as its value at x_i, $i = 1, 2, \cdots, n$;
- μ: a signed efficiency measure;
- a: a real-valued function defined on X which can be expressed as a shifting parameter vector $a = (a_1, a_2, \cdots, a_n)$;
- b: a real-valued function defined on X which can be expressed as a scaling parameter vector $b = (b_1, b_2, \cdots, b_n)$;
- $\bar{c}$: an interval-valued constant, $\bar{c} = [c_l \;\; c_r]$.

The introduction of parameters $a_1, a_2, \cdots, a_n$, and $b_1, b_2, \cdots, b_n$ attempts to balance the scales of the predictive attributes in case that they have different dimensions. They should satisfy constraints

$$\min_{1 \le i \le n} a_i = 0 \text{ and } \max_{1 \le i \le n} |b_i| = 1.$$

Under these constraints, of course, we have $a_i \ge 0$ and $-1 \le b_i \le 1$ for $i = 1, 2, \cdots, n$.

In this multiregression model, the regression coefficients are constant $\bar{c}$, all elements of vectors a and b, and $\mu(A)$ for every $A \in \mathscr{P}(X) - \{\varnothing\}$. Totally there are

$$2 + n + n + 2^n - 1 = 2n + 2^n + 1$$

unknown parameters. All unknown parameters should be optimally determined before the regression model is put into operation. The scheme now is to learn all these coefficients through a genetic algorithm by describing them as genes in the chromosome. As shown in Section 11.3.4, when CIII is with respect to a signed efficiency measure, its integration result is also calculated by a genetic approach. Obviously, during the process of learning coefficients for the CIII regression model, two genetic algorithms are involved.

11.4.2 *Double-GA optimization algorithm*

We propose a double-GA optimization algorithm to learn the unknown parameters in the CIII regression model. There are $2n + 2^n + 1$ parameters to be determined. All of them are represented by genes in a chromosome. Fig. 11.20 shows the structure of each individual chromosome represented in the double-GA optimization algorithm.

To evaluate the fitness value of an individual in the double-GA, we define the distance between two interval numbers $\bar{s}$ and $\bar{t}$ as

$$|\bar{s} - \bar{t}| = \sqrt{(\bar{s}_l - \bar{t}_l)^2 + (\bar{s}_r - \bar{t}_r)^2},$$

where $\bar{s}_l$, $\bar{s}_r$, $\bar{t}_l$, and $\bar{t}_r$ are the left and the right terminals of $\bar{s}$ and $\bar{t}$, respectively.

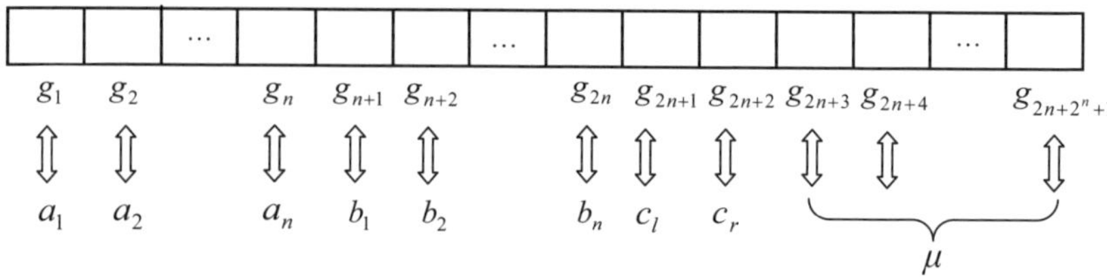

Fig. 11.20 Structure of an individual chromosome in the double-GA optimization algorithm.

Then the fitness value of an individual being considered in the population is

$$\hat{\sigma}^2 = \frac{1}{l}\sum_{j=1}^{l}(\bar{y}_j - \bar{y}_j^*)^2, \tag{11.13}$$

where $\bar{y}_j^*$ is the calculated integration result of the CIII regression model, which is identified by the parameters represented by the current individual, with respect to the j-th record of the predictive attributes, and $\bar{y}_j$ is the j-th record of the objective attribute in the training data set.

Now, the procedure of the double-GA is shown below.

(1) Choose a large prime p as the seed for the random number generator. Set the value for each genetic parameter as follows.

s :	The population size. It should be a large positive even integer. Its default is 100.
α, β :	The probabilities used in a random switch to control the choice of genetic operators for producing offspring from the selected parents. They should satisfy the condition that $\alpha \geq 0$, $\beta \geq 0$, and $\alpha + \beta \leq 1$. Their defaults are 0.4 and 0.4, respectively.
ε, δ :	Small positive numbers used in the stopping controller. Their defaults are 10^{-6} and 10^{-10}, respectively.

maxIC: The limit number of generations that have not significant improvement successively. Its default is 10.

maxGC: The limit number of generations. Its default is 10000.

(2) Read the number of the predictive attributes n, the number of training samples l, and the training samples.
Calculate

$$\hat{\sigma}_{\bar{y}}^2 = \frac{1}{l}\sum_{i=1}^{l}\left(\bar{y}_i - \frac{1}{l}\sum_{j=1}^{l}\bar{y}_j\right)^2 .$$

(3) Randomly create an initial population that consists of s individuals of chromosome. Initialize the Generation Counter (GC) and Improvement Counter (IC) by 0. Initialize $\hat{\sigma}_{\bar{y}}^2 \to m_0(\hat{\sigma}^2)$, where $m_0(\hat{\sigma}^2)$ stores the minimum fitness value of individuals in the closest previous generation.

(4) Decode each individual in the population to get its corresponding shifting parameters $a_1, a_2, \cdots, a_n$, scaling coefficients $b_1, b_2, \cdots, b_n$, interval-valued constant c_l, c_r, and values of signed efficiency measure $\mu_1, \mu_2, \cdots, \mu_{2^n-1}$.

(5) For each individual in current population, using the decoded regression coefficients, cooperated with each record in the training data set, to derive the calculated integration result of the CIII regression model represented by the current individual by

$$\bar{y}_j^* = \bar{c} + (\mathrm{C})\int (a + b\bar{f}_j)\,d\mu, \qquad j = 1, 2, \cdots, l .$$

If a monotone measure is considered, Theorem 11.2 is applied to derive the value of $(\mathrm{C})\int (a + b\bar{f}_j)d\mu$; otherwise, the GA-based optimization algorithm presented in Section 11.3.4 is performed. Then the fitness value of current individual is evaluated by Equation (11.13).

(6) Fitness value of the r-th chromosome is denoted by $\hat{\sigma}_r^2$. Set $m(\hat{\sigma}^2) = \min_{1 \le r \le s} \hat{\sigma}_r^2$, where $m(\hat{\sigma}^2)$ stores the minimum fitness value of individuals in current generation.

(7) If $m(\hat{\sigma}^2) < \varepsilon \hat{\sigma}_{\bar{y}}^2$ or $GC > GC_{\max}$, then go to (13); otherwise, take the next step.

(8) If $m_0(\hat{\sigma}^2) - m(\hat{\sigma}^2) < \delta \hat{\sigma}_{\bar{y}}^2$, then $IC + 1 \to IC$ and take the next step; otherwise, $0 \to IC$ and go to (10).

(9) If $IC > IC_{\max}$, divide the individuals in current population into three parts by ascending order on their fitness values. The individuals in the first part are kept, while those in the second part create new offspring by random mutation, and those in the third part are replaced by new randomly created individuals of chromosome. Evaluate the new created individuals, and update the population, go to (12); otherwise, take the next step.

(10) Do tournament selection (by tournament size as 2). Randomly select one operator among the non-uniform mutation (with probability α), the BLX crossover (with probability β), and the random mutation (with probability $1 - \alpha - \beta$) to produce new individuals of chromosome as the offspring.

(11) Repeat (10) until totally getting s new individuals. Evaluate this s new created individuals. Choose the best s individuals from the group of these s new created individuals and the original s individuals in current generation to form the population for the next generation.

(12) $GC + 1 \to GC$. Save $m(\hat{\sigma}^2)$ as $m_0(\hat{\sigma}^2)$. Then go to (6).

(13) Get the optimized regression coefficients from the best individual of the current generation.

(14) Stop.

11.4.3 *Explanatory examples*

In this part, two examples are implemented to verify the effectiveness and efficiency of the CIII regression model. These examples are conducted on synthetic data. Examples 11.14 and 11.15 are implemented on a CIII regression model with monotone measure and signed efficiency

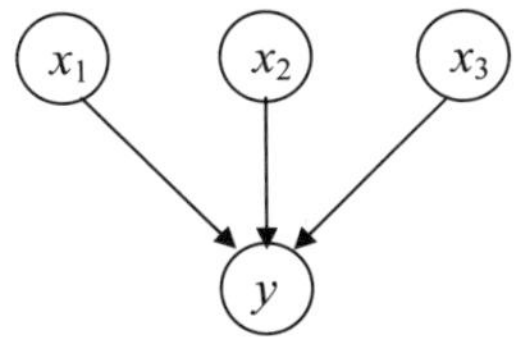

Fig. 11.21 Benchmark model in Examples 11.14 and 11.15.

measure, respectively. They all refer to a regression benchmark model with 3 predictive attributes and 1 objective attribute. Fig. 11.21 shows this benchmark model.

By presetting the shifting coefficients a_1, a_2, a_3 scaling coefficients b_1, b_2, b_3, constant c_l, c_r, and the values of monotone measure or signed efficiency measure $\mu_1, \mu_2, \cdots, \mu_7$, 10 training data sets, each of which consists of 200 observations, have been randomly generated for both experimental series, respectively.

Example 11.14 In this example, a CIII regression model with respect to a monotone measure is considered. The calculation of the CIII can be managed simply by Theorem 11.2. In this case, one genetic approach which is dedicated to the optimization of unknown parameters is involved.

10 randomly generated data sets, each of which consists of 200 observations, are applied to test the adaptability of our algorithm. The optimization results are recorded in Table 11.8. Here, among 10 randomly generated training data sets, five trials can converge to the global optimal before the maximum iteration time exceeds. For the remaining seven trials, they also reach the nearby space of the optimized solution. This shows that the proposed algorithm has satisfactory ability.

The comparisons of the preset and the estimated unknown parameters of the best one of 10 trials (the trial on Data set 2) are listed in Table 11.9. Here, all regression coefficients have been recovered well.

Table 11.8 Results of 10 trials in Example 11.14.

Data Set	Minimum fitness value
Set 1	2.15e–05 converge at generation 2003
Set 2	1.35e–04
Set 3	2.46e–05 converge at generation 2235
Set 4	2.17e–05 converge at generation 1701
Set 5	1.77e–03
Set 6	2.05e–05 converge at generation 1324
Set 7	7.49e–04
Set 8	1.18e–04
Set 9	2.48e–05 converge at generation 2321
Set 10	1.23e–04

Table 11.9 Comparisons of the preset and the estimated unknown parameters of the best trial in Example 11.14.

Coefficients	Preset value	Estimated value	Coefficients	Preset value	Estimated value
a_1	0.10	0.10012	$\mu(\emptyset)$	0.00	0.00000
a_2	0.20	0.20072	$\mu(\{x_1\})$	0.10	0.10141
a_3	0.30	0.30103	$\mu(\{x_2\})$	0.10	0.10268
b_1	0.20	0.19231	$\mu(\{x_1,x_2\})$	0.30	0.29987
b_2	0.50	0.50139	$\mu(\{x_3\})$	0.20	0.19921
b_3	0.90	0.91103	$\mu(\{x_1,x_3\})$	0.40	0.41001
c_l	0.10	0.10002	$\mu(\{x_2,x_3\})$	0.60	0.59623
c_r	0.50	0.49811	$\mu(X)$	1.00	1.00000

Example 11.15 In this example, a CIII regression model with respect to a signed efficiency measure is considered. Since Theorem 11.2 does not work for this case, the genetic approach presented in Section 11.4.2 is applied. Each of the 10 randomly generated data sets consists of 200 observations. The testing results on the ability of our algorithm are recorded in Table 11.10. Here, among 10 randomly generated training data sets, the trial on data set 3 gives the best optimization result. The optimization process stops at generation 4325 and converges to the

optimal solution. For the remaining trials on other data sets, the proposed double-GA can also reach into the nearby space of the optimized point. This shows that the algorithm still has satisfactory performance on the efficiency and effectiveness even double genetic approaches are involved. The comparisons of the preset and the estimated unknown parameters of the best trial are listed in Table 11.11. We can see the regression coefficients have been recovered well.

Table 11.10 Results of 10 trials in Example 11.15.

Data Set	Minimum fitness value
Set 1	1.45e–03
Set 2	2.56e–03
Set 3	2.43e–05 converge at generation 4325
Set 4	4.89e–04
Set 5	2.47e–05 converge at generation 5541
Set 6	2.86e–04
Set 7	1.67e–03
Set 8	2.89e–04
Set 9	4.98e–04
Set 10	1.62e–03

Table 11.11 Comparisons of the preset and the estimated unknown parameters of the best trial in Example 11.15.

Coefficients	Preset value	Estimated value	Coefficients	Preset value	Estimated value
a_1	0.10	0.10012	$\mu(\emptyset)$	0.00	0.00000
a_2	0.20	0.20072	$\mu(\{x_1\})$	0.10	0.99218
a_3	0.30	0.30103	$\mu(\{x_2\})$	-0.10	-0.10071
b_1	0.20	0.19231	$\mu(\{x_1, x_2\})$	0.30	0.29987
b_2	0.50	0.50139	$\mu(\{x_3\})$	0.70	0.71011
b_3	0.90	0.91103	$\mu(\{x_1, x_3\})$	0.40	0.39901
c_l	0.10	0.10002	$\mu(\{x_2, x_3\})$	0.60	0.60023
c_r	0.50	0.49811	$\mu(X)$	1.00	1.00000

Bibliography

Arslanov, M. Z. and Ismail, E. E. (2004). On the existence of possibility distribution function, *Fuzzy Sets and Systems*, 148(2), pp. 279–290.

Aubin, J. P. and Frankowska, H. (1990). Set-Valued Analysis, Birkhäuser, Boston.

Aumann, R. J. (1965). Integrals of set-valued functions, *J. of Mathematical Analysis and Applications*, 12(1), pp. 1–12.

Banon, G. (1981). Distinction between several subsets of fuzzy measures, *Fuzzy Sets and Systems*, 5, pp. 291–305.

Batle, N., and Trillas, E. (1979). Entropy and fuzzy integral, *Journal of Mathematical Analysis and Applications*, 69, pp. 469–474.

Bauer, H. (2001). Measure and Integration Theory. Walter de Gruyter, Berlin and New York.

Benvenuti, P. and Mesiar, R. (2000). Integrals with respect to a general fuzzy measure, *Fuzzy Measures and Integrals*. Springer-Verlag, New York, pp. 203–232.

Berberian, S. K. (1965). Measure and Integration. Macmillan, New York.

Berres, M. (1988). λ-additive measures on measure spaces, *Fuzzy Sets and Systems*, 27, pp. 159–169.

Billingsley, P. (1986). Probability and Measure (2nd Edition), John Wiley, New York.

Bouchon, B. and Yager, R. R., eds. (1987). Uncertainty in Knowledge-Based Systems, Springer-Verlag, New York.

Burk, F. (1998). Lebesgue Measure and Integration: An Introduction. Wiley-Interscience, New York.

Chae, S. B. (1995). Lebesgue Integration (2nd Edition), Springer-Verlag, New York.

Chen, T. Y., Wang, J. C., and Tzeng, G. H. (2000). Identification of general fuzzy measures by genetic algorithms based on partial information, *IEEE Trans. on Systems, Man, and Cybernetics* (*Part B*: *Cybernetics*), 30(4), pp. 517–528.

Chen, W., Cao, K., Jia, R., and Chen, K. (2009). An efficient algorithm for identification of real belief measures, *Proc. IEEE GrC 2009*, pp. 83-88.

Choquet, G. (1953-54). Theory of capacities, *Annales de l'Institut Fourier*, 5, pp. 131–295.

Choquet, G. (1969). Lectures on Analysis (3 volumes), W. A. Benjamin, Reading, MA.

Constantinescu, C., and Weber, K. (1985). Integration Theory, Vol. 1: Measure and Integration, Wiley-Interscience, New York.

De Campos, L. M., and Bolaños, M. J. (1989). Representation of fuzzy measures through probabilities, *Fuzzy Sets and Systems*, 31(1), pp. 23–36.

De Campos, L. M., and Bolaños, M. J. (1992). Characterization and comparison of Sugeno and Choquet integrals, *Fuzzy Sets and Systems*, 52(1), pp. 61–67.

Delgado, M., and Moral, S. (1989). Upper and lower fuzzy measures, *Fuzzy Sets and Systems*, 33, pp. 191–200.

Deng, X., and Wang, Z. (2005a). Learning probability distributions of signed fuzzy measures by genetic algorithm and multiregression, *Proc. IFSA 2005*, pp. 438-444.

Deng, X., and Wang, Z. (2005b). A fast iterative algorithm for identifying feature scales and signed fuzzy measures in generalized Choquet integrals, *Proc. FUZZ-IEEE 2005*, pp. 85-90.

Denneberg, D. (1994). Non-Additive Measure and Integral. Kluwer, Boston.

Denneberg, D. (2000a). Non-additive measure and integral, basic concepts and their role for applications, In: Grabisch, M. et al. (eds.), *Fuzzy Measures and Integrals*. Physica-Verlag, Heidelberg and New York, pp. 42–69.

Denneberg, D. (2000b). Totally monotone core and products of monotone measures, *International Journal of Approximate Reasoning*, 24(2-3), pp. 273–281.

Dubois, D., Nguyen, H. T., and Prade, H. (2000). Possibility theory, probability theory, and fuzzy sets: misunderstandings, bridges, and gaps, In: Dubois, D. and Prade, H. (eds.), *Fundamentals of Fuzzy Sets*, Kluwer, Boston, pp. 343–438.

Dubois, D., and Prade, H. (1980). Fuzzy Sets and Systems: Theory and Applications. Academic Press, New York.

Grabisch, M. (1995a). A new algorithm for identifying fuzzy measures and its application to pattern recognition, *Proc. FUZZ-IEEE/IFES'56*, Yokohama, Japan, pp.145-150.

Grabisch, M. (1995b). Fuzzy integral in multicriteria decision making, *Fuzzy Sets and Systems*, 69(3), pp. 279–298.

Grabisch, M. (1997a). k-order additive discrete fuzzy measures and their representation, *Fuzzy Sets and Systems*, 92(2), pp. 167–189.

Grabisch, M. (1997b). Alternative representations of discrete fuzzy measures for decision making, *Intern. J. of Uncertainty, Fuzziness, and Knowledge-Based Systems*, 5(5), pp. 587–607.

Grabisch, M. (1997c). Fuzzy measures and integrals: a survey of applications and recent issues, In: Dubois, D., Prade, H., and Yager, R. R. (eds.), *Fuzzy Information Engineering*, John Wiley, New York, pp. 507–529.

Grabisch, M. (2000a). The interaction and Möbius representations of fuzzy measures on finite spaces, *k*-additive measures: a survey. In: Grabisch, M., et al. (eds.), *Fuzzy Measures and Integrals: Theory and Applications*. Springer-Verlag, New York, pp. 70–93.

Grabisch, M. (2000b). Fuzzy measures and fuzzy integrals: theory and applications, In: Murofushi, T. and Sugeno, M. (eds.) .

Grabisch, M. and Nicolas, J. M. (1994). Classification by fuzzy integral: performance and tests, *Fuzzy Sets and Systems*, 65(2/3), pp. 255–271.

Guo, B., Chen, W., and Wang, Z. (2009). Pseudo gradient search for solving nonlinear multiregression based on the Choquet integral, *Proc. IEEE GrC 2009*, pp. 180-183.

Halmos, P. R. (1950). Measure Theory, Van Nostrand, New York.

Hawkins, T. (1975). Lebesgue's Theory of Integration: Its Origins and Development. Chelsea, New York.

Hui, J., and Wang, Z. (2005). Nonlinear multiregressions based on Choquet integral for data with both numerical and categorical attributes, *Proc. IFSA 2005*, pp. 445-449.

Ichihashi, H., Tanaka, H., and Asai, K. (1985). An application of the fuzzy integrals to multi-attribute decision problem, *Proc. First IFSA Congress*, Palma De Mallorca.

Ichihashi, H., Tanaka, H., and Asai, K. (1988). Fuzzy integrals based on pseudo-additions and multiplications, *Journal of Mathematical Analysis and Applications*, 130, pp. 354–364.

Ishi, K., and Sugeno, M. (1985). A model of human evaluation process using fuzzy measure, *International Journal of Man-Machine Studies*, 22, pp. 19–38.

Jumadinova, J. and Wang, Z. (2007). The pseudo gradient search and a penalty technique used in classifications, *Proc. 10th Joint Conference on Information Sciences*, pp. 1419-1426.

Keller, J., Qiu, H., and Tahani, H. (1986). Fuzzy integral and image segmentation, *Proc. North American Fuzzy Information Processing Soc.*, New Orleans, pp. 324–388.

Keller, J. M. et al. (1994). Advances in fuzzy integration for pattern recognition, *Fuzzy Sets and Systems*, 65(2/3), pp. 273–283.

Klement, E. P. and Weber, S. (1999). Fundamentals of generalized measure theory, In: Höhle, U. and Rodabaugh, S. E. (eds.), *Mathematics of Fuzzy Sets*. Kluwer, Boston and Dordrecht, pp. 633–651.

Klir, G. J., and Folger, T. A. (1988). Fuzzy Sets, Uncertainty, and Information, Prentice Hall, Englewood Cliffs, New Jersey.

Klir, G. J. and Yuan, B. (1995). Fuzzy Sets and Fuzzy Logic: Theory and Applications, Prentice Hall, Upper Saddle River, NJ.

Klir, G. J., Wang, Z., and Harmanec, D. (1997). Constructing fuzzy measures in expert systems, *Fuzzy Sets and Systems*, 92(2), pp. 251–264.

Klir, G. J., Wang, Z., and Wang, W. (1997). Constructing fuzzy measures by transformations, *J. of Fuzzy Mathematics*, 4(1), pp. 207–215.

Kruse, R. (1982a). A note on λ-additive fuzzy measures, *Fuzzy Sets and Systems*, 8, pp. 219–222.

Kruse, R. (1982b). On the construction of fuzzy measures, *Fuzzy Sets and Systems*, 8, pp. 323–327.

Lebesgue, H. (1966). Measure and the Integral. Holden-Day, San Francisco.

Leung, K. S. and Wang, Z. (1998). A new nonlinear integral used for information fusion, *Proc. of FUZZ-IEEE '98*, Anchorage, pp. 802–807.

Leung K. S., Wong M. L., Lam W., Wang Z. and Xu K. (2002), Learning nonlinear multiregression networks based on evolutionary computation, *IEEE Trans. SMC*, 32(5), pp. 630-644.

Li, W., Wang, Z., Lee, K.-H., and Leung, K.-S. (2005). Units scaling for generalized Choquet integral, *Proc. IFSA 2005*, pp. 121-125.

Liu, M., and Wang, Z. (2005). Classification using generalized Choquet integral projections, *Proc. IFSA 2005*, pp. 421-426.

Ma, Jifeng (1984). (IP)integrals, *Journal of Engineering Mathematics*, 2, pp. 169–170 (in Chinese).

Mahasukhon, M., Sharif, H., and Wang, Z. (2006). Using pseudo gradient search for solving nonlinear multiregression based on 2-additive measures, *Proc. IEEE IRI 2006*, pp. 410-413.

Murofushi, T. (2003). Duality and ordinality in fuzzy measure theory, *Fuzzy Sets and Systems*, 138(3), pp. 523–535.

Murofushi, T., and Sugeno, M. (1989). An interpretation of fuzzy measure and the Choquet integral as an integral with respect to a fuzzy measure, *Fuzzy Sets and Systems*, 29, pp. 201–227.

Murofushi, T. and Sugeno, M. (1991a). A theory of fuzzy measures: representations, the Choquet integral, and null set, *J. of Mathematical Analysis and Applications*, 159, pp. 532–549.

Murofushi, T. and Sugeno, M. (1991b). Fuzzy *t*-conorm integral with respect to fuzzy measures: Generalization of Sugeno integral and Choquet integral, *Fuzzy Sets and Systems*, 42, pp. 57–71.

Murofushi, T. and Sugeno, M. (1993). Some quantities represented by the Choquet integral, *Fuzzy Sets and Systems*, 56(2), pp. 229–235.

Murofushi, T., Sugeno, M., and Machida, M. (1994). Non-monotonic fuzzy measures and the Choquet integral, *Fuzzy Sets and Systems*, 64(1), pp. 73–86.

Pap, E. (1995). Null-Additive Set Functions, Kluwer, Boston.

Pap, E., ed. (2002a). Handbook of Measure Theory (2 volumes), Elsevier, Amsterdam.

Ralescu, D., and Adams, G. (1980). The fuzzy integral, *J. of Mathematical Analysis and applications*, 75(2), pp. 562–570.

Scott, B. L., and Wang, Z. (2006). Using 2-additive measures in nonlinear multiregressions, *Proc. IEEE GrC 2006*, pp. 639-642.

Shafer, G. (1976). A Mathematical Theory of Evidence, Princeton University Press, Princeton, New Jersey.

Shieh, C. –S and Lin, C. –T (2002). A vector neural network for emitter identification, *IEEE transaction on Antennas and Propagation*, 50(8), pp. 1120-1127.

Sims, J. R., and Wang, Z. (1990). Fuzzy measures and fuzzy integrals: An overview, *International Journal of General Systems*, 17, pp. 157–189.

Spilde, M., and Wang, Z. (2005). Solving nonlinear optimization problems based on Choquet integrals by using a soft computing technique, *Proc. IFSA 2005*, pp. 450-454.

Sugeno, M. (1974). Theory of Fuzzy Integrals and its Applications. Ph.D. dissertation, Tokyo Institute of Technology.

Sugeno, M. (1977). Fuzzy measures and fuzzy integrals: A survey, In: Gupta, Saridis, and Gaines (eds), *Fuzzy Automata and Decision Processes*. pp. 89–102.

Sugeno, M., and Murofushi, T. (1987). Pseudo-additive measures and integrals, *J. of Mathematical Analysis and Applications*, 122, pp. 197–222.

Tahani, H., and Keller, J. M. (1990). Information fusion in computer vision using the fuzzy integral, *IEEE Trans. on Systems, Man and Cybernetics*, 20, pp. 733–741.

Tanaka, H., and Hayashi, I. (1989). Possibilistic linear regression analysis for fuzzy data, *European Journal of Operations Research*, 40, pp. 389–396.

Tanaka, H., Sugihara, K., and Maeda, Y. (2004). Non-additive measures by interval probability functions, *Information Sciences*, 164, pp. 209–227.

Viertl, R. (1996). Statistical Methods for Non-Precise Data. CRC Press, Boca Raton, Florida.

Walley, P. (1991). Statistical Reasoning with Imprecise Probabilities. Chapman and Hall, London.

Wang, H., Sharif, H., and Wang, Z. (2006). A new classifier based on genetic algorithm, *Proc. IPMU 2006*, pp. 2479-2484.

Wang, H., Fang, H., Sharif, H., and Wang, Z. (2007). Nonlinear classification by genetic algorithm with signed fuzzy measure, *Proc. FUUZ/IEEE 2007*, pp. 1432-1437.

Wang, J. and Wang, Z. (1997). Using neural networks to determine Sugeno measures by statistics, *Neural Networks*, 10(1), pp. 183–195.

Wang, J.-C. and Chen, T.-Y. (2005). Experimental analysis of λ-fuzzy measure identification by evolutionary algorithms, *Intern. J. of Fuzzy Systems*, 7(1), pp. 1–10.

Wang, J.-F., Leung, K.-S., Lee, K.-H., and Wang, Z. (2008). Projection with Double Nonlinear Integrals for Classification, *Proc. ICDM 2008*, pp. 142-152.

Wang, P. (1982). Fuzzy contactability and fuzzy variables, *Fuzzy Sets and Systems*, 8, pp. 81–92.

Wang, R. and Ha, M. (2006). On Choquet integrals of fuzzy-valued functions, *J. of Fuzzy Mathematics*, 14(1), pp. 89–102.

Wang, R., Wang, L., and Ha, M. (2006). Choquet integrals on L-fuzzy sets, *J. of Fuzzy Mathematics*, 14(1), pp. 151–163.

Wang, W., Klir, G. J., and Wang, Z. (1996). Constructing fuzzy measures by rational transformations, *J. of Fuzzy Mathematics*, 4(3), pp. 665–675.

Wang, W., Wang, Z., and Klir, G. J. (1998). Genetic algorithms for determining fuzzy measures from data, *J. of Intelligent and Fuzzy Systems*, 6(2), pp. 171–183.

Wang, Xizhao, and Ha, Minghu (1990). Pan-fuzzy integral.,*BUSEFAL*, 43, pp. 37–41.

Wang, Z. (1981). Une class de mesures floues—les quasi-mesures, *BUSEFAL*, 6, pp. 28–37.

Wang, Z. (1984). The autocontinuity of set function and the fuzzy integral, *J. of Mathematical Analysis and Applications*, 99, pp. 195–218.

Wang, Z. (1985). Extension of possibility measures defined on an arbitrary nonempty class of sets, *Proc. of the 1st IFSA Congress*, Palma de Mallorca.

Wang, Z. (1986). Semi-lattice isomorphism of the extensions of possibility measure and the solutions of fuzzy relation equation, *Proc. of Cybernetics and Systems '86*, R. Trappl (ed), Kluwer, Boston, pp. 581–583.

Wang, Z. (1990). Absolute continuity and extension of fuzzy measures, *Fuzzy Sets and Systems*, 36, pp. 395–399.

Wang, Z. (1997). Convergence theorems for sequences of Choquet integrals, *Intern. J. of General Systems*, 26(1-2), pp. 133–143.

Wang, Z. (2002). A new model of nonlinear multiregression by projection pursuit based on generalized Choquet integrals, *Proc. of FUZZ-IEEE '02*, pp. 1240–1244.

Wang, Z. (2003). A new genetic algorithm for nonlinear multiregression based on generalized Choquet integrals, *Proc. of FUZZ-IEEE '03*, pp. 819–821.

Wang, Z. and Guo, H., Shi, Y., and Leung, K. S. (2004). A brief description of hybrid nonlinear classifier based on generalized Choquet integrals, *Lecture Notes in AI*, No. 3327, pp. 34–40.

Wang, Z. and Klir, G. J. (1992). Fuzzy Measure Theory, Plenum Press, New York.

Wang, Z. and Klir, G. J. (1997). Choquet integrals and natural extensions of lower probabilities, *Intern. J. of Approximate Reasoning*, 16(2), pp. 137–147.

Wang, Z. and Klir, G. J. (2007). Coordination uncertainty of belief measures in information fusion, in *Analysis and Design of Intelligent Systems Using Soft Computing Techniques* (Patricia Melin, Oscar Castillo, Eduardo Gomez Ramirez, Janusz Kacprzyk, and Witold Pedrycz eds.) —*Proc. IFSA 2007*, pp. 530-538.

Wang, Z. and Klir, G. J. (2008). Generalized Measure Theory, Springer, New York.

Wang, Z., Klir, G. J., and Wang, J. (1998). Neural networks used for determining belief measures and plausibility measures, *Intelligent Automation and Soft Computing*, 4(4), pp. 313–324.

Wang, Z., Klir, G. J., and Wang, W. (1996). Monotone set functions defined by Choquet integral, *Fuzzy Sets and Systems*, 81(2), pp. 241–250.

Wang, Z., Lee, K.-H., and Leung, K.-S. (2008). The Choquet integral with respect to fuzzy-valued signed efficiency measures, *Proc. WCCI 2008*, pp. 2143-2148.

Wang, Z. and Leung, K.-S. (2006). Uncertainty carried by fuzzy measures in aggregation, *Proc. IPMU 2006*, pp. 105-112.

Wang, Z., Leung, K.-S., and Klir, G. J. (2005). Applying fuzzy measures and nonlinear integrals in data mining, *Fuzzy Sets and Systems*, 156(3), pp. 371–380.

Wang, Z., Leung, K.-S., and Klir, G. J. (2006). Integration on finite sets. *Intern. J. of Intelligent Systems*, 21(10), pp. 1073–1092.

Wang, Z., Leung, K.-S., Wong, M.-L., Fang, J., and Xu, K. (2000). Nonlinear nonnegative multiregression based on Choquet integrals, *Intern. J. of Approximate Reasoning*, 25(2), pp. 71–87.

Wang, Z., and Li, F. (1985). Application of fuzzy integral in synthetical evaluations, *Fuzzy Mathematics*, 1, 109–114 (in Chinese).

Wang, Z. and Li, S. (1990). Fuzzy linear regression analysis of fuzzy valued variables, *Fuzzy Sets and Systems*, 36(1), pp. 125–136.

Wang, Z., Li, W., Lee, K.-H., and Leung, K.-S. (2008). Lower integrals and upper integrals with respect to nonadditive set functions, *Fuzzy Sets and Systems*, 159(3), pp. 646-660.

Wang, Z. Xu, K., Heng, P.-A., Leung, K.-S. (2003). Interdeterminate integrals with respect to nonadditive measures, *Fuzzy Sets and Systems*, 138(3), pp. 485–495.

Wang, Z., Xu, K., Wang, J., and Klir, G. J. (1999). Using genetic algorithms to determine nonnegative monotone set functions for information fusion in environments with random perturbation, *Intern. J. of Intelligent Sytems,* 14(10), pp. 949–962.

Wang, Z., Yang, R., and Leung, K.-S. (2005). On the Choquet integral with fuzzy-valued integrand, *Proc. IFSA 2005*, pp. 433-437.

Wang, Z., Yang, R., Heng, P.-A., and Leung, K.-S. (2006). Real-valued Choquet integrals with fuzzy-valued integrand, *Fuzzy Sets and Systems*, 157(2), pp. 256-269.

Wierzchon, S. T. (1983). An algorithm for identification of fuzzy measure, *Fuzzy Sets and Systems*, 9, pp. 69–78.

Wolkenhauer, O. (1998). Possibility Theory with Applications to Data Analysis. Research Studies Press, Taunton, UK.

Wu, Conxin, and Ma, Ming (1989). Some properties of fuzzy integrable function space $L^1(\mu)$, *Fuzzy Sets and Systems*, 31, pp. 397–400.

Wu, C. and Traore, M. (2003). An extension of Sugeno integral, *Fuzzy Sets and Systems*, 138(3), pp. 537–550.

Wu, Y. and Wang, Z. (2007). Using 2-interactive measures in nonlinear classifications, *Proc. NAFIPS'07*, pp. 248-353.

Xu, K., Wang, Z., Heng, P. A., and Leung, K. S. (2001). Using generalized Choquet integrals in projection pursuit based classification, *Proc. IFSA / NAFIPS*, pp. 506–511.

Xu, K., Wang, Z., Heng, P.-A., and Leung, K.-S. (2003). Classification by nonlinear integral projections, *IEEE Trans. on Fuzzy Systems*, 11(2), pp. 187–201.

Xu, K., Wang, Z., and Ke, Y. (2000). A fast algorithm for Choquet-integral-based nonlinear multiregression used in data mining, *J. of Fuzzy Mathematics*, 8(1), pp. 195–201.

Xu, K., Wang, Z., Wong, M.-L., and Leung, K.-S. (2001). Discover dependency pattern among attributes by using a new type of nonlinear multiregression, *Intern. J. of Intelligent Systems*, 16(8), pp. 949–962.

Yager, R. R. (2002). Uncertainty representation using fuzzy measures, *IEEE Trans. on Systems, Man, and Cybernetics*, Part B, 32(1), pp. 13–20.

Yager, R. R. and Kreinovich, V. (1999). Decision making under interval probabilities, *Intern. J. of Approximate Reasoning*, 22(3), pp. 195–215.

Yan, N., Wang, Z., Shi, Y., and Chen, Z. (2006). Nonlinear classification by linear programming with signed fuzzy measures, *Proc. FUZZIEEE 2006*, pp. 1484-1489.

Yang, Q. (1985). The pan-integral on the fuzzy measure space, *Fuzzy Mathematics*, 3, pp. 107–114 (in Chinese).

Yang, Q. and Song, R. (1985). A further discussion on the pan-integral. *Fuzzy Mathematics*, 4, pp. 27–36 (in Chinese).

Yang, R., Wang, Z., Heng, P.-A., and Leung, K.-S. (2005). Fuzzy numbers and fuzzification of the Choquet integral, *Fuzzy Sets and Systems*, 153(1), pp. 95–113.

Yang, R., Wang, Z., Heng, P.-A., and Leung, K.-S. (2007). Classification of heterogeneous fuzzy data by Choquet integral with fuzzy-valued integrand, *IEEE Trans.. Fuzzy Systems* 15(5), pp. 931-942.

Yang, R., Wang, Z., Heng, P.-A., and Leung, K.-S. (2008). Fuzzified Choquet integral with fuzzy-valued integrand and its application on temperature prediction, *IEEE Trans. SMCB*, 38(2), pp. 367-380.

Yuan, B. and Klir, G. J. (1996). Constructing fuzzy measures: a new method and its application to cluster analysis, *Proc. NAFIPS '96*, Berkeley, CA, pp. 567–571.

Yue S., Li P. and Yin Z. X. (2005), Parameter estimation for Choquet fuzzy integral based on Takagi-Sugeno fuzzy model. *Information Fusion* 6(2), pp. 175-182.

Zadeh, L. A. (1965). Fuzzy sets, *Information and Control*, 8, pp. 338–353.

Zadeh, L. A. (1968). Probability measures of fuzzy events, *J. of Mathematical Analysis and Applications*, 23, pp. 421–427.

Zadeh, L. A. (1975-76). The concept of a linguistic variable and its application to approximate reasoning, *Information Sciences*, 8, pp. 199–249, 301–357; 9, pp. 43–80.

Zhang, W., Chen, W., and Wang, Z. (2009). On the uniqueness of the expression for the Choquet integral with linear core in classification, *Proc. IEEE GrC 2009*, pp. 769-774.

Zong, T., Shi, P., and Wang, Z. (2006). Nonlinear integrals with respect to superadditive fuzzy measures on finite sets, *Proc. IPMU 2006*, pp. 2456-2463.

Index